Insect Biology

Insect Biology

Edited by **Christopher Fleming**

SYRAWOOD
PUBLISHING HOUSE

New York

Published by Syrawood Publishing House,
750 Third Avenue, 9th Floor,
New York, NY 10017, USA
www.syrawoodpublishinghouse.com

Insect Biology
Edited by Christopher Fleming

© 2016 Syrawood Publishing House

International Standard Book Number: 978-1-68286-066-3 (Hardback)

Contents

Preface

The world is advancing at a fast pace like never before. Therefore, the need is to keep up with the latest developments. This book was an idea that came to fruition when the specialists in the area realized the need to coordinate together and document essential themes in the subject. That's when I was requested to be the editor. Editing this book has been an honour as it brings together diverse authors researching on different streams of the field. The book collates essential materials contributed by veterans in the area which can be utilized by students and researchers alike.

The study of insect biology is of high importance for a number of fields like agriculture, chemistry, biology, health science, etc. This book on insect biology covers a diverse set of topics ranging from insect anatomy and physiology to topics like genetics, evolution, behavior of insects, etc. This text is a valuable compilation of researches, ranging from the basic to the most complex advancements in the field of insect biology. It provides significant information of this discipline to help develop a good understanding about the field among students and aid research scholars.

Each chapter is a sole-standing publication that reflects each author's interpretation. Thus, the book displays a multi-facetted picture of our current understanding of applications and diverse aspects of the field. I would like to thank the contributors of this book and my family for their endless support.

Editor

Congregation Sites and Sleeping Roost of Male Stingless Bees (Hymenoptera: Apidae: Meliponini)

CF Santos[12*], C Menezes[3,4], A Vollet-Neto[3], VL Imperatriz-Fonseca[1,5]

1 - Laboratório de abelhas, Instituto de Biociências, Universidade de São Paulo, São Paulo, SP, Brazil

2 - Laboratório de Entomologia, Pontifícia Universidade Católica do Rio Grande do Sul, Porto Alegre, RS, Brazil

3 - Embrapa Amazônia Oriental, Empresa Brasileira de Pesquisa Agropecuária, Belém, PA, Brazil

4 - Faculdade de Filosofia, Ciências e Letras de Ribeirão Preto, Universidade de São Paulo, Ribeirão Preto, SP, Brazil

5 - Universidade Federal Rural do Semi-Árido, Mossoró, RN, Brazil

Keywords

Males, mating swarm, roosting sites, social bees, reproductive strategies

Corresponding author

Charles Fernando dos Santos
Laboratório de Entomologia
Pontifícia Universidade Católica
do Rio Grande do Sul
90619-900 Porto Alegre, RS, Brazil
e-mail: charles.santos@pucrs.br

Abstract

Very little is known about stingless bee reproductive biology or male behaviour. In this note we provide the first observations on the male aggregations (congregation sites and roosting sites) of some stingless bee species. Our observations show that males of two stingless bee species can congregate on the same site. We also report for the first time the substrates used by stingless bee males for resting at night, that at least one species forms large sleeping roosts composed of hundreds of individuals, and that sleeping roost locations are not reused on subsequent nights.

Introduction

Male bees (Hymenoptera: Apoidea) patrol nests containing receptive females or aggregate at rendezvous sites in order to find sexual partners (Alcock et al., 1978; Eickwort & Ginsberg, 1980; Paxton, 2005). Particularly in the eusocial bees, *i.e.*, honeybees (Apidae: Apini) and stingless bees (Apidae: Meliponini), males can form mating swarms at specific areas, where they wait for virgin queens (Nogueira-Neto, 1954; Kerr et al., 1962; Zmarlicki & Morse 1963; Sommeijer & de Bruijn, 1995; Koeniger & Koeniger, 2000).

Male stingless bees reach sexual maturity about two to three weeks after eclosure, after which they permanently leave their nests (Engels & Engels, 1984; van Veen et al., 1997). They then join mating swarms close to nests containing virgin queens or nests in the process of being founded (Nogueira-Neto, 1954; Kerr et al., 1962; Nogueira-Ferreira & Soares,

1998; van Veen & Sommeijer, 2000). Still, in some stingless bee species (*Melipona*), males can congregate in non-nest associated congregation sites such as concrete retaining wall or beekeepers cabin which are periodically visited by workers and queens (Sommeijer & de Bruijn, 1995; Sommeijer et al., 2003; Cortopassi-Laurino, 2007). Apart from these facts little is known about the reproductive behaviour of male stingless bees. Nothing is known about where males rest overnight after visiting such reproductive aggregations, although males of solitary bees are known to frequently spend the night on plants, alone or in groups under leaves (Kaiser, 1995; Alcock, 1998; Alves-dos-Santos et al., 2009). Furthermore, we know that solitary bee males may exhibit site fidelity by returning to a particular sleeping roost on the successive nights (Kaiser, 1995; Alcock, 1998; Alves-dos-Santos, 1999; Oliveira & Castro, 2002; Alves-dos-Santos et al., 2009).

The present short note contributes to the knowledge

of reproductive behaviour (mating swarms) and sleeping roosts of males of some stingless bee species. As observations of such events are rare and have low predictability, we put together several sporadic observations we carried out during the course of other studies on stingless bee biology.

Non-nest associated congregation sites

In a meliponary located in the city of Pilar do Sul, São Paulo state, in October 2009, males of *Melipona quadrifasciata quadrifasciata* Lepeletier and *Plebeia droryana* (Friese) were observed aggregating at a congregation site, the leaves and twigs of an Oriental Raisin tree *Hovenia dulcis* (Rhamnaceae), approximately 4-5 m above ground. This congregation of males was observed on four consecutive days.

Males of both species periodically joined and departed the congregation site between about 08:00h and about 17:00 h. The congregation was formed of dozens of males of both *M. q. quadrifasciata* and *P. droryana*. However, it is unclear what the role of the frequent joining and leaving of individual males was. Near the congregation site several phorid flies (Diptera: Phoridae) were sitting on or near the branch used by *M. q. quadrifasciata* males. Such parasitic flies may have an important role in the mortality of stingless bee males (Simões et al., 1980; Brown, 1997; Sommeijer et al., 2003). The visit of some *M. quadrifasciata* workers carrying white resin on their corbiculae was noted, but no interactions between them and the males were observed. No virgin queens were observed at the congregation site. It remains uncertain why males of *M. q. quadrifasciata* and *P. droryana* chose to congregate at the same site.

Another male congregation site was observed in the city of Mossoró, state of Rio Grande do Norte, for one day in March 2009. *Melipona subnitida* (Ducke) males gathered on a branch of a cashew tree *Anacardium occidentale* (Anacardiaceae). This congregation occurred in a meliponary containing approximately 90 hives of this species, and was observed between 10:00 and 17:00h. The males were either sitting in the congregation site or flying close to it (Figure 1A) and received visits from a wasp (*Polybia* sp.) with which they performed trophallaxis (Figure 1B). One virgin queen with a distended abdomen was also observed at the congregation. However, this queen did not attract any males during her stay. The queen remained on the branch but in a very agitated manner, tightly circling a small point on the branch for approximately 20 minutes, and then flew away. The reason for the presence of virgin queens in non-nest associated congregation sites is unclear because their presence does not result in copulation attempts by the males (Sommeijer & de Bruijn, 1995; Sommeijer et al., 2003). Occasionally, worker bees were observed flying over the congregation site, landing briefly and disappearing afterwards. The presence of workers at some congregation sites of stingless bee males appears to be common; they perform trophallaxis with the males (Cortopassi-Laurino, 2007) or carry resins on their corbiculae to be distributed over the congregation site (Sommeijer et al., 2003). However, the adaptive significance of the behaviour of such workers at the congregation site remains unclear.

Sleeping roosts

Overnight aggregations of bee males have been described previously for solitary bee species that exhibit site fidelity by returning to a particular sleeping roost on the successive nights (Kaiser, 1995; Alcock, 1998; Alves-dos-Santos, 1999; Oliveira & Castro, 2002; Alves-dos-Santos et al., 2009). At Ribeirão Preto, the meliponary garden of the FFCLRP was daily inspected from June 03 to September 05 2010 for sleeping roosts between 17:00 h and 07:00 h. This meliponary contained approximately 150 colonies, including *Scaptotrigona depilis* (Moure), *Tetragonisca angustula* (Latreille), *Frieseomelitta varia* (Lepeletier), *M. scutellaris* (Latreille), *M. quadrifasciata* and *Nannotrigona testaceicornis* (Lepeletier). We searched for sleeping roosts on branches below 2 m. When a sleeping roost was found, the arrangement of the individuals was recorded and the location

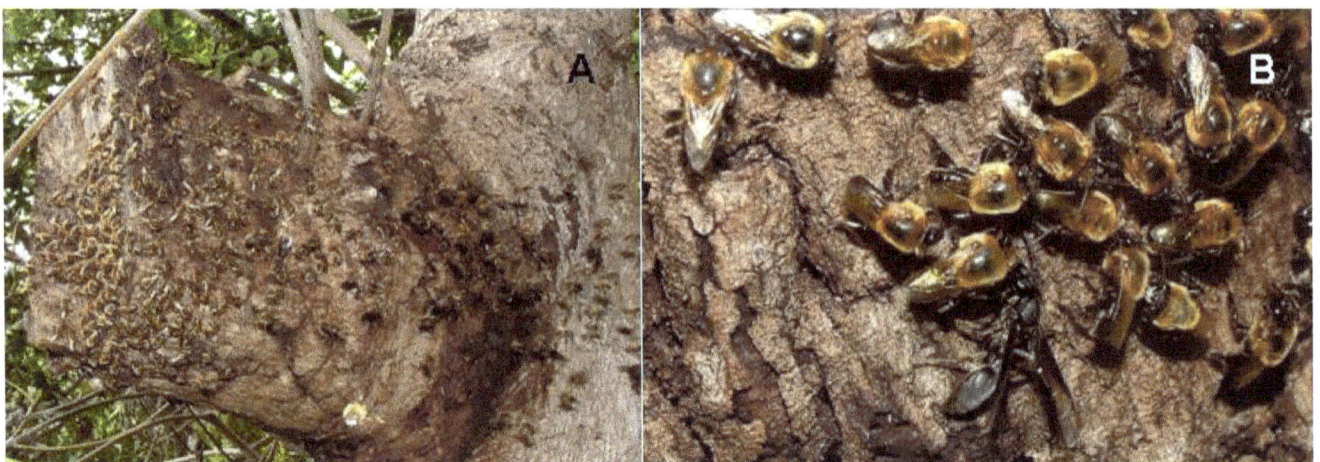

Fig. 1 (A) Congregation site of *Melipona subnitida* males on a branch of *Anacardium occidentale* (Anacardiaceae) at Mossoró, state of Rio Grande do Norte; (B) *Polybia* sp. wasp visiting the congregation.

Table 1. Features of the sleeping roosts for male stingless bees (Hymenoptera: Meliponini).

Stingless bee species	Substrates	plant species or locations	number of males	aggregation type
Scaptotrigona aff. depilis[1]	tree branch	Ocimum basilicum Lamiaceae)	3	solitary
	tree branch	Citrus limonia (Rutaceae)	5	solitary
	tree branch	Cosmos sp. (Asteraceae)	3	solitary
Melipona scutellaris[2]	flower petal	Montanoa pyramidata (Asteraceae)	1	solitary
Frieseomelitta varia[3]	metal wire	metal wire	± 250	Cluster

[1] - November 19, 2010; [2] - June 03, 2010; [3] - March 2011.

was marked and monitored during subsequent nights so as to observe whether the males returned to it. We found *S. depilis* males at FFCLRP sleeping on branches of *Ocimum basilicum* (Lamiaceae), *Citrus limonia* (Rutaceae) and *Cosmos* sp. (Asteraceae) (Figure 2A). Each of these plants was used as sleeping roosts by three to five males, but they remained separated from each other (Table 1). None of roosts were re-used in subsequent nights. A single *M. scutellaris* (Latreille) male was also found sleeping on a daisy flower (*Montanoa pyramidata*: Asteraceae) (Figure 2B).

Additionally, in a private meliponary approximately 1.2 km from the FFCLRP-USP campus, a sleeping roost of *F. varia* males was observed in March 2011. There were three hives of this species on the site. The sleeping roost comprised approximately 250 *F. varia* males on any one night. The males were sitting on a metal wire (Table 1, Figure 2C) approximately 2 m from one of the three colonies of this species at the site. They were first observed at 22:00 h, and again at 01:30 h.

The *F. varia* males remained densely grouped. Occasionally, some males moved their hind legs or walked slowly on the wire and even over other males. They did not react to contact with entomological tweezers during the collection of some individuals. On the following morning at approximately 08:00 h, some males were still at the sleeping roost site, but they gradually left the site over the course of the morning. Neither additional aggregations nor other sleeping roosts were reported thereafter.

To our knowledge these are the first report of sleeping roosts of stingless bee males. The causes that led males of *S. depilis*, *M. scutellaris* and *F. varia* to choose their sleeping roosts remain unclear. There is still much to learn about the biology of stingless bee males and this report will help other researchers to find and study them.

Acknowledgements

The authors thank Dr. Silvia R. M. Pedro and Dr. Sidnei Mateus (Faculdade de Filosofia, Ciências e Letras de Ribeirão Preto, USP) for the identification of *Plebeia droryana* and and *Polybia* sp., respectively. Financial support received: CFS and AVN from the Coordenação de Aperfeiçoamento de Pessoal de Nível Superior (CAPES); CM from the Fundação de Amparo à Pesquisa do Estado de São Paulo (FAPESP, process 07–50218-1); VLIF from the PVNS scholarship from CAPES. The authors also thank the CAPES – PROAP 2010 program for field trip financial support for CFS. Authors gratefully acknowledge Rodolfo R. Jaffé (BeeLab USP) for his critical reading of the manuscript and suggestions.

References

Alcock, J., Barrows, E.M., Gordh, G., Hubbard, L.J., Kirkendall, L., Pyle, D.W., Ponder, T.L. & Zalom, K.G. (1978). The ecology and evolution of male reproductive behaviour in the bees and wasps. Zoological Journal of the Linnean. Society, 64: 293-326. doi: 10.1111/j.1096-3642.1978.tb01075.x

Alcock, J. (1998). Sleeping aggregations of the bee

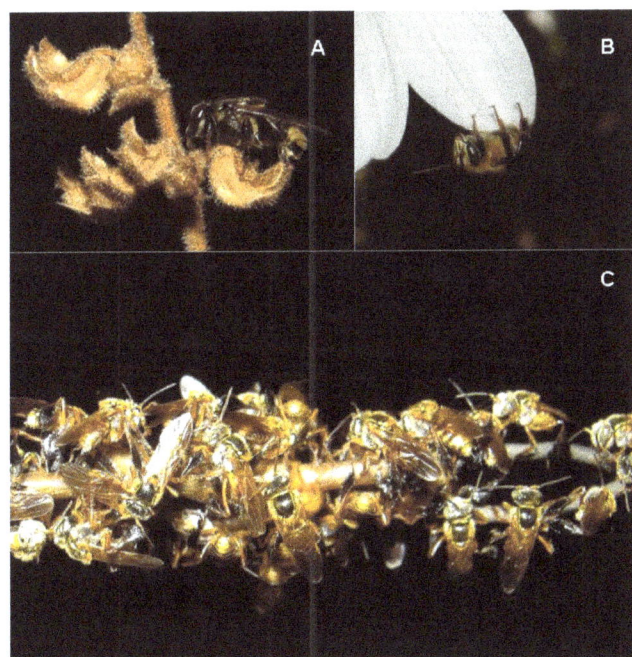

Fig. 2 Sleeping roost of males of: (A) *Scaptotrigona* aff. *depilis*; (B) *Melipona scutellaris*; (C) *Frieseomelitta varia*.

Idiomelissodes duplocincta (Cockerell) (Hymenoptera: Anthophorini) and their possible function. Journal of the Kansas Entomological Society, 71: 74–84.

Alves-dos-Santos, I. (1999). Aspectos morfológicos e comportamentais dos machos de *Ancyloscelis* Latreille (Anthophoridae, Apoidea). Revista Brasileira de Zoologia, 16, (supl. 2): 37–43. doi: 10.1590/S0101-81751999000600005

Alves-dos-Santos, I., Gaglianone, M.C., Naxara, S.R.C., Engel, M.S. (2009). Male sleeping aggregations of solitary oil-collecting bees in Brazil (Centridini, Tapinotaspidini, and Tetrapediini; Hymenoptera: Apidae). Genetics and Molecular Research, 8: 515-524.

Brown, B.V. (1997). Parasitic phorid flies: a previously unrecognized cost to aggregation behavior of male stingless bees. Biotropica, 29: 370-372. doi: 10.1111/j.1744-7429.1997.tb00439.x

Cortopassi-Laurino, M. (2007). Drone congregations in Meliponini: what do they tell us? Bioscience Journal, 23: (supppl. 1): 153–160.

Eickwort, G.C., Ginsberg, H.S. (1980). Foraging and mating behavior in Apoidea. Annual Review of Entomology, 25: 421-446. doi: 10.1146/annurev.en.25.010180.002225

Engels, E. & Engels, W. (1984). Drohnen-Ansammlungen bei Nestern der stachellosen Biene *Scaptotrigona postica*. Apidologie, 15: 315–328. doi: 10.1051/apido:19840304

Kaiser, W. (1995). Rest at night in some solitary bees – a comparison with the sleep-like state of honey bees. Apidologie, 26: 213–230. doi: 10.1051/apido:19950304

Kerr, W.E., Zucchi R., Nakadaira, J.T., Butolo, J.E. (1962). Reproduction in the social bees (Hymenoptera: Apidae). Journal of the New York Entomological Society, 70: 265-276. jstor.org/stable/25005835

Koeniger, N. & Koeniger, G. (2000). Reproductive isolation among species of the genus *Apis*. Apidologie, 31: 313-339. doi: 10.1051/apido:2000125

Nogueira-Ferreira, F.H. & Soares, A.E.E. (1998). Male aggregations and mating flight in *Tetragonisca angustula* (Hymenoptera, Apidae, Meliponinae). Iheringia, Série Zoologia, 84: 141-144.

Nogueira-Neto, P. (1954). Notas bionômicas sobre meliponíneos III – sobre a enxameagem. Arquivos do Museu Nacional, 42: 419-452.

Oliveira, F.F. & Castro, M.S. (2002). Nota sobre o comportamento de agregação de machos de *Oxaea austera* Gerstaecker (Hymenoptera, Apoidea, Oxaeinae) na caatinga do Estado da Bahia. Brasil. Revista Brasileira de Zoologia, 19: 301–303.

Paxton, R.J. (2005). Male mating behaviour and mating systems of bees: an overview. Apidologie, 36: 145-156. doi: 10.1051/apido:2005007

Simões, D., Bego, L.R., Zucchi, R., Sakagami, S.F. 1980. *Melaloncha sinistra* Borgmeier, an endoparasitic phorid fly attacking *Nannotrigona* (*Scaptotrigona*) *postica* Latr. (Hym., Meliponinae). Revista Brasileira de Entomologia, 24: 137-142.

Sommeijer, M.J. & de Bruijn, L.L.M. (1995). Drone congregations apart from the nest in *Melipona favosa*. Insectes Sociaux, 42: 123-127. doi: 10.1007/BF01242448

Sommeijer, M.J., de Bruijn, L.L.M., Meeuwsen, F.J.A.J. (2003). Behaviour of males, gynes and workers at drone congregation sites of the stingless bee *Melipona favosa* (Apidae: Meliponini). Entomologische Berichten, 64: 10-15.

van Veen, J.W., Sommeijer, M.J., Meeuwsen, F. (1997). Behaviour of drones in *Melipona* (Apidae, Meliponinae). Insectes Sociaux, 44: 435–447. doi: 10.1007/s000400050063

van Veen, J.W. & Sommeijer, M.J. (2000). Observations on gynes and drones around nuptial flights in the stingless bees *Tetragonisca angustula* and *Melipona beecheii* (Hymenoptera, Apidae, Meliponinae). Apidologie, 31: 47–54. doi: 10.1051/apido:2000105

Zmarlicki, C. & Morse, R.A. (1963). Drone congregation areas. Journal of Apicultural Research., 2: 64-66.

Species Composition of Termites (Isoptera) in Different Cerrado Vegetation Physiognomies

DE Oliveira[1], TF Carrijo[2], D Brandão[3]

1 - Universidade de Brasília, Brasília – DF, Brazil

2 - Universidade de São Paulo, Ribeirão Preto – SP, Brazil

3 - Universidade Federal de Goiás, Goiânia – GO, Brazil

Keywords

termite assemblage, biodiversity, forest, savanna, feeding guilds

Corresponding author

Danilo Elias de Oliveira
Laboratório de Termitologia
Departamento de Zoologia
Instituto de Biologia
Universidade de Brasília
Brasília, DF, Brazil
70910-900
E-Mail: daniloelo@gmail.com

Abstract

Little is known about the termite fauna of the different vegetation physiognomies in the Cerrado biome. It is suggested that the species compositions in grassland and savanna areas are closely related to each other, and quite distinct from those of forests. This study compared the species composition from five different physiognomies of Cerrado, and tested the hypothesis that the termite faunas of savannas and grasslands form a distinct group from that of forests. The study was conducted in the Parque Estadual da Serra de Jaraguá, state of Goiás, Brazil. Termites were sampled from two physiognomies of savanna, one natural grassland, one pasture, and one gallery forest. A transect with 10 parcels of 5x2 m was established in each physiognomy. The relative abundance was inferred by the number of encounters, termites were classified in feeding guilds, and the dissimilarity in the species composition between the physiognomies was calculated. A total of 219 encounters, of 42 species of two families were recorded. The most abundant feeding guilds were the humivores (98) and xylophages (55). The physiognomies with the largest number of species were rupestrian cerrado (23 species) and cerrado *sensu stricto* (21). The physiognomies had a similar species composition (less than 55% dissimilarity), mainly the natural open areas. The hypothesis of a distinct fauna of termites in forest vegetation was refuted. The termite fauna of gallery forest is very different from that of pasture, but most species also occur in natural open areas. The impact of pasture on the diversity and composition of termites seems to be significant, but the impact is even greater on the proportion of the feeding guilds, reducing the proportion of xylophages and intermediates.

Introduction

Termites (Isoptera) are one of the most abundant soil invertebrates in tropical ecosystems (Wilson, 1971; Wood & Sands, 1978; Eggleton et al., 1996) and the most important soil ecosystem engineers of these environments (Bignell, 2006; Jouquet et al., 2011). In some arid and semi-arid tropical savannas, during the dry season, termites are the only active group of invertebrates able to decompose organic matter (Jouquet et al., 2011) and provide ecosystem services such as soil formation and aeration (Lavelle et al., 2006). Most species of termites are tropical, and among more than 2800 described

species, approximately 500 occur in the Neotropical region (Kambhampati & Eggleton, 2000; Constantino, 2012). With about 150 species, 50% of them endemic (Constantino, 2005), the Cerrado biome in central Brazil probably harbors the most diverse and highly endemic savanna termite fauna (Domingos et al., 1986).

The Cerrado is the largest and richest tropical savanna in the world, and is also probably the most threatened one (Silva & Bates, 2002). Ranked among the 25 most important biodiversity hotspots (Myers et al., 2000), the Cerrado is the second most extensive biome in Brazil (Klink & Machado, 2005), and the most strongly affected by human disturbance.

During the past 50-60 years, with the evolution of agricultural techniques, this savanna also has become the largest agricultural frontier in Brazil (Sano et al., 2010). The latest analysis of the conservation status of the Cerrado showed that only 20% of the original area remained relatively undisturbed, and only 6.2% was preserved in protected areas (Myers et al., 2000).

The Cerrado biome is a mosaic of 10 different kinds of vegetation formations (physiognomies) (Ribeiro & Walter, 1998). The development of these physiognomies depends on the soil, groundwater, and other environmental variables. The physiognomies can be distinguished from one another mainly by the species composition, richness pattern, and relative abundance of plant species (Eiten, 1983). The termite faunas of most of these habitats are still little known, and for some of them no published information is available.

The majority of termite surveys in the Cerrado biome have been conducted in the cerrado *sensu stricto* physiognomy (Mathews, 1977; Coles, 1980; Brandão & Souza, 1998; Constantino & Schlemmermeyer, 2000; Constantino, 2005; Cunha et al., 2006; Carrijo et al., 2009). Some surveys have also described local faunas in the cerradão (Mathews, 1977; Constantino & Schlemmermeyer, 2000; Cunha et al., 2006), gallery forest (Mathews, 1977; Constantino & Schlemmermeyer, 2000; Cunha et al., 2006), and interfluve mesophytic forest (Brandão & Souza, 1998; Cunha et al., 2006).

The diversity of animals is often linked to the type of vegetation where they live or breed, and often differs between forests and grassy (grasslands and savannas) environments (Bond & Parr, 2010); termites can be expected to follow these patterns. Constantino (2005) remarked that in the Cerrado biome there are two termite faunas, one closely related to those of grasslands and savannas, and another one, quite distinct, related to forests. Constantino (2005) also suggested that the species composition of termites from Cerrado forests resembles faunas of the Amazon and Atlantic Forest biomes. Also, Mathews (1977) observed a gradient in the species composition of termites that build epigeal (above-ground) nests, from grasslands and savannas to forests.

The present study aimed to 1) compare the species composition, richness, and abundance of termites in five Cerrado physiognomies including three types of savanna (ranging from shortgrass savanna to woodland), one forest, and one pasture, all situated in the Parque Estadual da Serra de Jaraguá, Goiás, Brazil; and 2) test if there are two different groups of termite species related to the vegetation in the Cerrado biome: one in open areas (grasslands and savannas), and another in forests.

Material and Methods

This study was carried out at the Parque Estadual da Serra de Jaraguá, municipality of Jaraguá, state of Goiás,

Brazil (15.75° to 15.85° S and 49.27° to 49.37° W). The park covers an area of about 2,860 ha, ranging in altitude from 640 to 1,140 m. The predominant soil type is Litolic Neosol, and the Köppen-Geiger climate type is "Aw" (Agência Ambiental, 2004). The park is surrounded by cattle grazing land and is mostly covered by cerrado *sensu stricto* (a vegetation type composed of tropical xeromorphic semideciduous broadleaf trees and scrub woodland of cerrado), which is more or less dense depending on the location. At higher altitudes (above 1,000 m) some rupestrian cerrado is present (tropical rupestrian semideciduous broadleaf trees and scrub woodland on rocky soil). Also, gallery forests (tropical mesophytic semideciduous broadleaf gallery forest) extend along the permanent water courses. Near the border of the park are some open shortgrass savannas (tropical xeromorphic, semideciduous broadleaf scrub with seasonal shortgrass).

Termites were sampled in these four physiognomies, and in a cattle pasture that was described by Carrijo et al. (2009) as "a 7-year-old pasture cultivated with *Brachiaria brizantha* Stapf. (Poaceae), originally covered by cerrado *sensu stricto* and cleared and turned into pasture approximately 50 years ago". We used the same methods as Carrijo et al. (2009), that is: in each physiognomy we established one linear transect with 10 plots of 5 x 2 m and 2 m height. Each plot was 30 m distant from the next one, and the transect was at least 50 m distant from the border of the particular physiognomy. The plots were carefully searched for termites in all possible places where these insects can be found: epigean nests, litter, plant stalks, fallen branches, soil surface, and to a depth of 20 cm below ground. Each plot was sampled by two collectors during 30 min. A total of 50 plots were sampled, comprising an area of 500 m².

Because of the difficulty of estimating the number of colonies (relative abundance) with this protocol, all individuals of the same species that were collected in the same plot were considered as one encounter. Thus, the maximum abundance of each species in each plot is 1 (one), and for all plots combined, 50 encounters (following Bignell and Eggleton, 2000).

Samples from all colonies were identified using appropriate literature and/or comparing with the Termitological Collection of the Universidade Federal de Goiás (UFG), where the vouchers were deposited. Species were classified in four functional groups (feeding guilds) according to field observations and literature information (Mathews, 1977; Gontijo & Domingos, 1991; DeSouza & Brown, 1994): xylophages, humivores, grass/litter feeders and intermediates.

The statistical analyses were performed with the software R (R Development Core Team, 2010), using the packages 'vegan' (Oksanen et al., 2011) and 'vcd' (Meyer et al., 2006, 2011). The sampling effort for each physiognomy was evaluated using a species accumulation curve, constructed using the function 'specaccum'. The number of species of termites was estimated by the estimators Jackknife 1 and Bootstrap, which are two of the best-known richness estimators (Colwell & Coddington, 1994; Magurran, 2004); for this, we used the function 'poolaccum'. A Pearson Chi-square test was performed between the number of encounters observed for each feeding guild, and that estimated based on a Poisson distribution; the results of this test were presented in a Cohen-Friendly association plot, generated by the function 'assoc' (Meyer et al., 2006). The dissimilarity in species composition between the five physiognomies was assessed using the Chao-Jaccard index (Chao et al., 2005), with the function 'vegdist', 1,000 permutations and method 'chao'. The cluster was performed with the function 'hclust' and method 'average'.

Results

In the five physiognomies, a total of 42 species and 27 genera of termites from the families Termitidae and Rhinotermitidae were collected (Table 1). Rhinotermitidae was represented by two species, *Coptotermes* sp. and *Heterotermes tenuis*. Of the four subfamilies of Termitidae, Apicotermitinae was the richest, with 15 species, followed by Nasutitermitinae, with nine species. A slight tendency toward stabilization can be observed in the species accumulation curve (Fig 1). The number of observed species was closer to the Bootstrap estimate (48 species) than to the Jackknife 1 estimate (58). The sampling completeness was 87.5% if we use the first estimator, and 72.4% for the latter (Fig 1).

The savannas (cerrado *sensu stricto* and rupestrian cerrado) were the physiognomies with higher numbers of species, followed by the grasslands (shortgrass savanna and pasture), and the gallery forest, respectively (Fig 2). None of

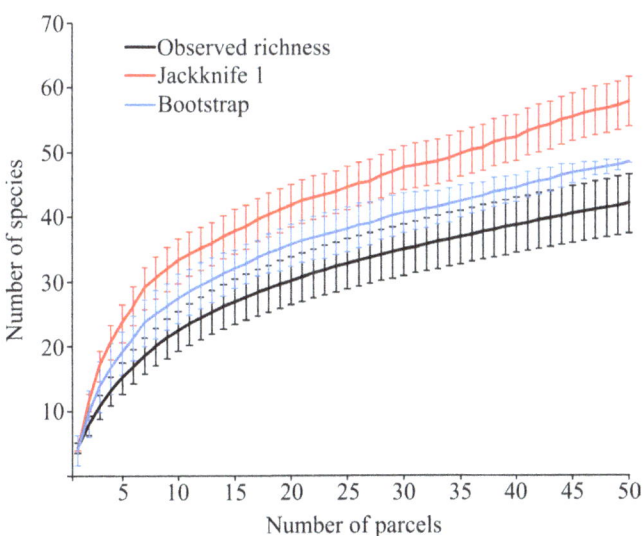

Figure 1 Accumulation curves with the observed termite richness and the estimates from Bootstrap and Jackknife 1 of five physiognomies pooled together from the Parque Estadual da Serra de Jaraguá. Vertical lines are the standard deviations.

the accumulation curves from the five physiognomies showed a stabilization tendency, which means that the numbers of species are certainly higher (Fig 2).

The relative abundance of termites in all physiognomies was 219 encounters, with 5.214 ± 5.542 (average $\pm$ standard deviation) encounters per species (Table 1). However, a strong dominance of some species was found. The most common species, *Amitermes* sp., *Anoplotermes* sp. 3 and *Nasutitermes* sp. 1, had 23, 21 and 17 encounters, respectively, while 16 species were found only once, comprising 38% of the total pool of species surveyed. The subfamily Apicotermitinae was the most abundant both in absolute number (90 encounters) and in number of encounters per species (six encounters per species). Rhinotermitidae was the least abundant by both measurements: six in absolute number, and three encounters per species.

The cerrado *sensu stricto* showed the highest abundance of termites, both absolute (61), and per species (2.9); while the gallery forest had the lowest absolute abundance (25), and the shortgrass savanna had the lowest abundance per species (1.8). The most abundant species in the savannas (cerrado *sensu stricto* and rupestrian cerrado) was *Armitermes* sp., while *Anoplotermes* sp. 3 was the most abundant in the pasture and gallery forest, and *Nasutitermes* sp. 1 in the shortgrass savanna (Table 1).

Considering all physiognomies together, the most abundant and diverse guild was the humivores (98 encounters and 20 species) followed, in number of encounters, by the xylophages (55 encounters and six species) and, in number of species, by the intermediates (25 encounters and nine species). The grass/litter feeders had 41 encounters and seven species. Humivorous and xylophagous species were the most abundant in all the physiognomies considered separately; except in the pasture, where the grass/litter feeders had 18 encounters, against three of xylophagous species. The proportion of feeding guilds in cerrado *sensu stricto* and rupestrian cerrado was similar, with more humivores and xylophages, but also with a high proportion of intermediates, compared to the other physiognomies (Fig 3).

The Pearson Chi-square test showed that the shortgrass savanna had a lower proportion of grass/litter feeders than that expected by chance (Fig 3). The gallery forest showed a higher proportion of humivores, while the proportions of the other guilds were lower than those expected by chance. The pasture was the most singular physiognomy; the proportions of humivores and, mainly of grass/litter feeders were much higher, while those of intermediates and xylophages were much lower than those expected by chance (Fig 3).

Regarding the species composition, all physiognomies were very similar to each other (Fig 4). All the open areas had a dissimilarity index less than 0.5. The gallery forest and pasture were the most dissimilar areas (Chao-Jaccard Index = 0.8426). In fact, only three species were shared between gallery forest and pasture, and two of them (*Anoplotermes*

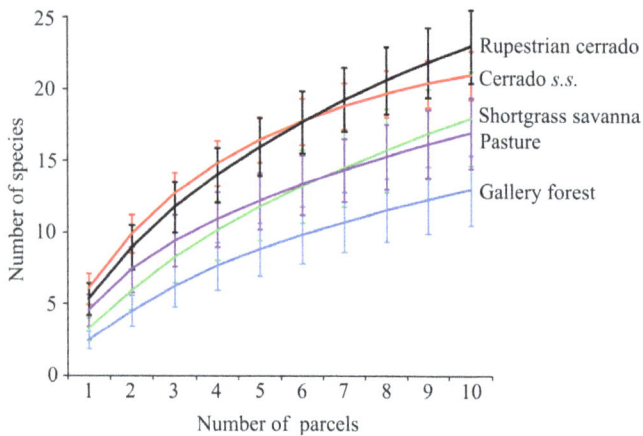

Fig 2 Accumulation curves with the observed termite richness for each physiognomy present in the Parque Estadual da Serra de Jaraguá. Vertical lines are the standard deviations.

sp. 3 and *Nasutitermes* sp. 1) were the only species that were present in all physiognomies.

Discussion

The species richness at the Parque Estadual da Serra de Jaraguá was as expected, as the local richness of termites in the Cerrado biome is about 40 to 60 species (Constantino, 2005), ranging from 15 (Cunha et al., 2006) to 114 (Mathews, 1977). In previous surveys in the Cerrado, the subfamily Nasutitermitinae was frequently the richest and most common group, but in these studies the species of Apicotermitinae were frequently pooled together due to limitations of identification (Constantino, 2005; Cunha et al., 2006). The true richness and abundance of Apicotermitinae is expected to be very high (Eggleton, 1999), which was the case in the present study. However, most previous studies in the Cerrado cannot be properly compared, since the protocols used are not standardized.

The gallery forest had the second smallest proportion of xylophages (6 species of a total of 25). This was an unexpected result, since the amount of wood is expected to be higher in forests than in the other physiognomies, and, consequently, more xylophagous species were expected in this environment. The reason is probably that, in the savannas, most of the wood needed by the xylophages is in shrubs, below two meters (the maximum height collectable by the protocol used), while in the forest the xylophages are situated much higher, in the canopy, and out of reach of our collections. In forests, a vertical gradient of termite species occurs and the xylophagous species found on the ground are only a subsample of those on the canopy (Roisin et al., 2006).

The grass/litter feeders were the most abundant termites in the pasture. This was expected, since grass is the most abundant food resource for termites in this environment. Moreover, litter-feeding species occur in higher proportions in

Cerrado than in forest biomes (Constantino & Acioli, 2008). However, the high proportion of grass/litter feeders led to a skewed pattern in the feeding-guild proportions. Compared to the other physiognomies, the pasture had fewer xylophages and intermediates, and many more grass/litter feeders. This pattern was not as evident for the shortgrass savanna, the other

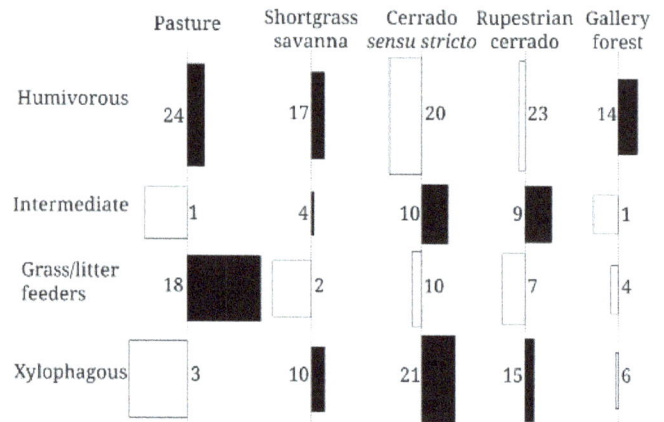

Fig 3 Cohen-Friendly association plot for the alimentary guilds of termites for each physiognomy of the Parque Estadual da Serra de Jaraguá. The bar length is proportional to the contribution to the Pearson Chi-Squared; and the bar area is proportional to the difference between the observed and expected frequencies. The independence values for each physiognomy are the vertical dashed lines. The black bars, to the right, are the values higher than expected by chance, and the white bars, to the left, are the values lower than the expected.

grassland studied, which suggests that this is linked to human activity, turning cerrado *sensu stricto* into pasture, and not only the coverage of grass. The effects of pasture formation on termite communities are well known and include a reduction of richness and abundance of humivores and mostly xylophages, leading to high rates of local extinctions and an increase in the abundance of grass-feeders, which transforms some species into aesthetic pests (Brandão & Souza, 1998; Constantino, 2002; Cunha & Orlando, 2011).

The termite fauna studied here had a very similar composition between all physiognomies. Although previous studies have concluded that each physiognomy has a different termite species composition (e.g., Constantino & Acioli, 2008), the natural open areas (rupestrian cerrado + cerrado *sensu stricto* + shortgrass savanna) investigated here have about the same termite species pool. Mathews (1977) also found a gradient of termite species from open areas to forests, with a high number of shared species of two or more physiognomies. Eiten (1983) observed a continuum of plant species between cerrado *sensu stricto* and rupestrian cerrado. In an inventory of plant species of the Parque Estadual da Serra de Jaraguá (unpublished data), the present authors found a continuum of plant species between all the natural open areas, with some exclusive species in each physiognomy.

The gallery forest and the pasture each had a distinct

Table 1 List of the number of termite encounters in different physiognomies of Cerrado in Parque Estadual da Serra de Jaraguá, state of Goiás, Brazil. (Feeding group X – xylophagous; H – humivorous; G – grass/ litter feeder; I – intermediate).

Family/subfamily/species	Feeding group	Pasture	Shortgrass savanna	Cerrado *sensu stricto*	Rupestrian cerrado	Gallery forest
Rhinotermitidae						
Coptotermes sp.	X	0	0	1	0	0
Heterotermes tenuis	X	0	2	1	0	2
Termitidae: Apicotermitinae						
Anoplotermes sp.1	H	0	3	5	5	3
Anoplotermes sp.2	H	1	1	2	1	0
Anoplotermes sp.3	H	7	4	4	2	4
Anoplotermes sp.4	H	0	3	3	2	4
Anoplotermes sp.5	H	0	1	2	1	1
Anoplotermes sp.6	H	6	1	0	2	0
Anoplotermes sp.7	H	3	2	0	0	0
Anoplotermes sp.8	H	1	0	0	0	0
Anoplotermes sp.9	H	0	1	0	0	0
Aparatermes sp.	H	1	0	1	0	1
Grigiotermes sp.1	H	1	0	0	0	0
Grigiotermes sp.2	H	1	0	0	0	0
Grigiotermes sp.3	H	0	0	0	1	0
Ruptitermes sp.	G	6	0	1	0	0
Tetimatermes sp.	H	0	0	1	0	0
Termitidae: Nasutitermitinae						
Agnathotermes sp.	H	1	0	0	0	0
Anhangatermes sp.	H	0	0	0	1	0
Atlantitermes stercophilus	I	0	1	0	0	0
Nasutitermes sp.1	X	1	5	4	6	1
Nasutitermes sp.2	X	0	0	5	1	2
Nasutitermes sp.3	X	0	0	0	0	1
Parvitermes bacchanalis	G	0	0	1	0	0
Subulitermes sp.	H	0	0	0	0	1
Velocitermes heteropterus	G	0	1	5	1	1
Termitidae: Syntermitinae						
Silvestriermes euamignathus	I	1	0	2	3	0
Cornitermes silvestrii	G	2	0	3	3	0
Cornitermes villosus	G	0	0	0	0	3
Curvitermes minor	I	0	1	0	1	0
Cyrilliotermes cupim	H	0	1	0	0	0
Embiratermes festivellus	I	0	0	3	2	0
Labiotermes emersoni	H	1	0	2	6	0
Procornitermes araujoi	G	4	0	0	2	0
Syntermes nanus	G	6	1	0	1	0
Termitidae: Termitinae						
Amitermes sp.	X	2	3	10	8	0
Dihoplotermes cf. *inusitatus*	I	0	0	0	1	0
Neocapritermes araguaia	I	0	1	2	1	0
Neocapritermes opacus	I	0	1	0	0	0
Neocapritermes talpa	I	0	0	0	0	1
Spinitermes trispinosus	H	0	0	0	2	0
Termes sp.	I	0	0	3	1	0
Encounters (total = 219)		45	33	61	54	26
Species richness (total = 42)		17	18	21	23	13

subsample of termite species from the natural open areas, plus some exclusive species. For example, of all the termite species found in the gallery forest, four were exclusive to this physiognomy and the other nine species were also found in cerrado *sensu stricto*. In spite of their high overall similarity, the pasture and gallery forest had very different species compositions, with only three shared species. This is an expected result, since the vegetation structure in gallery forest is very different from pasture. In gallery forest there is an abundant amount of wood and a comparatively small stock of grass, and the opposite in pasture. In fact, two species of xylophages were found for each species of grass/litter feeders termites in the gallery forest. In pasture the opposite occurred, with two species of grass/litter feeders found for each species of xylophagous termites.

Although the gallery forest had a different species composition pool, there is no evidence in the present study to support the hypothesis that this physiognomy hosts species from the Amazon and Atlantic Forest biomes. As mentioned above, the species pool of the gallery forest is a subsample of that in cerrado *sensu stricto* plus some exclusive species (*Nasutitermes* sp. 1, *Subulitermes* sp. 1, *Cornitermes villosus* and *Neocapritermes talpa*). Of these species, *Cornitermes villosus* only occurs in forests of the Cerrado biome; *Nasutitermes* and *Subulitermes* have species occurring in open areas of Cerrado; and only *Neocapritermes talpa* occurs in both the Amazon and Cerrado forests.

In the last decade, different physiognomies have undergone different rates of deforestation, and those with denser vegetation (i.e., forests and savannas) have been most affected (Rocha et al., 2011). It is alarming that gallery forests, which have a relatively depauperate but little-known termite fauna, are among the physiognomies that are most affected by deforestation.

This study showed that termites have a high number of species and abundance in Cerrado areas, mainly in the savanna physiognomies. The physiognomies have a similar species composition, mainly the natural open areas, and there is no support in the present study for the hypothesis of a distinct fauna of termites in forest. As expected, the termite fauna of gallery forest is very different from pasture, but most of the species also occur in natural open areas. The implantation of pasture seems to impact the termite diversity and, mainly, the species composition and proportions of the feeding guilds. However, this statement should be reinforced by future studies, with a greater sampling effort.

Acknowledgements

We thank our friends Diogo A. Costa and Thiago Santos for their help in discussing the project and also for their essential help in the fieldwork. This study was supported by the Conselho Nacional de Desenvolvimento Científico e Tecnológico, Brazil, and the Coordenação de Aperfeiçoamento de Pessoal de Nível Superior, CAPES, Brazil. The Secretaria do Meio Ambiente e dos Recursos Hídricos, SEMARH, Brazil, for granting a collecting permit. We also thank two anonymous reviewers for valuable comments that improved this article.

References

Agência Ambiental, G. (2004). Proposta de delimitação e reavaliação da categoria de Parque Ecológico da Serra de Jaraguá. Goiânia.

Bignell, D.E. (2006). Termites as Soil Engineers and Soil Processors. In H. König & A. Varma (Eds.), Intestinal Microorganisms of Termites and Other Invertebrates (pp. 183–220). Berlin: Springer.

Bignell, D.E., & Eggleton, P. (2000). Termites in ecosystems. In T. Abe, D.E. Bignell, & M. Higashi (Eds.), Termites: Evolution, Sociality, Symbioses, Ecology (pp. 363–387). Dordrecht: Kluwer Academic Publishers.

Bond, W.J., & Parr, C.L. (2010). Beyond the forest edge: Ecology, diversity and conservation of the grassy biomes. Biol. Conserv., 143: 2395–2404. doi: 10.1016/j.biocon.2009.12.012

Brandão, D., & Souza, R.F. (1998). Effects of deforestation and implantation of pastures on the fauna in the Brazilian "Cerrado" region. J. Trop. Ecol., 39: 19–22.

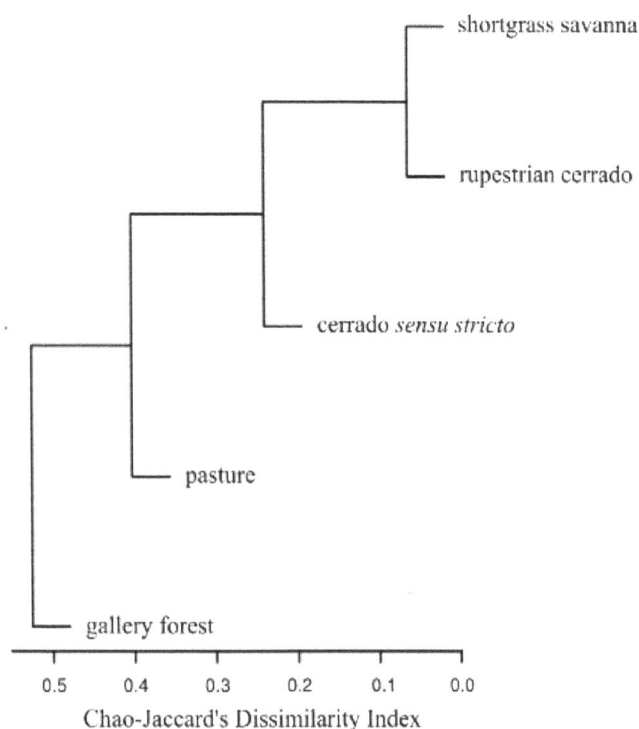

Fig. 4 Dendrogram with the dissimilarity (Chao-Jaccard's Index) between termite species composition for each physiognomy present in the Parque Estadual da Serra de Jaraguá.

Carrijo, T.F., Brandão, D., Oliveira, D.E., Costa, D.A., & Santos, T. (2009). Effects of pasture implantation on the termite (Isoptera) fauna in the Central Brazilian Savanna (Cerrado). J. Insect Conserv., 13: 575–581. doi:10.1007/s10841-008-9205-y

Chao, A., Chazdon, R.L., Colwell, R.K., & Shen, T.J. (2005). A new statistical approach for assessing similarity of species composition with incidence and abundance data. Ecol. Lett., 8: 148-159. doi: 10.1111/j.1461-0248.2004.00707.x

Coles, H.R. (1980). Defensive strategies in the ecology of neotropical termites. University of Southampton.

Colwell, R.K., & Coddington, J.A. (1994). Estimating terrestrial biodiversity through extrapolation. Philos. T. Roy. Soc. B, 345: 101–118. doi: 10.1098/rstb.1994.0091

Constantino, R. (2002). The pest termites of South America: taxonomy, distribution and status. J. Appl. Ent., 126: 355–365. doi: 10.1046/j.1439-0418.2002.00670.x

Constantino, R. (2005). Padrões de diversidade e endemismo de térmitas no bioma Cerrado. In A.O. Scariot, J.C.S. Silva, & J.M. Felfili (Eds.), Cerrado: Ecologia, Biodiversidade e Conservação (pp. 319–333). Brasília: Ministério do Meio Ambiente.

Constantino, R. (2012). On Line Termite Database. Retrieved from http://vsites.unb.br/ib/zoo/catalog.html

Constantino, R., & Acioli, A.N.S. (2008). Diversidade de Cupins (Insecta: Isoptera) no Brasil. In F.M.S. Moreira, J.O. Siqueira, & L. Brussaard (Eds.), Biodiversidade do Solo em Ecossistemas Brasileiros (pp. 278–297). Lavras: UFLA.

Constantino, R., & Schlemmermeyer, T. (2000). Cupins (Insecta: Isoptera). In Fauna Silvestre da Região do Rio Manso - MT (pp. 129–151). Brasília: IBAMA / ELETRONORTE.

Cunha, H.F., Costa, D.A., & Brandão, D. (2006). Termite (Isoptera) assemblages in some regions of the Goiás State, Brazil. Sociobiology, 47: 505–518.

Cunha, H.F., & Orlando, T.Y.S. (2011). Functional composition of termite species in areas of abandoned pasture and in secondary succession of the Parque Estadual Altamiro de Moura Pacheco, Goiás, Brazil. Biosci. J., 27:986–992.

DeSouza, O., & Brown, W.L. (1994). Effects of habitat fragmentation on Amazonian termite communities. J. Trop. Ecol., 10: 197–206.

Domingos, D.J., Cavenaghi, T.M.C.M., & Gontijo, T.A. (1986). Composição em espécies, densidade e aspectos biológicos da fauna de térmitas de cerrado em Sete Lagoas-MG. Ciência e Cultura, 38:199–207.

Eggleton, P. (1999). Termite species description rates and the state of termite taxonomy. Insectes Soc., 46: 1–5. doi: 10.1007/s000400050105

Eggleton, P., Bignell, D.E., Sands, W.A., Mawdsley, N.A., Lawton, J.H., Wood, T.G., & Bignell, N.C. (1996). The Diversity, Abundance and Biomass of Termites under Differing Levels of Disturbance in the Mbalmayo Forest Reserve, Southern Cameroon. Philos. T. Roy. Soc. B, 351:51–68. doi: 10.1098/rstb.1996.0004

Eiten, G. (1983). Classificação da vegetação do Brasil (p. 305). Brasília, DF: CNPq/Coordenação Editorial.

Gontijo, T.A., & Domingos, D.J. (1991). Guild Distribution of Some Termites from Cerrado Vegetation in South-East Brazil. J. Trop. Ecol., 7: 543–529.

Jouquet, P., Traoré, S., Choosai, C., Hartmann, C., & Bignell, D. (2011). Influence of termites on ecosystem functioning. Ecosystem services provided by termites. Eur. J. Soil Biol., 47: 215–222. doi:10.1016/j.ejsobi.2011.05.005

Kambhampati, S., & Eggleton, P. (2000). Taxonomy and phylogeny of termites. In T. Abe, D. Bignell, & M. Higashi (Eds.), Termites: Evolution, Sociality, Symbioses, Ecology (pp. 1–23). Dordrecht: Kluwer Academic Publishers.

Klink, C.A., & Machado, R.B. (2005). A conservação do Cerrado Brasileiro. Megadiversidade, 1, 147–155.

Lavelle, P., Decaëns, T., Aubert, M., Barot, S., Blouin, M., Bureau, F., Rossi, J.-P. (2006). Soil invertebrates and ecosystem services. Eur. J. Soil Biol., 42: S3–S15. doi:10.1016/j.ejsobi.2006.10.002

Magurran, A.E. (2004). Measuring Biological Diversity (p. 256). Oxford: Blackwell Publishing Inc.

Mathews, A.G.A. (1977). Studies of termites from Mato Grosso State, Brazil (p. 264). Rio de Janeiro, RJ: Academia Brasileira de Ciências.

Meyer, D., Zeileis, A., & Hornik, K. (2006). The strucplot framework: visualizing multi-way contingency tables with vcd. J. Stat. Soft., 17:1–8. Retrieved from http://www.jstatsoft.org/v17/i03/

Meyer, D., Zeileis, A., & Hornik, K. (2011). vcd: visualizing categorical data. R package

Myers, N., Mittermeier, R.A., Mittermeier, C.G., Da Fonseca, G.A.B., & Kent, J. (2000). Biodiversity hotspots for conservation priorities. Nature, 403: 853–858. doi:10.1038/35002501

Oksanen, J., Blanchet, F.G., Kindt, R., Legendre, P., Minchin, P.R., O'Hara, R.B., ... Wagner, H. (2011). vegan: community ecology package. Retrieved from http://cran.r-project.org/package=vegan

R Development Core Team. (2010). R: A Language and Environment for Statistical Computing. Vienna, Austria. Retrieved from http://www.r-project.org/

Ribeiro, J.F., & Walter, B.M.T. (1998). Fitofisionomias do

bioma Cerrado. In S.M. Sano & S.P. Almeida (Eds.), Cerrado: Ambiente e Flora (pp. 89–166). Planaltina: Embrapa-CPAC Planaltina.

Rocha, G.F., Ferreira, L.G., Ferreira, N.C., & Ferreira, M.E. (2011). Detecção de desmatamentos no bioma Cerrado entre 2002 e 2009: padrões, tendências e impactos. Rev. Bras. Cart., 63: 341–349.

Roisin, Y., Dejean, A., Corbara, B., Orivel, J., Samaniego, M., & Leponce, M. (2006). Vertical stratification of the termite assemblage in a neotropical rainforest. Oecologia, 149: 301–311. doi:10.1007/s00442-006-0449-5

Sano, E.E., Rosa, R., Brito, J.L.S., & Ferreira, L.G. (2010). Land cover mapping of the tropical savanna region in Brazil.

Environ. Monit. Assess., 166: 113–124. doi: 10.1007/s10661-009-0988-4

Silva, J.M.C., & Bates, J.M. (2002). Biogeographic patterns and conservation in the South American Cerrado: a tropical savanna hotspot. BioScience, 52: 225–234. doi: 10.1641/0006-3568(2002)052[0225:BPACIT]2.0.CO;2

Wilson, E.O. (1971). The Insect Societies (p. 548). Cambridge: Belknap Press.

Wood, T.G., & Sands, W.A. (1978). The role of termites in ecosystems. In M.V. Brian (Ed.), Production Ecology of Ants and Termites (pp. 245–292). Cambridge: Cambridge University Press.

Age Polyethism in the Swarm-founding Wasp *Metapolybia miltoni* (Andena & Carpenter) (Hymenoptera: Vespidae; Polistinae, Epiponini)

L Chavarría[1]; FB Noll[1,2]

1 - Universidade de São Paulo, Ribeirão Preto, São Paulo, Brazil

2 - Universidade Estadual Paulista Júlio de Mesquita Filho, São José do Rio Preto, São Paulo, Brazil

Keywords

Social wasps, workers, division of labor

Corresponding author

Laura Chavarría
FFCLRP-USP
Av. Bandeirantes 3900 Bloco 9
Ribeirão Preto, São Paulo, Brazil
14040-901
E-Mail: laurachp@usp.br

Abstract

In Epiponini division of labor is associated with age polyethism and individual task specialization. We observed worker activities in three colonies of *Metapoybia miltoni* in Brazil. We analyzed differences of task allocation among age groups. Old workers tend to forage more than young, but age polyethism was less evident in other tasks. Age composition of population could be a determinant factor in task allocation. Workers are probably allocating to perform tasks according to colony needs, and not to individual's age. Considering age population in studies of division of labor could help to understand how colonies respond to different situations.

Introduction

Division of labor characterizes insect societies. Workers allocate tasks responding to internal and external conditions, and to individual workers decisions (Sendova-Franks & Franks, 1999). There are two general patterns: worker age polyethism (task allocation correlated with age) and morphological polymorphisms (changes in size or shape related to task performance) (Beshers & Fewell, 2001). Organization of labor in wasps of the tribe Epiponini seems not to be related with morphological polymorphism because there is no morphological specialization or subcastes. Division of labor could be associated with worker's age, or with individual task specialization (Jeanne, 1996; Karsai & Wenzel, 2000).

According to Jeanne (1991) epiponines have the most evident worker age polyethism of social Vespidae. Previous studies in *Polybia* (Lepeletier), *Protopolybia* (Ducke) and *Agelaia* (Ducke) (Simões, 1977; Forsyth, 1978; Jeanne et al., 1988; Jeanne et al., 1992) found that young and middle aged workers perform nest tasks (building, brood care, nest main-tenance, defense), while old workers forage. Nevertheless, the presence of workers specialized in a particular task is rare in most of social Hymenoptera (Robinson, 1992; Sendova-Franks & Franks, 1999; O' Donnell, 1998; Karsai & Wenzel, 2000; Beshers & Fewell, 2001; Johnson, 2003). Within Epiponini, Karsai and Wenzel (2000) did not find specialization in colonies of *Metapolybia aztecoides* (Richards) and *M. mesoamericana* (Smethurst & Carpenter).

Because organization and regulation of work is complex, conventional patterns are insufficient to explain the division of labor of several insect societies (Beshers & Fewell, 2001). For these reasons, we studied task allocation according to age in three colonies of *Metapolybia miltoni* (Andena & Carpenter) in Maranhão, Brazil.

We observed colonies for three days in 2008: colonies N-1 and N-2 on February, in Reserva Merck (S 02° 39' 7.8" and W 44° 09' 04.0"); and colony N-3 on March, in Reserva das Paineiras (S 03° 14' 35.4" W 43° 25' 28.7"). Nest envelope was removed in order to perform video recordings (Sony Handycam DCR-SR42). We took a random sample of workers to perform individual observations (N=15 for N-1, N=11 for N-2 and N=18 for N-3). All individuals were marked with quick-drying paint. Video recordings included 197, 194 and 370 minutes for N-1, N-2 and N-3, respectively. We observed allocation of work in three tasks: cell inspection, construction (envelope and cells) and forage. In these cases, forager's activities were directly observed due to difficulties to follow them in video.

Females were classified according to age, based on three categories for the coloration of the transverse apodeme (Richards, 1971): light for younger, brown for middle age, and black for older females. To verify if the amount of workers that performs a determinate task varied according to age, and within colonies N-1 and N-3 (all observed workers of N-2 were young) a Chi-square test was applied.

Our results indicate that colony cycle stage did not affect workers frequency performing tasks among colonies (cell inspection $x^2=0,854$ df= 2 p>0.5; construction $x^2=0,381$ df= 2 p>0.5; foraging $x^2=3,641$ df= 2 p>0.5). Colony N-1 was in a mature stage of colony cycle (eggs, larvae, pupae and low queen proportion), and most of workers were young (old= 1% middle= 8% young= 90%). Colony N-2 was also in a mature stage, but in male production and most of the workers were young (young= 90% middle= 10%). Colony N-3 was in a pre-emergence phase (only eggs and higher queen proportion), and most of workers were old (middle= 17% old= 83%).

We observed a similar number of old and young workers perfoming nest tasks (cell inspection and construction) and foraging (young $x^2= 4.588$ df= 2 p>0.05; middle $x^2= 1.333$ df= 2 p>0.5; old $x^2= 0.839$ df= 2 p>0.5). Nevertheless, it is clear that young workers tended to forage less than olders (Fig 1 colonies N-1, N-3), even in colony N-2 young workers tended to forage less. Also, the amount of workers that perform tasks across different ages inside colonies N-1 and N-3 was similar (N-1 $x^2= 2,798$ df= 2 p>0.1; N-3 $x^2= 0,802$ df= 2 p>0.5).

As mencioned before, previous studies of division of labor in the Epiponini found that as workers get old they switch to perform tasks less related with brood care (Jeanne, 1991). Similarly, we observed that old workers tend to forage more than young individuals (Fig 1). We did not find significant differences probably because of the small sample of foragers (N=24). Nevertheless, even when old workers tend to forage more, young workers can also forage (Fig 1).

Age polyethism was less evident in the other tasks: old workers as young ones can build and inspect cells (Fig 1). Age composition of worker population within colonies may be a determinant factor in task allocation. The observed colonies did not include individuals of different ages; workers of colonies N-1 and N-2 were mostly young, and workers of colony N-3 were mostly old. Because colonies population has little variation in terms of age, workers must be allocated to perform tasks according to colony needs and not to individual's age. On the other hand, in colonies with differently aged workers, polyethism could be more important to delimit

tasks, as demonstrated in previous studies (Jeanne et al., 1988; Jeanne et al., 1992).

As observed by Karsai and Wenzel (2000), we also found flexibility in activities performed by workers; young, middle and old individuals perform different tasks. Workers of all insect societies retain some behavioral flexibility that helps to respond to changing conditions (Robinson, 1992). Caste flexibility is decisive for colony survival in swarm wasps because it allows colonies to respond efficiently to different situations that may arise.

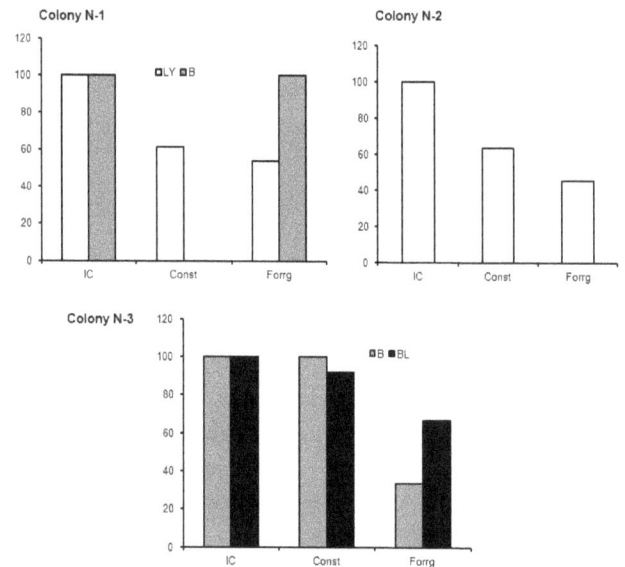

Fig 1. Percentage of workers of different age (LY= light yellow for younger, B= brown for middle age, BL= black for older females) that perform a task (IC= cell inspection, Const= construction, Forrg= foraging) throught the studied colonies of *Metapolybia miltoni*.

Workers decisions are dependent of colony context, and workers would be allocated to perform certain tasks when necessary (Karsai & Wenzel, 2000).

In conclusion, considering age composition of population and studying colonies exposed to different situations would help to understand how colonies allocate tasks. The division of labor of swarm wasps is more complex than previously thought; colonies do not organize labor in the same manner. In fact, as evidenced by Noll & Wenzel (2008), caste dimorphism may have evolved at least eight times and social organization probably derived directly from an ancestor with incipient caste dimorphism in most taxa. For this reason general patterns are not enough to understand different strategies across the tribe.

Acknowledgments

We especially thank to Organization of America States (OAS), the Incentive Commission of Ministerio de Ciencia y Tecnología (MICIT) and Consejo Nacional para Investigaciones Científicas y Tecnológicas (CONICIT) from Costa Rica, for financial support to do this investigation. To Conselho Nacional de Desenvolvimento Científico e Tecnológico (CNPq) and Fundação de Amparo à Pesquisa do Estado de São Paulo (FAPESP) (2007/08633-1) from Brazil to support part of the investigation. To Gisele Garcia for all the help and friendship offered during the stay in Maranhão. To Otávio Augusto Lima de Oliveira for field assistance. To John Wenzel and Sergio Jansen for their comments, suggestions and helpful ideas.

References

Andena, SR. & Carpenter, JM. (2011). A new species of *Metapolybia* (Hymenoptera: Vespidae; Polistinae, Epiponini). Entomol. Am. 117(3/4): 117-120.

Beshers, SN. & Fewell, JH. (2001). Models of Division of Labor in Social Insects. Annu. Rev. Entomol. 46: 413-440. DOI: 10.1146/annurev.ento.46.1.413

Forsyth, AB. (1978). Studies on the behavioral ecology of polygynous social wasps. Dissertation, Harvard University.

Jeanne, RL., Downing, HA. & Post, DC. (1988). Age polyethism and individual variation in *Polybia ocidentalis*, and advance eusocial wasp. In RL. Jeanne (Ed). Interindividual behavioral variability in social insects (pp. 323-357). Boulder, Colorado: Westview Press.

Jeanne, RL. (1991). Polyethism. In KG. Ross & RW. Matthews (Eds). The social biology of wasps. (pp. 389-425). Ithaca, New York: Cornell University Press.

Jeanne, RL., William, NM. & Yandell, BS. (1992). Age polyethism and defense in tropical social wasps (Hymenoptera: Vespidae). J. Insect Behav. 5(2): 211-227. DOI: 10.1007/BF01049290

Jeanne, RL. (1996). Regulation of nest construction behavior in *Polybia occidentalis*. An. Behav. 52: 473-488. DOI: 10.1006/anbe.1996.0191

Johnson, BR. (2003). Organization of work in the honeybee: a compromise between division of labor and behavioral flexibility. Proc. R. Soc. Lond. 270: 147-152. DOI: 10.1098/rspb.2002.2207

Karsai, I. & Wenzel, JW. (2000). Organization and Regulation of Nest Construction behaviour in *Metapolybia* wasp. J. Insect Behav. 13(1): 111-140. DOI: 10.1023/A:1007771727503

Noll, F.B. & Wenzel, J.W. (2008). Caste in the swarming wasps: "queenless" societies in highly social insects. Biol. J. Lin. Soc. 93: 509-522. DOI: 10.1111/j.1095-8312.2007.00899.x

O'Donnell, S. (1998). Reproductive Caste Determination in Eusocial Wasps (Hymenoptera: Vespidae). Annu. Rev. Entomol. 43: 323-346. DOI: 10.1146/annurev.ento.43.1.323

Richards, OW. (1971). The biology of the social wasps (Hymenoptera, Vespidae). Biol. Rev. 46: 483-528. DOI: 10.1111/j.1469-185X.1971.tb01054.x

Robinson, GE. (1992). Regulation of Division of Labor in Insect Societies. Annu. Rev. Entomol. 37: 637-665. DOI: 10.1146/annurev.en.37.010192.003225

Sendova- Franks, AB & Franks, NR. (1999). Self- assembly, self- organization and division of labour. Phil. Trans. R. Soc. Lond. 1388 (354): 1395-1405. DOI: 10.1098/rstb.1999.0487

Simões, D. (1977). Etologia e diferenciação de casta em algumas vespas sociais (Hymenoptera, Vespidae). Tese de doutorado. Ribeirão Preto- USP. 169pp.

Toxicity of Hydramethylnon to the Leaf-cutting Ant *Atta sexdens rubropilosa* Forel (Hymenoptera: Formicidae)

FC Bueno[1], LC Forti[2], OC Bueno[1]

1 - *Universidade Estadual Paulista, Campus de Rio Claro, SP, Brazil*

2 - *Universidade Estadual Paulista, Campus de Botucatu, SP, Brazil*

Key words

Leaf-cutting ant control, inhibitor of cellular respiration, toxicological bioassays.

Corresponding author:

Odair Correa Bueno
UNESP - Departamento de Biologia
CEIS – Centro de Estudos de Insetos
Sociais, Av. 24-A, 1515, HI 13506-900
Rio Claro / SP, Brazil
E-Mail: odaircb@rc.unesp.br

Abstract

Since 2009, when sulfluramid was listed in annex B of the Stockholm Convention's Persistent Organic Pollutants, effort has been made to search for other active ingredients to use in baits for controlling leaf-cutting ants in Brazil. Considering that active ingredients that inhibit insect cellular respiration have been shown to be effective in controlling ants, the current work aimed at assessing the toxicity of hydramethylnon to *Atta sexdens rubropilosa* workers. Hydramethylnon was dissolved in acetone and in a solution of acetone + soy oil then incorporated in artificial diet at concentrations of 1 µg/mL, 5 µg/mL, 10 µg/mL, 100 µg/mL, 200 µg/mL and 1000 µg/mL. The treatments where ants were daily fed on the diet containing hydramethylnon at 100 µg/mL, 200 µg/mL and 1000 µg/mL, especially those dissolved in soy oil, exhibited high mortality in comparison to the controls. The data presented here confirms the insecticidal activity of hydramethylnon and highlights the importance of employing soy oil in the formulation of baits to control leaf-cutting ants because it enhances hydramethylnon efficiency.

Introduction

Chemicals used to control pests of cultivated plants, which include leaf-cutting ants, have been always one of the main ecological concerns because of their harmful effects on the environment, human health and other animals (Williams, 1990). As a consequence, the number of studies has greatly increased aiming to replace traditional pesticides for those of rapid degradation, high specificity and less noxious to the environment (Morini et al., 2005).

Most of the strategies of chemical control are based on killing leaf-cutting ants by contact, but it is usually not enough to control populations in a certain area. Efficient control involves exterminating the whole colony, not only some individuals. Currently, the most appropriate method for controlling leaf-cutting ants is the use of toxic baits because they are incorporated into the colony feeding cycle and the insecticide acts through ingestion (Loeck & Nakano, 1984).

Since dodecachlor (organochlorine pesticide - Mirex) was prohibited in 1993, the chemical sulfluramid became the most used active ingredient in toxic baits in Brazil. Nevertheless, this compound was included in annex B of the Stockholm Convention's Persistent Organic Pollutants in 2009 with restrictions of only being used for controlling leaf-cutting ants in Brazil until a novel compound is found to replace it (Stockholm Convention, 2009).

To develop efficient and economically viable toxic baits for ant control, it is essential that the active ingredient acts slowly, so that workers live long enough to spread the chemical among other ants, is toxic by ingestion, does not repel workers, is lethal at low concentrations and environmentally acceptable (Etheridge & Phillips, 1976; Forti et al., 1993; Bueno & Campos-Farinha, 1999). Recent toxicological analysis of several active ingredients used for pest control reveal that in general inhibitors of cellular respiration meet the requirements for use in baits to control leaf-cutting ants (Nagamoto et al., 2004; Decio et al., 2013).

Hydramethylnon acts on insect cellular respiration by inhibiting electron transport system and consequently blocking ATP production and decreasing mitochondrial oxygen consumption. Metabolism disruption and subsequent decrease in ATP result in delayed mortality by this active ingredient (Bloomquist, 2010; Irac, 2010).

In view of this, the aim of the current work was to assess the toxicity of hydramethylnon to workers of *Atta sexdens rubropilosa* Forel.

Material and Methods

The *A. sexdens rubropilosa* workers used in the assays, whose body mass was about 20-25 mg, were randomly picked from a laboratory nest kept at Centro de Estudos de Insetos Sociais (Instituto de Biociências, UNESP – Univ. Estadual Paulista, Campus de Rio Claro, SP) and some specimens were deposited in the Coleção Entomológica Adolph Hempel (Instituto Biológico, São Paulo – SP, Brazil). Before the assays, nests were daily supplied with leaves of *Eucalyptus* sp., oat seeds and occasionally with leaves of other plants such as *Hibiscus* sp., *Ligustrum* sp. or rose petals.

Fifty ants were put into five Petri dishes (ten ants per dish) for each treatment. During the assays the ants were maintained on an artificial diet prepared with glucose (50 g/L), bacto-peptone (10 g/L), yeast extract (1.0 g/L) and agar (15 g/L) in distilled water (0.1 L) (Bueno et al. 1997). The diet (0.5 g per dish) with hydramethylnon (experimental) or without (control) was offered daily on a small plastic cap.

Hydramethylnon dissolved either in acetone (HA) or in acetone and soy oil (9 mL of acetone per 1 mL of oil) (HAO) was added to the artificial diet at concentrations of 1 µg/mL, 5 µg/mL, 10 µg/mL, 100 µg/mL, 200 µg/mL and 1000 µg/mL. Three controls were established to verify that the hydramethylnon toxicity results were not biased by the chosen solvent: one group received the artificial diet ('diet control'), another group received the artificial diet with acetone ('acetone control'), and a third group received the artificial diet with acetone and soy oil ('acetone + oil control'). Acetone or the combination of acetone and soy oil were added at the same proportions as those used for the hydramethylnon-treated groups.

During the assays, ants were maintained in an incubator at temperature of 25 ± 1°C and relative humidity ranging between 70-80% for maximum length of 25 d and the number of dead was registered daily. The survival average 50% (S_{50}) was calculated and survival curves were compared by the computer-assisted software Graph-Pad ™ using the log-rank test (Elandt-Johnson & Johnson 1980).

Results and Discussion

Mortality rates of both controls 'acetone' and 'acetone + oil' did not significantly differ comparing to the group that was only fed on artificial diet, no solvent added (diet control), indicating that solvents did not affect the mortality of *A. sexdens rubropilosa* workers (Tables 1 and 2).

Hydramethylnon dissolved in acetone resulted in decreased worker survival which was more drastic at concentrations of 200 µg/mL and 1000 µg/mL. Ant survival median time was reduced from 14 d (acetone control) to 10 d (200 µg/mL) and 9 d (1000 µg/mL). All ants were dead on the 19th day for the concentration of 100 µg/mL, 16th for 200 µg/mL and 15th for 1000 µg/mL (Table 1).

Hydramethylnon dissolved in acetone + soy oil also resulted in decreased worker survival being more drastic at 100 µg/mL, 200 µg/mL and 1000 µg/mL. Ant survival median was reduced from 13 d (acetone + oil control) to only 6 d (100 µg/mL, 200 µg/mL and 1000 µg/mL). Mortality of all ants occurred on the 13th day for concentration 100 µg/mL and on the 9th day for concentrations of 200 µg/mL and 1000 µg/µL (Table 2).

Soy oil in the diet made the toxicity of hydramethylnon to leaf-cutting ants more potent, a fact that was shown by the higher ant mortality rate in the treatment with hydramethylnon dissolved in acetone + oil in comparison with the ants treated with hydramethylnon dissolved only in acetone. Few studies have revealed the mode of action of oils in insects. Hewlett (1947) suggested that intoxication by oil was due to the mechanical action, interfering with breathing by blocking the spiracles. On the other hand, there are authors who believe that toxicity of oils is attributed to chemical action (Singh et al. 1978) or may act both physically and chemically in the insect (Obeng-Ofori 1995). For Taverner et al. (2001), oils can act in the insect nervous system by increasing the permeability of neuron membrane and thus affecting ion change and stressing excitability of neuron cells. However, in leaf-cutting ants, the oil allows an alternative via for the ingestion of oil-soluble active ingredients due to the feeding behavior of workers (Bueno et al., 2008; Decio et al., 2013).

Adult ants feed primarily on liquid. The ingested food moves into the infrabuccal cavity of workers, wich remain for 24h. Solid parts of the food are retained in the cuticle folds and spines present on infrabuccal cavity and are subsequently discared in the trash (Fowler et al., 1991; Moreira et al., 2011). Nevertheless, the liquid portion of the food passes after the opening post-pharyngeal gland, where occurs the separation of water-soluble compounds and oil-soluble compounds. The water-soluble compounds moving into the crop and oil-soluble compound enter the ducts of post-pharyngeal gland, where they are absorbed and transferred to the hemolymph and subsequently to the whole body (Bueno et al., 2008). Thus, recently Decio et al. (2013) observed major part of soy oil from the diet is likely allocated in the lumen of post-pharyngeal glands where it will be metabolized and suggested that ant post-pharyngeal glands are involved in the metabolism of lipids.

The current work demonstrated that hydramethylnon has a great potential as active ingredient to be incorporated in baits for leaf-cutting ant control especially because of its slow mode of action. This characteristic is fundamental for

Table 1. Mortality (%) of *A. sexdens rubropilosa* workers fed on hydramethylnon dissolved in acetone * S_{50}: Survival median 50%. Different letters after S_{50} values show a significant difference according to the log-rank test (P < 0.05).

Treatment	% Daily accumulated mortality										S_{50} *
	1	2	3	6	8	10	14	17	21	25	
Diet control	0	0	0	8	16	30	54	84	100	—	13 *a*
Acetone control	0	0	0	10	18	30	58	84	100	—	14 *a*
Hydramethylnon 1 µg/ml	0	0	0	8	16	30	54	72	92	98	13 *a*
Hydramethylnon 5 µg/ml	0	0	0	10	28	38	56	78	100	—	13 *a*
Hydramethylnon 10 µg/ml	0	0	0	14	32	48	82	96	100	—	11 *b*
Hydramethylnon 100 µg/ml	0	0	0	6	18	48	92	98	100	—	10 *b*
Hydramethylnon 200 µg/ml	0	0	0	10	22	54	90	100	—	—	10 *b*
Hydramethylnon 1000 µg/ml	0	1	2	20	38	62	94	100	—	—	9 *b*

Table 2. Mortality (%) of *A. sexdens rubropilosa* workers fed on hydramethylnon dissolved in acetone and soy oil* S_{50}: Survival median 50%. Different letters after S_{50} values show a significant difference according to the log-rank test (P < 0.05).

Treatment	% Daily accumulated mortality										S_{50} *
	1	2	3	6	8	10	14	17	21	25	
Diet control	0	0	0	6	16	30	54	74	82	94	13 *a*
Acetone + oil control	0	0	0	6	18	32	56	76	86	94	13 *a*
Hydramethylnon 1 µg/mL	0	0	0	10	20	36	62	88	96	100	13 *a*
Hydramethylnon 5 µg/mL	0	0	2	10	28	46	90	100	—	—	11 *b*
Hydramethylnon 10 µg/mL	0	0	0	4	6	60	96	100	—	—	10 *b*
Hydramethylnon 100 µg/mL	0	0	2	56	90	92	100	—	—	—	6 *b*
Hydramethylnon 200 µg/mL	0	0	4	52	88	100	—	—	—	—	6 *b*
Hydramethylnon 1000 µg/mL	0	0	2	50	98	100	—	—	—	—	6 *b*

obtaining ant-insecticidal baits of high efficiency since it allows that the workers live long enough to spread the active ingredient inside the colony. The concept of slow toxicity was firstly used to select active ingredients for controlling fire ants (Stringer et al., 1964; Williams, 1983; Vander Meer et al., 1985) and more recently it was adapted by Nagamoto et al. (2004) for leaf-cutting ants. The selection based on this concept favour compounds that cause low mortality on the first 24 h after application (lower than 15%), but high mortality at the end of the experiment (above 90%).

Acknowledgements

We thank FAPESP (Fundação de Amparo a Pesquisa do Estado de São Paulo) and CNPq (Conselho Nacional de Desenvolvimento Científico e Tecnológico) for providing financial support for this research. We also thank two anonymous reviewers for valuable comments that improved this article.

References

Bloomquist, J.R. Insecticides: chemistries and characteristics. In: <www.ipmworld.umn.edu/chapters/bloomq.htm. Acess: 20 de dezembro de 2010.

Bueno, O.C., Morini, M.S.C., Pagnocca, F.C., Hebling, M.J.A. & Silva, O.A. (1997). Sobrevivência de operárias de *Atta sexdens rubropilosa* Forel (Hymenoptera: Formicidae) isoladas do formigueiro e alimentadas com dietas artificiais. An. Soc. Entom. Bras., 26: 107-12.

Bueno, O.C. & Campos-Farinha A.E.C. (1999). As formigas domésticas. In: Mariconi, F.A.M. (Ed.). Insetos e outros invasores de residência. Piracicaba: FEALQ 135-180.

Bueno, O.C., Bueno, F.C., Diniz, E.A. & Schneider, M.O. (2008). Utilização de alimento pelas formigas-cortadeiras. In: Vilela, F.C. et al. (eds). Insetos Sociais: Da Biologia à Aplicação. Viçosa: Editora UFV 96-114.

Decio, P., Silva-Zacarin, E.C.M., Bueno, F.C. & Bueno O.C. (2013). Toxicological and histopathological effects of hydramethylnon on *Atta sexdens rubropilosa* (Hymenoptera: Formicidae) workers. Micron, 45: 22-31. doi: 10.1016/j.micron.201210.008.

Elandt-Johnson, R. & Johnson, N.L. (1980). Survival models and data analysis. John Wiley and Sons, New York.

Etheridge, P. & Phillips, F.T. (1976). Laboratory evaluation of new insecticides and bait matrices for the control of leaf-cutting ants (Hymenoptera, Formicidae). Bul. Entom. Res., 66: 569-78.

Forti, L.C., Della Lucia, T.M.C., Yassu, W.K., Bento, J.M.S. & Pinhão, M.A.S. (1993). Metodologias para experimentos com iscas granuladas para formigas cortadeiras. In: Della Lucia, T.M.C. (ed.). As formigas cortadeiras. Viçosa: Folha de

Viçosa 191-211.

Fowler, H.G., Forti, L.C., Brandão, C.R.F, Delabie, J.H.C. & Vasconcelos, H.L. (1991). Ecologia nutricional de formigas. In: Pannizi, A.R.; Parra, J.R.P. (eds). Ecologia nutricional de insetos e suas implicações no manejo de pragas. São Paulo: Manoele 131-223.

Hewlett, P.S. (1947). The toxicities of three petroleum oils to the grains weevils. Ann. Appl. Biol., 34: 575-585.

Irac. Irac Moa Classification Scheme v.7.0. Disponível em: <http://www.irac-online.org/wp-content/uploads/2009/09/MoA-classification_v7.0.4-5Oct10.pdf>. Acesso em: 03 mar. 2011.

Loeck, A.E. & Nakano, O. (1984). Efeito de novas substâncias visando o controle de sauveiros novos de Atta laevigata (Smith, 1858) (Hymenoptera: Formicidade). Solo, 1: 25-30.

Moreira, D.D.O., Erthal Jr., M. & Samuels, R.I. (2011). Alimentação e digestão em formigas cortadeiras. In: In: DELLA LUCIA, T.M.C. (ed). Formigas-cortadeiras: da biologia ao manejo. Viçosa: UFV 204-225.

Morini, M.S.C., Bueno, O.C., Bueno, F.C., Leite, A.C., Hebling, M.J.A., Pagnocca, F.C., Fernandes, J.B., Vieira, P.C. & Silva, M.F.G.F. (2005). Toxicity of sesame seed to leaf-cutting ant Atta sexdens rubropilosa (Hymenoptera: Formicidae). Sociobiology, 45(1): 195-204.

Nagamoto, N.S., Forti, L.C., Andrade, A.P.P., Boaretto, M.A.C. & Wilcken, C.F. (2004). Method for the evaluation of insecticidal activity over time in Atta sexdens rubropilosa workers (Hymenoptera: Formicidae). Sociobiology, 44: 413-431.

Obeng-Ofori, D. (1995). Plant oils as grain protectants against infestations of Cryptolestes pusillus and Rhyzopertha dominica in stored grain. Entomol. Exp. Appl., 77: 133-139.

Singh, S.R., Luse, R.A., Leuschner, K. & Nangju, D. (1978). Groundnut oil treatment for the control of Callosobruchus maculates during cowpea storage. J. Stored Prod. Res., 14: 77-80.

Stockholm Convention. The new POPs under the Stockholm Convention. [2009]. Disponível em: <http://chm.pops.int/Convention/ThePOPs/TheNewPOPs/tabid/2511/Default.aspx>. Acesso em: 14 fev. 2013.

Stringer, C.E., Lofgren, C.S. & Bartelett, F.J. (1964) Imported fire ant toxic bait studies: evaluation of toxicants. J. Econ. Entomol., 57 (6): 941-945.

Taverner, P.D., Gunning, R.V., Kolesik, P., Bailey, P.T., Inceoglu, A.B., Hammock, B. & Roush, R.T (2001). Evidence for direct toxicity of a "light" oil on the peripheral nerves of lightbrown apple moth. Pest. Biochem. Physiol. 69:153-165.

Vander Meer, R.K., Lofgren, C.S. & Williams, D.F. (1985). Fluoroaliphatic sulfones: a new class of delayed-action insecticides for control of Solenopsis invicta (Hymenoptera: Formicidae). J. Econ. Entomol., 78: 1190-1197.

Williams, D.F. (1983). The development of toxic baits for the red important fire ant. Fla. Entomol., 66: 162-171.

Williams, D.F. (1990). Overview. In: Applied Myrmecology: a world perspective. Vander Meer, R.K., K. Jaffe & A. Cedeno (eds) San Francisco: Westview Press 493-495.

5

Temporal Activity Patterns and Foraging Behavior by Social Wasps (Hymenoptera, Polistinae) on Fruits of *Mangifera indica* L. (Anacardiaceae)

BC Barbosa, MF Paschoalini, F Prezoto

Laboratório de Ecologia Comportamental e Bioacústica (LABEC), Universidade Federal de Juiz de Fora, Juiz de Fora, MG, Brazil.

Keywords
Fruit trees, Harvest, Hymenoptera, Vespidae

Corresponding autor
Bruno Corrêa Barbosa
Lab. de Ecologia Comportamental e Bioacústica (LABEC), Universidade Federal de Juiz de Fora 36036-900, Juiz de Fora, MG, Brazil.
E-mail: brunobarbosabiologo@hotmail.com

Abstract

This study had as objective to determine which species of social wasps visit mango fruits, to record the behaviors displayed by them while foraging and to verify which the species of wasps visitors offer risk of accidents to farmers. The studied area was monitored during February 2012, from 8:00 to 17:00, in a 144 hour effort, and the data collected included the time of activity, wasps diversity, aggressiveness and the general behavior of social wasps around the fruits. There were registered a total of 175 individuals of 12 different species. Social wasps damaged the healthy fruits, and we registered the abundance and richness peaks during the hot period of the day. This study indicated the need for special care during the harvest, as aggressive wasps are indeed present and abundant, resulting in a possible increase of accident risk for the workers.

Foraging of social wasps comprises a collection of the following resources: carbohydrates (used mainly for adult diet), animal protein (used for immature diet), plant fiber (used for nest building), and water (used for cooling and building the nest) (Hunt, 2007, Prezoto et al. 2008, Elisei et al. 2010, Clemente et al. 2012).

While foraging for these resources, social wasps show a generalistic and opportunistic behavior, and evidence of foraging optimization. This behavior has been documented in wasps foraging on different fruit species, such as grapes (Hickel & Schuck 1995), cacti (Santos et al 2007b), jabuticaba trees (De Souza et al. 2010), cashew trees (Santos & Presley 2010), guava trees (Brugger et al. 2011), pitanga trees (Souza et al 2013), and Spanish prune (Prezoto & Braga, 2013). In these studies, the authors reported that the wasps might prey on crop pests that damage fruits to collect carbohydrates.

Despite the growing number of studies on social wasps in the past decade, information on the role of these insects in orchards is still scarce, despite these insects having their highest diversity in the Neotropics (Rafael et al. 2012). Hence, there is a need for studies that answer questions such as: Which species of social wasps forage on fruits? What types of behavior do they display? Do wasps offer risk of accidents to fruit farmers?

This study aims to increase the knowledge about the occurrence of social wasps on mango tree plantations and to describe the richness and abundance of social wasps that forage on fruits throughout the day, as well as to describe the behaviors displayed by them while foraging.

The study was conducted in a farm in the municipality of Juiz de Fora, Zona da Mata Mineira (21°43'55"S, 43°22'16"W, 800 m a.s.l.). Observations were made in February 2012, during the fruiting of mango trees, from 8:00 to 17:00, in a total of 144 h of observation. For each observation event, we established 4-m² quadrants on the base of trees to record visiting wasp and their behavior (*ad libitum sensu* Altmann 1974) during foraging on fallen fruits. All were record by direct observation.

To record behavioral information we defined four types of arrival behavior of wasps on fruits: (I) direct landing on fruit, (II) hovering before landing on fruit, (III) hovering over other fruits before landing, and (IV) landing elsewhere and moving to the fruit. After observation, wasps were collected with an insect net for identification, using for genera and species keys proposed by Richards (1978) and Carpenter & Marques (2001), vouchers were deposited in the Laboratory of Behavioral Ecology and Bioacustics at the Federal University of Juiz de Fora (LABEC).

We also classified behavioral displays in the presence of other insects as: no aggressive behavior (the wasp remained on the fruit even after being touched by other insects); aggressive behavior (attack or threat to other insects landed on the fruit). For the correlation test between richness and abundance of social wasps we used the Pearson coefficient test (r). We also determined the Berger-Parker index of dominance in R 3.0 (Freeware).

We recorded 175 individual wasps of four genera and 12 species foraging on mango fruits (Table 1). Most species (87.8%; n = 9) were swarm-founding wasps (whose nests are founded by a swarm composed of tens of queens and hundreds of workers), which have large biomass, and, therefore, need a large amount of food. These species form large colonies, which makes their local abundance higher than that of species of independent foundation (whose colonies may be founded by one or a few wasps) and may determine resource consumption. These studies corroborate studies carried out in areas of eucalyptus plantations (Ribeiro Junior 2008), silvopastoral systems (Auad et al. 2010), rainforests (Souza & Prezoto 2006), and arid (Santos et al 2009) and island environments (Santos et al 2007a) in which swarming species were more abundant than species of independent foundation.

The abundance peak occurred from 10:00 to 14:00h (Fig. 1). There was a positive correlation between abundance and the warmest times of the day (r = 0.7635; $P = 0.0062$). These results corroborate studies on social wasp foraging, in which the activity peak of wasps was observed in the warmest times of the day (Rezende et al. 2001; Elisei et al. 2010; Bichara Filho et al. 2010; Castro et al., 2011). There was no wasp visit from 14:30 to 17:00h. We believe that this may have occurred, because after 14:00h there was shading on fallen fruits, causing a decrease in the temperature in this environment and probably interfering in the foraging of wasps. The species *Polybia*

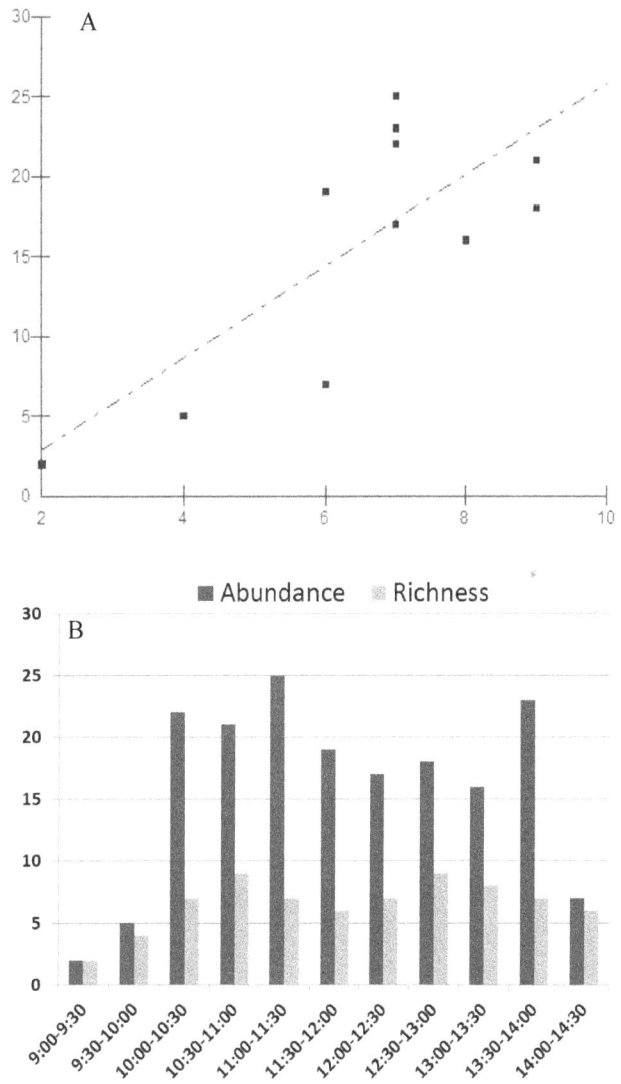

Fig 1. Relationship between abundance and richness (A), and variations in abundance and richness (B) throughout the day.

Table 1: Frequency of social wasps, arrival behavior, aggressiveness, and dominance while foraging on fruits of *Mangifera indica* L. (AB - aggressive behavior, NA - did not exhibited aggressiveness).

Species	Abundance	Arrival behavior of wasps on fruits				Dominance	Agressiveness
		I	II	III	IV		
Agelaia vicina	3			X		0.017	NA
Polybia bifasciata	8	X				0.045	NA
Polybia ignobilis	28	X				0.160	AB
Polybia jurinei	18	X			X	0.102	NA
Polybia sp	8	X				0.045	AB
Polybia platycephala	62	X	X			0.354	NA
Polybia scutellaris	1	X				0.005	NA
Polybia fastidiosuscula	15		X			0.085	NA
Synoeca cyanea	3	X				0.017	NA
Mischocyttarus araujo	3		X			0.017	NA
Mischocyttarus cassununga	22			X		0.125	NA
Polistes versicolor	4	X				0.022	NA
Total	**175**						

platycephala Richards, 1951, *Polybia ignobilis* (Haliday, 1836), and *Mischocyttarus cassununga* (Von Ihering, 1903) presented the highest dominance indices (d = 0.354; d = 0.160, and d = 0.120, respectively) and *Polybia scutellaris* (Write, 1841) was recorded only once (Table 1).

All species exploited fruits with pre-existent orifices (mainly caused by other insects such as *Atta* ants and the bee *Trigona spinipes* Fabricius, 1973. The only exception was the species *Synoeca cyanea* (Fabricius, 1775), which was always observed breaking the skin of fruits. This behavior suggests that this species may become a pest in some environments due to its potential to damage fruits. The same behavior was observed by De Souza et al. (2010) in jabuticaba trees and by Brugger et al. (2011) in guava trees. However, Prezoto & Braga (2013) recorded that this behavior of *S. cyanea* in Spanish prune results from wasp predation on larvae of the fruit fly *Zaprionus indianus* Gupta, 1970, which qualifies this wasp as a natural enemy of this pest.

Most wasp species landed directly on the fruit (I). Only the species *Polybia jurinei* Saussure, 1854 and *P. platycephala* displayed more than one arrival behavior (Table I). The wasps *P. ignobilis* and *Polybia sp.* were the only species that displayed aggressive behavior (Table I). All the other species were recorded using the same fruit without displaying aggressive behavior (Table I).

Although the species *P. ignobilis* and *Polybia sp.* represent only 20% (n = 36) of the wasps recorded in the study, the aggressive potential of these species should be taken into account to avoid accidents, since they are swarming species, whose colony population easily surpass hundreds of individuals. We also emphasize that *P. ignobilis* was also described by Hermes & Kohler (2004) as an aggressive species.

Based on our results, we suggest that in the period from 10:00 to 14:00, characterized as the activity peak of wasps on fruits, the collectors have an extra care during their activities, as for example the use of personal protective equipment, or even the interruption of the collection activity to reduce the risk of accidents by stings.

References

Altmann, J. (1974). Observation study of behavior: sampling methods. Behaviour, 49: 223-265.

Auad, A.M.; Carvalho C.A.; Clemente M.A. & Prezoto, F. (2010). Diversity of social wasps in a silvipastoral system. Sociobiology, 55: 627-636.

Bichara Filho, C. C., Santos, G. M. M., Santos Filho, A.B., Santana-Reis, V.P., Cruz, J. D. da, & Gobbi, N. (2010). Foraging Behavior of the Swarm-founding Wasp *Polybia* (*Trichothorax*) *sericea* (Hymenoptera, Vespidae): Daily Resource Collection Activity and Flight Capacity. Sociobiology, 55: 899-907.

Brugger, B.P., Souza, L.S.A., Souza, A.R. & Prezoto, F. (2011). Social wasps (*Synoeca cyanea*) damaging *Psidium sp.* (Myrtaceae) fruits in Minas Gerais State, Brazil. Sociobiology, 57: 533-535

Carpenter, J. M. & Marques, O. M. (2001). Contribuição ao estudo dos vespídeos do Brasil (Insecta, Hymenoptera, Vespoidea), Cruz das Almas. Universidade Federal da Bahia. Serie Publicações Digitais, 2: 147.

Castro, M. M., Guimaraes, D. L., Prezoto, F. (2011). Influence of Environmental Factors on the Foraging Activity of *Mischocyttarus cassununga* (Hymenoptera, Vespidae). Sociobiology, 58: 133-141.

Clemente, M.A., Lange, D., Del-Claro K., Prezoto F., Campos, N.R. & Barbosa, B. C. (2012). Flower-visiting social wasps and plants interaction: Network pattern and environmental complexity. Psyche: A Journal of Entomology. doi: 10.1155/2012/478431

De Souza, A.R., Venancio D. & Prezoto, F. (2010). Social wasps (Hymenoptera: Vespidae: Polistinae) damaging fruits of *Myrciaria sp.* (Myrtaceae). Sociobiology, 55: 297-299.

Elisei, T.; Nunes, J.V.E.; Ribeiro Junior, C.; Fernandes Junior, A.J. & Prezoto, F. (2010). Uso da vespa social *Polistes versicolor* no controle de desfolhadores de eucalipto. Pesquisa Agropecuária Brasileira, 45: 958-964.

Hermes, M.G. & Köhler, A. (2004). Chave ilustrada para as espécies de Vespidae (Insecta, Hymenoptera) ocorrentes no Cinturão Verde de Santa Cruz do Sul, RS, Brasil. EDUNISC. Caderno de Pesquisa-Série Biologia, 16(2): 65-115.

Hickel, E.R. & Schuck, E. (1995). Vespas e abelhas atacando a uva no Alto Vale do Rio do Peixe. Agropecuária Catarinense, 8: 38-40.

Hunt, J. H. (2007). *The Evolution of Social Wasps*. Oxford University Press, New York, 259 p.

Prezoto, F. & Braga, N. (2013). Predation of *Zaprinus indianus* (Diptera: Drosophilidae) by the Social Wasp *Synoeca cyanea* (Hymenoptera: Vespidae). Florida Entomologist, 96: 670-672. doi: 10.1653/024.096.0243

Prezoto, F., Cortes, S.A.O. & Melo, A.C. (2008). Vespas: de vilãs a parceiras. Ciência Hoje, 48: 70-73.

Rafael, J.A.; Melo, G.A.R.; DE Carvalho, C.J.B.; Casari, S.A. & Constantino R. (2012). *Insetos do Brasil: Diversidade e Taxonomia*. Ribeirão Preto. Holos Editora, 810 p.

Resende, J.J., Santos, G. M. M., Bichara Filho C.C. & Gimenes M. (2001). Atividade de busca de recursos pela vespa social *Polybia occidentalis occidentalis* (Olivier, 1791) (Hymenoptera, Vespidae). Revista Brasileira de Zoociências, 3: 105-115

Ribeiro Junior C. 2008. Levantamento de vespas sociais (Hymenoptera: Vespidae) em uma cultura de Eucalipto. Dissertação de Mestrado. Universidade Federal de Juiz de Fora, Juiz de Fora, MG, Brasil. 68p.

Richards, O. W. (1978). *The social wasps of the Americas excluding the Vespinae*. London, British Museum (Natural History), 580 p.

Santos, G. M. M. & Presley, S. J. (2010). Niche Overlap and Temporal Activity Patterns of Social Wasps (Hymenoptera: Vespidae) in a Brazilian Cashew Orchard. Sociobiology, 56: 121-131.

Santos, G. M. M., Cruz, J. D. da, Marques, O. M. & Gobbi, N. (2009). Diversidade de Vespas Sociais (Hymenoptera: Vespidae) em Áreas de Cerrado na Bahia. Neotropical Entomology, 38: 317-320. doi: 10.1590/S1519-566X2009000300003

Santos, G. M. M., Bichara Filho, C. C., Resende, J. J., Cruz, J. D. da, & Marques, O. M. (2007)a. Diversity and community structure of social wasps (Hymenoptera: Vespidae) in three ecosystems in itaparica island, Bahia State, Brazil. Neotropical Entomology, 36: 180-185. doi: 10.1590/S1519-566X2007000200002

Santos, G. M. M., Cruz, J.D. da, Bichara Filho, C.C., Marques, O.M. & Aguiar, C.M.L. (2007)b. Utilização de frutos de cactos (Cactaceae) como recurso alimentar por vespas sociais (Hymenoptera, Vespidae, Polistinae) em uma área de Caatinga (Ipirá, Bahia, Brasil). Revista Brasileira de Zoologia, 24: 1052-1056. doi: 10.1590/S0101-81752007000400023

Souza, M.M. & Prezoto, F. (2006). Diversity of social wasps (Hymenoptera: Vespidae) in semideciduous forest and cerrado (Savanna) regions in Brazil. Sociobiology, 47: 135-147.

Souza, G. K.; Pikart, T.G.; Jacques, G. C.; Castro, A. A.; Souza, M. M.; Serrao, J. E. & Zanuncio, J. C. 2013. Social Wasps on *Eugenia uniflora* Linnaeus (Myrtaceae) Plants in an Urban Area. Sociobiology, 60: 204-209. doi: 10.13102/sociobiology.v60i2.204-209

6

Does Forest Phisiognomy affect the Structure of Orchid Bee (Hymenoptera, Apidae, Euglossini) Communities? A Study in the Atlantic Forest of Rio de Janeiro state, Brazil

WM Aguiar[1], GAR Melo[2], MC Gaglianone[3]

1 - Universidade Estadual de Feira de Santana, Departamento de Ciências Exatas, Feira de Santana, BA, Brazil

2 - Universidade Federal do Paraná, Departamento de Zoologia, Curitiba, PR, Brazil

3 - Universidade Estadual do Norte Fluminense , Laboratório de Ciências Ambientais, Campos dos Goytacazes, RJ, Brazil

Keywords

Biodiversity, geographic transition conservation, Neotropics, pollinators

Corresponding author

Willian Moura de Aguiar
Universidade Estadual de Feira de Santana
Departamento de Ciências Exatas
Av. Transnordestina s/n Novo Horizonte
Feira de Santana, BA, Brazil 44036-900.
Email: wmag26@yahoo.com.br

Abstract

We describe and discuss the composition, abundance and diversity of euglossine in three vegetation types of the Atlantic Forest (Lowland Seasonal Semideciduous, Submontane Seasonal and Dense Montane Ombrophilous Forest) in Rio de Janeiro state, Brazil, compare them to previous studies in the region and investigate the importance of the vegetation types, climatic and geomorphological factors on the species composition. Male euglossine bees attracted by fragrances were sampled monthly from August/2008 to July/2009 using entomological nets and traps. Euglossine bee communities exhibited differences in their species composition and abundance along the year and in the vegetation types. The precipitation, altitude and vegetation types demonstrated a significant influence on the ordination of the euglossine communities. Our study found differences in the composition of euglossine bee communities as well as in their patterns of abundance and dominance among different vegetation formations, stressing the importance of the conservation of landscape mosaics in the region.

Introduction

Orchid bee communities have been widely sampled in different neotropical ecosystems in recent decades, including the Atlantic Forest (Tonhasca et al., 2002a; Sofia & Suzuki, 2004; Nemésio & Silveira, 2006a; Ramalho et al., 2009; Mattozo et al., 2011; Silveira et al., 2011; Aguiar & Gaglianone, 2012; Ramalho et al., 2013), the Amazonian Forest (e.g., 1985; Powell & Powell, 1987; Storck-Tonon et al., 2009; Abrahamczyk et al., 2011), Central American forests (e.g., Ackerman, 1983; Brosi 2009), the savannas of central Brazil (Cerrado) (Souza et al., 2005; Alvarenga et al., 2007; Faria & Silveira, 2011), and the dry forest of northeastern Brazil (Caatinga) (Souza et al., 2005; Alvarenga et al., 2007; Andrade-Silva et al., 2012). These studies have shown structural differences in bee communities from distinct biogeographical regions, particularly in relation to composition, richness, patterns of dominant species and numbers of endemic species. The differences have usually been attributed mainly to historical factors, although regional differences in community structure at less encompassing spatial scales can be analyzed based on current ecological characteristics related to climatic, geomorphological, and/or vegetational parameters (Silveira et al., 2002; Sydney et al., 2010). Recently, Nemésio and Vasconcelos (2013) evaluated the beta diversity of Euglossina in the Atlantic Forest and noted that climate variations explain twice as much variation in the species data than the spatial variation in species distribution. Nevertheless, part of the observed latitudinal changes in community composition appears to be explained by a concomitant seasonal gradient of precipitation. Similarly, low temperatures and a seasonal rainfall may help explain the relative specificity of the fauna of some of the most western Atlantic Forest. This Tropical Forest extends along almost the entire eastern coast of Brazil and it is composed of a mosaic of rainforest, "restinga" (coastal vegetation on Quaternary sandy soils) and mangrove swamp ecosystems (Galindo-Leal & Câmara, 2003).

The wide latitudinal extension and significant longitudinal width of this biome, associated with altitudinal variations throughout the region, constitute important factors that define different vegetation types (Galindo-Leal & Câmara,

2003; Ribeiro et al., 2009) and likewise favor the high diversity and endemism of animals found there (Myers et al., 2000).

Comparative studies of orchid-bee communities between different vegetational formations have been carried out, and, in the Atlantic Forest, include those by Nemésio and Silveira (2007), Sydney et al. (2010), Mattozo et al. (2011) and Nemésio and Vasconcelos (2013). These authors reported differences between euglossine bee species composition in Dense Ombrophilous and Semideciduous forests, along the northern and the southern coast of Brazil. Additionally, Nemésio (2008) demonstrated the strong influence of altitude on spatial distributions and species composition of communities living in the same geographical region.

Dense ombrophilous forest areas within the Atlantic Forest biome generally occur from the scarps of the coastal mountains, which are directly influenced by masses of humid air moving in from the sea, to the coastline inland. Forested regions in the northern part of Rio de Janeiro State, however, exhibit somewhat unusual features due to the occurrence of the "Campos dos Goytacazes Gap", which designates a geographic transition between the Serra do Mar forest domain (which extends from north of Paraná to Rio de Janeiro) and the Central Atlantic forest domain (which covers the state of Espírito Santo, small areas of eastern Minas Gerais and southern Bahia).

The low humidity associated with this geographical transition (gap) between the two Atlantic forest domains favors the appearance of seasonal forests that extend to the coast (Oliveira-Filho & Fontes, 2000; Oliveira-Filho et al., 2005). This incursion of semideciduous forest to the coast in the northern region of the state of Rio de Janeiro makes the vegetation show limited number of species common to both areas of dense Ombrophilous forest in the state of Rio de Janeiro (Silva & Nascimento 2001).

Oliveira-Filho and Fontes (2000) consider the weather and especially the temperature as the factor most strongly related to the floristic variation observed. This interposition of a seasonal forest between two large belts of ombrophilous forests might be expected to result in differences in the composition, richness and abundance of the orchid bee communities, as these bees are highly dependent on local floral resources for food, nesting, and reproductive resources (see Roubik & Hanson, 2004).

The present study was therefore designed to: (1) Describe and discuss the community structure of euglossine bees in three distinct vegetation types of the Atlantic Forest in the northern portion of Rio de Janeiro State; (2) Comparatively analyze the structure of those bee communities with other communities previously studied in the same region (Tonhasca et al., 2002a; Aguiar & Gaglianone, 2008, 2011, 2012; Ramalho et al., 2009;); (3) Investigate the importance of the different vegetation types and different climatic conditions (temperature, humidity, precipitation) and geomorphological factors (altitude) on the species composition of those bee communities.

Materials and Methods

Study Sites

We selected three forest fragments, from the three different Atlantic Forest vegetation types that originally covered the northern portion of the Rio de Janeiro state (Veloso et al., 1991), based on their large area, the advanced stage of regeneration, and their degree of conservation, considering that they have not been completely devastated in the past.

1- "Mata do Carvão" (21°24'S - 41°04'W), located in the conservation unity "Estação Ecológica Estadual Guaxindiba", in the municipality of São Francisco de Itabapoana, is a fragment of Lowland Seasonal Semideciduous Forest (LSSF) covering approximately 1,200ha, and situated at 40 m.a.s.l. (Silva & Nascimento, 2001).

2- "Mata da Prosperidade" (21°24'S - 42°02'W), located within a private farm in the municipality of São José de Ubá, is a fragment of Semideciduous Submontane Seasonal Forest (SSSF) covering approximately 900 ha, and situated between 350 and 500 m.a.s.l. (Dan et al., 2010).

3- "Mata da Cabecinha" (22°05'S - 42°05'W), located in the municipality of Trajano de Moraes, is a fragment of Dense Montane Ombrophilous Forest (DMOF) covering approximately 900ha, at 750 to 1,000 m.a.s.l. (RioRural/GEF, 2007).

The climate in the first two areas (LSSF and SSSF) is classified as Aw (Köppen & Geiger, 1928), with an average total annual rainfall of approximately 1,100 mm (RadamBrasil, 1983) and a well-defined dry season from May to September (Radam Brasil, 1983). During the study year, the total precipitation was 1,600mm, with average temperatures of 25.3°C and 24.8°C in the two areas respectively (Data: Instituto Nacional de Meteorologia).

In the DMOF area, the climate is classified as Cwa (Köppen & Geiger, 1928), with an average total annual rainfall of approximately 1,300 mm and a predominantly humid climate with no (or only small) water deficits during the year. During the study year, the precipitation was 1,600mm, with an average temperature of 21.4°C (Data: Instituto Nacional de Meteorologia).

Data Collection

Male euglossine bees were sampled once a month from August, 2008 to July, 2009. At each collecting day, fragrance baits (methyl cinnamate, vanilla, eucalyptol, benzyl acetate, and methyl salicylate) were exposed from 09:00 to 15:00h, using two quantitative sampling methods: insect nets and traps, totaling 144 sampling hours (72 hours with an insect net and 72 with traps). The traps and sampling methodologies were the same described in Aguiar and Gaglianone (2011); baits were applied to cotton balls and inserted in the traps or hung on bushes for direct capture using the net. In the capture using an insect net, a single collector inspected the cotton balls throughout the sampling period and all bees that landed on cotton were collected. The attractors were placed 1.5m from the ground and 2m apart from each other. The minimum distance between collection stations was approximately 500m. Two different capture methods were used to maximize the number of captured species. The data were used together to characterize the euglossine bee community. Previous studies in the region (Aguiar & Gaglianone, 2011, 2012) suggested that the methodologies can be complementary, as traps collect more individuals of *Eulaema* while more *Euglossa* are captured with nets.

Sampling was consistently undertaken on sunny days, and never during periods of atypical low temperatures. All voucher specimens are deposited in the Zoological Collection of the Laboratory of Environmental Sciences at the Universidade Estadual do Norte Fluminense.

Data analysis

Community descriptions

Three descriptive indices were estimated, Shannon-Wiener diversity (H′), Berger-Parker dominance (d), and Pielou's evenness (J) (Magurran, 2003), using the software Past version 1.91 (Hammer et al., 2001).

In order to compare the diversity among the three euglossine communities in the study area, we generated a 95% confidence interval for the Shannon-Wiener diversity index using the Jackknife method (Zahl, 1977) and sites with non-overlapping confidence intervals were considered significantly different. Diversity estimates were generated using the software Spade (Chao & Shen, 2005).

Species abundance distribution patterns were determined using the Rank-Abundance Plot, with their relative abundance plotted in descending order (Whittaker, 1965). All species whose relative abundance was larger than 10% were considered dominant for that area.

Rarefaction curves for species richness of each study area were obtained using 999 randomizations, following Magurran (2003). This procedure was undertaken to evaluate the sampling effort based on the species richness in the study areas. The analyses were carried out with the software EcoSim 7 (Gotelli & Entsminger, 2001) and the results plotted using Statistica 8.0 (StatSoft, 2007). The nonparametric richness estimators (Chao1, Jack1 and Bootstrap) were also calculated using EstimateS 8.0 (Colwell, 2006).

Comparative analyses with other Atlantic Forest communities

The structure of the euglossine communities sampled were compared with those reported in studies previously undertaken in Rio de Janeiro State, based on the following criteria: (1) We considered surveys that used baited traps or direct capture with entomological net in baits, that undertook sampling at least on a monthly basis for at least one year, and used at least five fragrance baits (and at least one of which was cineol or eucalyptol); (2) In the case of surveys on a quarterly basis, we considered only those that lasted for at least two years. The study areas examined in the present work, as well as those selected for comparisons (according to the criteria delimited above), are presented in Table S1 (available online as supplemenatry material).

Detrended Correspondence Analysis (DCA) was used to evaluate the ordination of the communities being compared and a data matrix was built to that end containing abundance data for the individuals in each area (considering the methodological criteria described above for comparisons between areas).

The first two axes of the DCA were correlated with climatic (temperature, humidity, precipitation) and geomorphological (altitude) parameters using Pearson's linear correlations. The same axes were utilized to evaluate the influence of the forest vegetation types on the ordination of the euglossine bee communities using nonparametric variance analysis (Kruskal-Wallis); the vegetation types were categorized according to Table S1. These analyses provided representations of the patterns of gradual species substitutions along environmental gradients, as suggested by Ter Braak (1995), and were carried out in Statistica 8 (StatSoft, 2007). Some of the species names used in the original publications were updated to correct identification, or in response to taxonomic changes.

Results

Description of the studied communities

A total of 1,710 male euglossine bees were collected, belonging to four genera and 15 species. *Euglossa cordata* (Linnaeus), *Eulaema nigrita* Lepeletier and *Eulaema cingulata* (Fabricius) were the dominant species, representing more than 80% of the individuals collected (Table 1).

In the SSSF fragment 978 individuals belonging to four genera and 11 species were collected; 444 individuals belonging to four genera and 10 species were collected in the LSSF fragment and 288 individuals belonging to three genera and 12 species were collected in the DMOF forest fragment (Table 1).

The estimated Shannon diversity in SSSF (H′ = 1.52, d.f = 0.024) significantly differed from the LSSF and DMOF sites (H′ = 1.23, d.f = 0.05 and H′ = 1.36, d.f = 0.08, respectively). Similarly, the evenness in the SSSF area showed a higher value than those in LSSF or DMOF (Table 1).

Species composition differed in the three forest fragments: *Eulaema atleticana* Nemésio was sampled only in the LSSF site, *Eufriesea violacea* (Blanchard) only in SSSF, and *Euglossa annectans* Dressler, *E. bembei* Nemésio, and *E. truncata* Rebêlo & Moure were restricted to the DMOF site; *Euglossa clausi* Nemésio & Engel was found only in SSSF and DMOF (Table 1).

Euglossa cordata was dominant in LSSF (52% of the total), while *Eulaema cingulata* was dominant in both SSSF and DMOF (31% and 59%, respectively) (Table 1). Four species were dominant in SSSF, three species in LSSF, and two species in DMOF (Fig 1). In spite of the differences in species dominance, the curves obtained for species importance did not demonstrate significant differences among the different vegetation types (Kolmogorov-Smirnov test, $P > 0.05$), suggesting essentially the same distribution patterns in all three studied sites (Fig 1).

The rarefaction curves based on species richness as a function of abundance revealed that the DMOF site requires a higher sampling effort (Fig 2). This result was also corroborated by the richness estimators (Table 2). The rarefaction curves in the other two areas demonstrated a tendency toward stabilization beyond the abundance of 400 individuals, confirming the results of the richness estimators, which likewise indicated values very close to those actually observed (Table 2). Comparisons of the richness curves calculated for the study areas indicated that species richness was significantly greater in DMOF than in the other two areas (Fig 2).

The studied euglossine communities exhibited differences in their species composition and abundance along the year and among the different vegetation types (Fig 3). In LSSF and SSSF, the communities showed a peak of abundance during the dry season, as well as another smaller peak during the rainy season (although less prolonged in SSSF). The highest abundance in the DMOF area was observed between December and February, during the rainy season (Fig 3B).

Temporal variation in abundance was largely determined by the dominant species. *Eulaema cingulata* was more abundant during the dry season in SSSF and LSSF, but did not demonstrate

a noticeable temporal pattern in the DMOF site (Fig 3D). The abundance peak of *Eulaema nigrita* occurred during the rainy season (January to April) in LSSF and DMOF, while its abundance in SSSF varied very little along the year (Fig 3E).

Euglossa cordata showed a peak of abundance in the LSSF site only during the dry period; this species showed two similar abundance peaks in the SSSF site, one during the dry season and one in the rainy season, while a single peak was observed in DMOF during the rainy season (Fig 3F). *Euglossa securigera* Dressler demonstrated two peaks of abundance in SSSF (Fig 3C) that were similar to those of *Euglossa cordata*; a few individuals were collected in the other two areas, making interpretations of its abundance more difficult. *Eufriesea violacea* and *E. surinamensis* (Linnaeus) were sampled only during the rainy period (between November and January).

Comparative analyses with other Atlantic Forest communities and correlation with abiotic factors

The correspondence analysis (DCA) of euglossine communities resulted in four major groupings (Fig 4): (1) one group clustered all of the Lowland Seasonal Semideciduous Forest areas (LSSF), between 0.0 and 0.4 on axis 1 and between -0.2 and -0.4 on axis 2, plus one of the Seasonal Semideciduous Submontane Forest sites (SSSF-5); (2) the second group encompassed the areas of Dense Lowland Ombrophilous Forests (DLOF) between -0.3 and -0.6 on axis 1 and between 0.0 and -0.4 on axis 2; (3) the third group, represented by the ar-

Table 1: Composition, abundance (Ab.), relative frequency (fr), richness, diversity, dominance, and evenness of the bee communities of the subtribe Euglossina in different vegetation types of the Atlantic Forest in Rio de Janeiro State, Brazil. LSSF: Lowland Seasonal Semideciduous Forests; SSSF: Submontane Seasonal Semideciduous Forest; DLOF: Dense Lowland Ombrophilous Forest; DSOF: Dense Submontane Ombrophilous Forest; and DMOF: Dense Montane Ombrophilous Forest.

Species	LSSF		SSSF		DMOF		TOTAL	
	Ab.	fr	Ab.	fr	Ab.	fr	Ab.	fr
Eufriesea surinamensis (Linnaeus)	1	0.2	1	0.1	4	1.4	6	0.4
E. violacea (Blanchard)	0	0	3	0.3	0	0	3	0.2
Euglossa annectans Dressler	0	0	0	0	9	3.2	9	0.5
E. bembei Nemésio	0	0	0	0	1	0.3	1	0.1
E. cordata (Linnaeus)	230	51.8	171	17.5	20	7	421	24.6
E. clausi Nemésio & Engel	0	0	6	0.6	1	0.3	7	0.4
E. despecta Moure	5	1.1	2	0.2	1	0.3	8	0.5
E. fimbriata Moure	2	0.5	22	2.3	4	1.4	28	1.6
E. pleosticta Dressler	23	5.2	10	1.0	5	1.7	38	2.2
E. securigera Dressler	12	2.7	213	21.8	9	3.1	234	13.7
E. truncata Rebêlo & Moure	0	0	0	0	1	0.3	1	0.1
Eulaema atleticana Nemésio	1	0.2	0	0	0	0	1	0.1
E. cingulata (Fabricius)	48	10.8	306	31.3	170	59	524	30.5
E. nigrita Lepeletier	120	27.0	242	24.7	63	22	425	24.9
Exaerete smaragdina (Guérin-Méneville)	2	0.5	2	0.2	0	0	4	0.2
TOTAL	444	100	978	100	288	100	1710	100
Richness	10		11		12		15	
Diversity (H')	1.26		1.52		1.32		1.61	
Dominance (D)	0.518		0.313		0.590		0.306	
Evenness (J)	0.570		0.650		0.528		0.595	

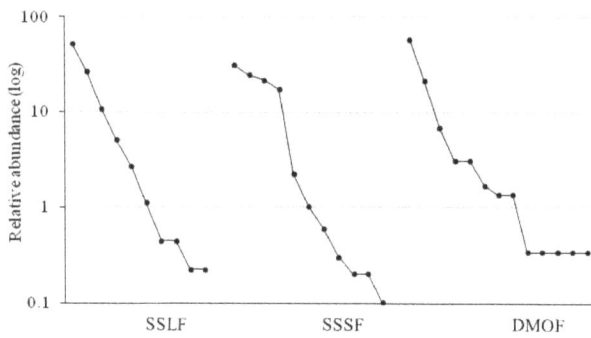

Fig 1. Abundance of orchid bee species in three forest communities in Rio de Janeiro State. LSSF: Lowland Seasonal Semideciduous Forests; SSSF: Submontane Seasonal Semideciduous Forest; and DMOF: Dense Montane Ombrophilous Forest. For descriptions of the areas, refer to Table S1.

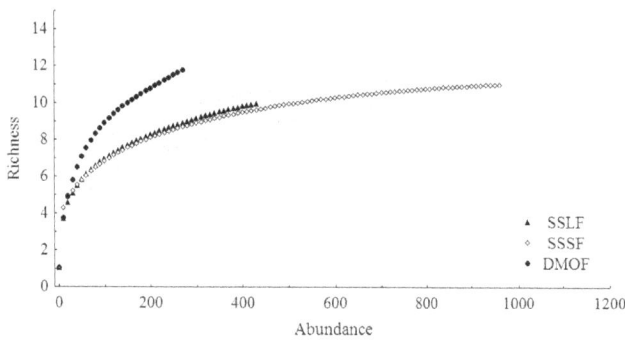

Fig 2. Rarefaction curves (1,000 simulations) for the species richness of orchid bees as a function of their abundance in areas with different vegetation types of forest in Rio de Janeiro State. Dotted lines indicate the upper and lower limits (95%) of each curve. The abbreviations follow those of Table S1.

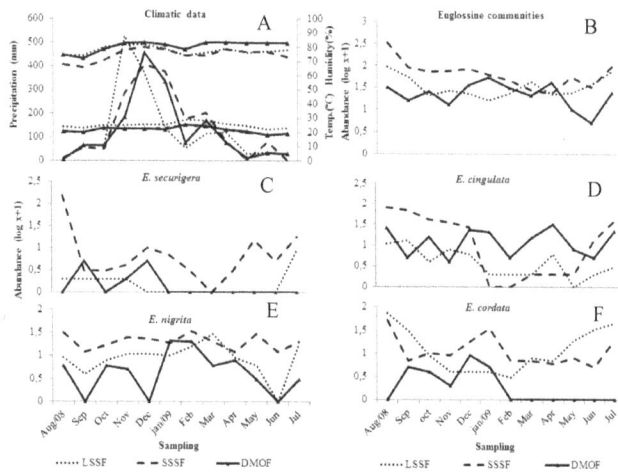

Fig 3. Climatic data (A) and temporal variations of Euglossine communities (B) in three vegetation types of forest in northern and northwestern Rio de Janeiro State, between August/2008 and July/2009. 3C to 3F represent the temporal variations of the most abundant species.

eas of SSSF and DMOF, was situated between 0.6 and 1.3 on axis 1 and between 0.0 and 0.4 on axis 2; (4) the fourth group, which encompassed the areas of DSOF, was more scattered in the diagram, with clear indication of subgroups within it. The eigenvalues of axis 1 and axis 2 were 0.20 and 0.19, respectively, while the percentage of variance explained by axis 1 was 24.9%, and 23.7% by axis 2.

The correlation between axis 1 of the DCA with the abiotic variables demonstrated a significant negative influence of precipitation (Person's r = - 0.56, N = 21, p = 0.001) and a positive influence of altitude (Pearson's r = 0.523, N = 21, P = 0.003) on the ordination of the euglossine communities. Only altitude had a significant positive influence on the second axis of the DCA (Pearson's r = 0.61, N = 21, P = 0.0004) (Table 3). The vegetation type likewise demonstrated an influence on the ordination of the euglossine communities on both axes 1 and 2 (Kruskal-Wallis test: H= 7.85, P = 0.049 and H= 15.39, P= 0.0015 respectively).

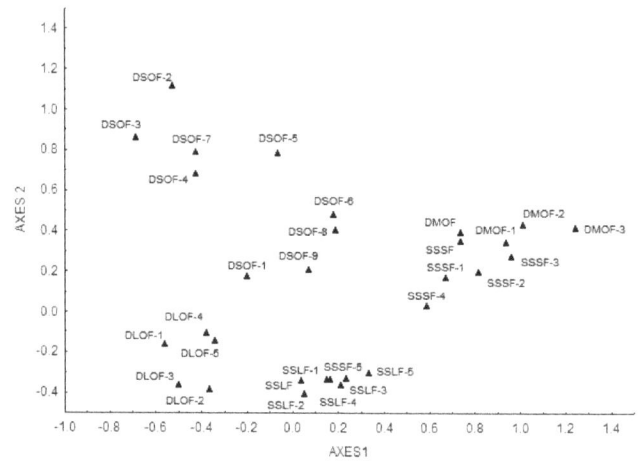

Fig 4: Diagram of the Detrended Correspondence Analyses (DCA) based on the composition of the orchid bee fauna found in areas in northern and northwestern Rio de Janeiro State. LSSF: Lowland Seasonal Semideciduous Forests; SSSF: Submontane Seasonal Semideciduous Forest; DLOF: Dense Lowland Ombrophilous Forest; DSOF: Dense Submontane Ombrophilous Forest; and DMOF: Dense Montane Ombrophilous Forest. The abbreviations follow those of Table S1.

Table 2 - Richness estimators for euglossine bee communities sampled between August/2008 and July/2009 in three different vegetation types of forest in Rio de Janeiro State, Brazil. The abbreviations follow those of Table S1.

Richness Estimator	LSSF	SSSF	DMOF
Jack1	12.75 ± 1.43	13.75 ± 1.97	17.6 ± 2.85
Chao1	10.33 ± 0.92	11 ± 0.16	23 ± 10.2
Bootstrap	11.17 ± 0	12.2 ± 0	14.93 ± 0

Table 3 - Pearson correlations between axes 1 and 2 of the Detrended Correspondence Analysis (DCA) and the variables of precipitation, temperature, humidity, and altitude in study areas in the northern and northwestern regions of Rio de Janeiro State, Brazil, relative to the compositions of their orchid bee fauna.

	Precipitation	Temperature	Humidity	Altitude
Axis 1	r = -0.56; P = 0.001	r = 0.014; P = 0.94	r = 0.24; P = 0.20	r = 0.52; P = 0.003
Axis 2	r =- 0.009; P = 0.96	r = -0.20; P = 0.291	r = 0.08; P = 0.632	r = 0.61; P = 0.0004

Discussion

The results of the present study, plus those obtained by Tonhasca et al. (2002a), Aguiar and Gaglianone (2008, 2011, 2012), and Ramalho et al. (2009) in other areas in the northern portion of the Rio de Janeiro State, revealed a total of 32 euglossine bee species for this region. This number corresponds to about 60% of the total orchid-bee fauna reported by Nemésio (2009) for the Atlantic Forest biome. Considering the result of the rarefaction curves and richness estimators, the orchid bee fauna in Dense Montane Ombrophilous Forest was underestimated and the real number of species is probably higher in the region.

Studies in other ombrophilous forests in Rio de Janeiro State (Tonhasca et al., 2002a; Ramalho et al., 2009) indicated higher number of species. In addition to insufficient sampling in the studied area, its regeneration time (approximately 60 years, according to local inhabitants), may not have been sufficient for more restrictive species to recolonize it. Species such as *Euglossa marianae* Nemésio and *Euglossa iopoecila* Dressler are known from areas of well-preserved Ombrophilous Forest (Tonhasca et al., 2002a; Ramalho et al., 2009).

These results point out to a great richness of orchid bee species in Dense Montane Ombrophilous Forest of the Atlantic Forest biome, even at the small spatial scales examined. This is possibly due to the great diversity of habitats, influenced by the wide geomorphological and climatic variation, besides the "Campos dos Goytacazes Gap", extending the Lowland Seasonal Semideciduous Forest to the coastline. This configuration forms a mosaic of landscapes probably favoring the occurrence of orchid bees in the region. The euglossine bee communities sampled in Dense Montane Ombrophilous Forest demonstrated the highest species richness among the studied physiognomies. The greatest average relative humidity values (up to 80%) seem to be the most important abiotic factor for this result. It was expected given the fact that other studies have indicated preference for humid forests by orchid bees (see Roubik & Hanson, 2004).

Eufriesea surinamensis and *Eulaema atleticana* were not found in Lowland Seasonal Semideciduous Forest in a previous study (Aguiar & Gaglianone, 2008) using the same methodology. Also four other species previously found there (*Eulaema niveofasciata* (Friese), *Euglossa gaianii* Dressler, *Euglossa leucotricha* Rebêlo & Moure, and *Euglossa truncata*) were not present in the current study. This reflects the dynamics of the bee communities in a short time scale, which might result in smaller chances of resampling rare species (Roubik, 2001).

Euglossa annectans (treated as a junior synonym of *Euglossa stellfeldi* Moure in Nemésio, 2009) had not been cited in previous surveys undertaken in Rio de Janeiro State although the species has been originally described from specimens collected at the Tijuca Forest, RJ (Dressler, 1982b). This species is endemic to the Atlantic forest domain, being distributed from the states of Bahia to Santa Catarina, and then into Argentina (Faria & Melo, 2007; Andrade-Silva et al., 2012). In this study *E. annectans* along with *E. bembei* were found only in Dense Montane Ombrophilous Forest, which, together with data from other studies (Tonhasca et al., 2002a; Darrault et al., 2006; Moure et al., 2007; Ramalho et al., 2009; Cortopassi-Laurino et al., 2009), indicates their close association with dense ombrophilous forests in Brazil. *Eulaema atleticana,* also endemic to the Atlantic Forest, has been considered restricted to coastal areas in northeastern and southeastern Brazil (Nemésio, 2009). However, more recent surveys have indicated that this species is common at altitudes above 700m (A. Nemesio, personal communication). The occurrence of *Eulaema atleticana* in the Lowland Seasonal Semideciduous Forests in Campos dos Goytacazes Gap may represent the southern limit of its distribution.

Eufriesea violacea was collected only in Submontane Seasonal Semideciduous Forest. This species has been previously found in other vegetation type of the Atlantic Forest domain (Tonhasca et al., 2002a; Nemésio & Silveira, 2006a; Giangarelli et al., 2009), but always in well-preserved areas. Giangarelli et al. (2009) reported that it was sensitive to habitat fragmentation and that its abundance becomes significantly reduced in small forest fragments. This can explain its reduced abundance in the present study.

Effects of fragmentation in the orchid bee communities were detected by Aguiar and Gaglianone (2012), who found changes in the pattern of species dominance in small fragments of Atlantic Forest. Originally the Lowland Seasonal Semideciduous Forests extended from southern Rio Grande do Norte to northern of Rio de Janeiro state. This forest was severely fragmented and the studied fragment represents only a very small portion of its original cover in the region (Silva & Nascimento, 2001). The number of species recorded by Bonilla-Gómez (1999) for this vegetation type was much higher and some relatively common species in this physiognomy have become rare in small forest fragments.

The dominance of *Euglossa cordata, Eulaema nigrita* and *Eulaema cingulata* in the present study was likewise observed in other surveys in the Atlantic Forest (Peruquetti et al.,1999; Aguiar & Gaglianone, 2008; Farias et al., 2008; Ramalho et al., 2009; Nemésio & Silveira, 2010). Ramalho et al. (2009) and Aguiar and Gaglianone (2012) demonstrated that these species were abundant in areas at different stages of conservation and do not confirm that they are indicators of disturbed areas, as suggested previously (e.g. Morato et al., 1992; Peruquetti et al., 1999). However, their tolerance to disturbance areas is unquestionable. Furthermore, the communal nesting or social behavior could favor their dominance in many of the communities studied (Zucchi et al., 1969; Garófalo, 1992; 1994; Augusto & Garófalo, 2004; Roubik & Hanson, 2004), as well as the great flight distance of Eulaema nigrita and E. cingulata (Dressler, 1982).

Despite exhibiting high relative abundances in all of the vegetation types studied, *Euglossa cordata* and *Eulaema cingulata* showed variations in their dominance patterns. *Euglossa cordata*, for example, was not dominant in Dense Montane Ombrophilous Forest, although its abundance was greater than 50% in Lowland Seasonal Semideciduous Forest. *Eulaema cingulata* was the most abundant species in Dense Montane Ombrophilous Forest but the third one in Lowland Seasonal Semideciduous Forests. *Euglossa securigera* also showed great variation in their dominance patterns, represented 23% of the individuals collected in Submontane Seasonal Semideciduous Forest but only about 3% of the individuals sampled in Lowland Seasonal Semideciduous Forests and

Dense Montane Ombrophilous Forest. Vegetation types and differences in the availability of key resources can change population patterns of pollinators (Lázaro & Totland, 2010), including euglossine communities (Ramirez et al., 2002; Souza et al., 2005; Nemésio & Silveira, 2007); detailed analyses of resource availability in the forests would be important to clarify this issue. According to Smith et al. (2012) the floral landscape in the neotropical forest is spatially and temporally heterogeneous for foraging bees, promoting considerable change in the abundance pattern and amount of brood in the nests for bees.

The abundance distribution patterns observed in the present study were similar to those described by other authors in different Atlantic Forest areas: few dominant species; usually two to four species, each one with more than 10% of abundance and many rare species (Rebêlo & Garófalo, 1997; Sofia et al., 2004; Aguiar & Gaglianone, 2008; Nemésio, 2007; Ramalho et al., 2009). However, this pattern differs from that found by Aguiar and Gaglianone (2012) for small fragments of Atlantic Forest, where most species had each more than 10% of abundance, suggesting the loss of rare species in small forest fragments.

The euglossine communities in Lowland Seasonal Semideciduous Forests and Submontane Seasonal Semideciduous Forest demonstrated seasonal peaks of abundance, with one conspicuous peak during the dry period that was influenced by the abundances of *Euglossa cordata*, *E. securigera* and *Eulaema cingulata*, as well as another peak in the rainy season with a predominance of *Eulaema nigrita* (thus similar to data presented by Aguiar and Gaglianone [2008, 2011]). The peak of abundance in Dense Montane Ombrophilous Forest occurred during the rainy season but the same was not demonstrated for the studied semideciduous forests. Long-term studies are necessary to confirm the different seasonal pattern observed in these areas.

Short periods of adult activity were observed for *Eufriesea violacea* and *E. surinamensis*. Seasonality of *Eufriesea* species was observed in different regions in Brazil, for example, semideciduous forest in Paraná (Giangarelli et al., 2009), ombrophilous forest in the Rio de Janeiro (Tonhasca et al., 2002a; Ramalho et al., 2009) and sandbanks in Maranhão state (Silva et al. 2009). Our data corroborate the seasonality of these bees in different physiognomies of Atlantic Forest.

Three of the four main groups detected in the Detrended Correspondence Analysis (DCA) (groups 1, 2 and 4 in the Results) showed a close correspondence of the euglossine communities with the predominant vegetation types. A single group combined fragments from distinct vegetation types, Seasonal Semideciduous Submontane Forest and Dense Montane Ombrophilous Forest. This grouping may be partly explained by their uniquely sharing of *Euglossa clausi*, as well as a higher relative abundance of *Eulaema cingulata*.

As stressed by Linsley (1958) and Abrahamczyk et al. (2011), different distribution patterns along natural gradients reflect different responses to changes in the biotic and abiotic factors acting along those gradients. In the case of the present study, the original Atlantic Forest was a continuum of forest, "restinga", and mangrove swamp ecosystems stretching along the entire eastern coast of Brazil (Galindo-Leal & Câmara, 2003) and included diverse natural gradients within its latitudinal and longitudinal extensions, as well as altitudinal variation going from the coast to inland (Galindo-Leal & Câmara, 2003).

Rainfall and altitude, and also vegetation types, were the main factors influencing the ordination of the euglossine communities in the study areas. According to Abrahamczyk et al. (2011), these factors were found to have a significant effect on the ordination of euglossine communities in the western Amazon region. Altitudinal variation, can drastically reduce abundance patterns and alter the species compositions of euglossine communities. Our data suggest that altitude is important and can modify species composition and abundance of euglossine bees, even at small scales of variation. This result has to be tested in future studies. Temperature is influenced by altitudinal variations, and it is an important factor in the regulation of flowering time in the Atlantic Forest (Talora & Morellato, 2000; Pereira et al., 2008), altering plant resource availability for bees. According to Hegland and Boeke (2006) and Lázaro and Totland (2010) the diversity and density of plants affect the foraging behavior and community of pollinators. Indeed, the incursion of the Lowland Seasonal Semideciduous Forests, due to the Campos of Goytacazes Gap, significantly modified the structure of vegetation in the region and can represent a great influence on the community structure of euglossine bees.

Precipitation has been found to be significantly related to the abundance and species richness of euglossine bees in the western Amazon region (Abrahamczyk et al., 2011). The Lowland Seasonal Semideciduous Forests area has relatively high average temperatures (25.3°C) associated with lower relative humidity levels (70%), which influenced the ordination of euglossine communities.

In addition to the climate and type of vegetation identified as strong influences on euglossine bee communities, other components, such as competition with similar species, historical occurrences, and habitat homogeneity (see Armbruster, 1993; Rosenzweig, 1995; Roubik, 2001; Tonhasca et al., 2002b; Roubik & Hanson, 2004) are also determinant for these bees. Also cleptoparasitic euglossine bees, such as those of the genus *Exaerete*, require additional biotic factors for their occurrence (such as the occurrence of host species in the genera *Eulaema* and *Eufriesea*) (Wcislo & Cane, 1996; Nemésio & Silveira, 2006b); however these factors have not been evaluated in this study, but should be considered in future studies.

Our data confirmed that a set of climatic, geomorphological and vegetational factors act to strongly influence euglossine species richness and composition (Ramirez et al., 2002; Souza et al., 2005; Nemésio & Silveira, 2007). *Euglossa annectans* and *E. bembei*, for example, were found only in Dense Montane Ombrophilous Forest, the survey region with the highest altitude (850 m) and humidity and the lowest average temperature (21 °C). Altitude, however, does not appear to be a determinant factor in the geographical distribution of these species, as they have been also found in areas of dense ombrophilous forest at lower altitudes in southern Brazil (with warmer temperatures) (Cortopassi-Laurino et al., 2009).

Our study reinforces the importance of the conservation of landscape mosaics that include various vegetation types, as the composition of euglossine bee communities as well as their patterns of abundance and species dominance have been found to differ among these distinct sites. Furthermore, the occurrence of rare species, such as *Eulaema atleticana* in the Lowland Seasonal Semideciduous Forest, demonstrates the need to preserve these areas, as was similarly observed for *Euglossa annectans* and *E. bembei* in Dense Montane Ombrophilous Forest and for *Eufriesea violacea* in Seasonal Semideciduous Submontane Forest. The conservation of this important group of neotropical pollinators is essential for the maintenance of ecological services and the genetic diversity of their host plant populations, and attention should be given to studies focusing on these points.

Acknowledgements

We thank Instituto Nacional de Meteorologia (INMET) for climatic data; Frederico Machado Teixeira, Marcelita França Marques, Mariana Scaramussa Deprá and Giselle Braga Menezes for their help in the field; FAPERJ for fellowships to the first author. We also thank the financial support provided by Procad/Capes (158/07), Rio Rural/SEAPPA/GEF, FAPERJ, and CNPq for the PQ scholarships to MCG and GARM.

References

Abrahamczyk, S., Gottleuber, P., Matauschek, C. & Kessler, M. (2011). Diversity and community composition of euglossine bee assemblages (Hymenoptera: Apidae) in western Amazonia. Biodiversity and Conservation, 20: 2981-3001. doi: 10.1007/s10531-011-0105-1

Ackerman, J.D. (1983). Specificity and mutual dependency of the orchid-euglossine bee interaction. Biological Journal of the Linnean Society, 20: 301-314. doi: 10.1111/j.1095-8312.1983.tb01878.x

Aguiar, W.M. & Gaglianone, M.C. (2008). Comunidade de Abelhas Euglossina (Hymenoptera: Apidae) em Remanescentes de Mata Estacional Semidecidual sobre Tabuleiro no Estado do Rio de Janeiro. Neotropical Entomology, 37: 118-125. Doi 10.1590/S1519-566X2008000200002

Aguiar, W.M. & Gaglianone, M.C. (2011). Euglossine bees (Hymenoptera, Apidae, Euglossina) on an inselberg in the Atlantic Forest domain of southeastern Brazil. Tropical Zoology, 24: 107-125.

Aguiar, W.M. & Gaglianone, M.C. (2012). Euglossine bee communities in small forest fragments of the Atlantic Forest, Rio de Janeiro state, southeastern Brazil (Hymenoptera, Apidae). Revista Brasileira de Entomologia, 56: 210-219. doi 10.1590/S0085-56262012005000018

Alvarenga, P.E.F., Freitas, R.F. & Augusto, S.C. (2007). Diversidade de Euglossini (Hymenoptera: Apidae) em áreas de cerrado do Triângulo Mineiro, MG. Bioscience, 23 Supplement 1: 30-37.

Andrade-Silva, A.C.R., Nemésio, A., Oliveira, F.F., & Nascimento, F.S. (2012). Spatial-Temporal Variation in Orchid Bee Communities (Hymenoptera: Apidae) in Remnants of Arboreal Caatinga in the Chapada Diamantina Region, State of Bahia, Brazil. Neotropical Entomology, 41: 296-305. doi 10.1007/s13744-012-0053-9.

Armbuster, W.S. (1993). Within-habitat heterogeneity in baiting samples of male Euglossine bees: possible causes and implications. Biotropica, 25: 122-128.

Augusto, S.C. & Garófalo, C.A. (2004). Nesting biology and social structure of *Euglossa* (*Euglossa*) *townsendi* Cockerell (Hymenoptera, Apidae, Euglossini). Insectes Sociaux, 51: 400-409.

Becker, P., Moure, J.S. & Peralta, F.J.A. (1991). More about euglossine bees in Amazonian forest fragments. Biotropica, 23: 586-591. doi: 10.2307/2388396

Bonilla-Gómez M.A. (1999). Caracterização da estrutura espaço-temporal da comunidade de abelhas euglossinas (Hymenoptera, Apidae) na Hiléia Baiana. Tese de Doutorado. Universidade Estadual de Campinas, 153p.

Brosi, B.J. (2009). The effects of forest fragmentation on euglossine bee communities (Hymenoptera: Apidae: Euglossini). Biological Conservation, 142: 414-423.

Chao, A. & Shen, T.J. (2005). Program SPADE (Species Prediction And Diversity Estimation). Programand user's guide at http://chao.stat.nthu.edu.tw (accessed on 18 October 12).

Colwell, R.K. (2006). EstimateS: statistical estimation of species richness and shared species from samples. Version 8.0. User's guide and application. University of Connecticut, USA. Available at: http://purl.oclc.org/estimates (accessed on 18 October 12)

Cortopassi-Laurino, M., Zillikens, A. & Steiner, J. (2009). Pollen sources of the orchid bee *Euglossa annectans* Dressler 1982 (Hymenoptera: Apidae, Euglossini) analyzed from larval provisions. Genetics and Molecular Research, 8: 546-556.

Dan, M.L., Braga, J.M.A. & Nascimento, M.T. (2010). Estrutura da comunidade arbórea de fragmentos de floresta estacional semidecidual na bacia hidrográfica do rio São Domingos, Rio de Janeiro, Brasil. Rodriguésia, 61: 1-18.

Darrault, R., Medeiros, P.C.R., Locatelli, E., Lopes, A. V., Machado, I. C. & Schlindwein, C. (2006). Abelhas Euglossini, in: Porto, K. Almeida Cortez, J. & Tabarelli (eds), Diversidade biológica e conservação da floresta Atlântica ao norte do rio São Francisco, Ministério do Meio Ambiente, Brasília, pp.352-354.

Dressler, R.L. (1982a). New species of Euglossa II. (Hymenoptera: Apidae). Revista de Biologia Tropical, 30: 121-129.

Dressler, R.L. (1982b). Biology of the orchid bee (Euglossini). Annual Review of Ecology and Systematics, 13: 373-394.

Faria, L.R.R. & Silveira, F.A. (2011). The orchid bee fauna (Hymenoptera, Apidae) of a core area of the Cerrado, Brazil: the role of riparian forests as corridors for forest-associated bees. Biota Neotropica, 11: 87-94 . doi 10.1590/S1676-06032011000400009

Farias, R.C.A.P., Madeira-da-Silva, M.C., Pereira-Peixoto, M.H. & Martins, C.F. (2008). Composição e Sazonalidade de Espécies de Euglossina (Hymenoptera: Apidae) em Mata e Duna na Área de Proteção Ambiental da Barra do Rio Mamanguape, Rio Tinto, PB. Neotropical Entomology, 37: 253-258. doi: 10.1590/S1519-566X2008000300003

Galindo-Leal, C. & Câmara, I.G. (2003). Atlantic forest hotspots status: an overview. In: C. Galindo-Leal & I.G. Câmara (Eds). The Atlantic Forest of South America: biodiversity status, threats, and outlook (pp. 3-11).Center for Applied Biodiversity Science e Island Press, Washington, D.C.

Garófalo, C.A. (1992). Comportamento de nidificação e estrutura de ninhos de *Euglossa cordata* (Hymenoptera: Apidae: Euglossini). Revista Brasileira de Biologia, 52: 187-198.

Giangarelli, D.C., Freiria, G.A., Colatreli, O.P., Suzuki, K.M. & Sofia, S.H. (2009). *Eufriesea violacea* (Blanchard) (Hymenoptera: Apidae): An orchid bee apparently sensitive to size reduction in forest patches. Neotropical Entomology, 38: 1-6.

Gotelli, N.J. & Entsminger, G.L. (2001). EcoSim: Null models programa for ecology. Version 7.0. Acquired Intelligence Inc. & Kesey-Bear (accessed on 23 September 12)

Hammer, O., Harper, D.A.T. & Ryan, P.D. (2001). Past. Paleontological Statistics Programa Package for Education and Data Analysis. Paleontologia Eletronica 4: 1-9.

Hegland, S.J. & Boeke, L. (2006). Relationships between the density and diversity of floral resources and flower visitor activity in a temperate grassland community. Ecological Entomology, 31: 532–538.

Janzen, D.H., de Vries, P.J., Higgins, M.L. & Kimsey, L.S. (1982). Seasonal and site variation in Costa Rican euglossine bees at chemical baits in lowland deciduous and evergreen forests. Ecology , 63: 6-74.

Köppen, W. & Geiger, R. (1928). Klimate der Erde. Gotha: Verlag Justus Perthes. Wall-map 150cmx200cm.

Lázaro, A. & Totland, O. (2010). Local floral composition and the behaviour of pollinators: attraction to and foraging within experimental patches. Ecological Entomology, 35: 652 - 661.

Linsley, E.G. (1958). The ecology of solitary bees. Hilgardia 27: 543-597.

Magurran, A.E. (2003). Measuring biological diversity. Blackwell Publishing, Oxford. 215p.

Mattozo, V.C., Faria, L.R.R. & Melo, G.A.R. (2011). Orchid bees (Hymenoptera: Apidae) in the coastal forests of southern Brazil: Diversity, efficiency of sampling methods and comparison with other Atlantic forest surveys. Papéis Avulsos de Zoologia, 51: 505-515. 10.1590/S0031-10492011003300001

Michener, C.D. (1979). Biogeography of the bees. Annals of the Missouri Botanical Garden, 66: 277–347.

Morato, E.F., Campos, L.A.O. & Moure, J.S. (1992). Abelhas Euglossini (Hymenoptera, Apidae) coletadas na Amazônia Central. Revista Brasileira de Entomologia, 36: 767-771.

Moure, J.S., Melo, G.A.R. & Faria Jr., L.R.R. (2007). Euglossini Latreille, 1802. In: J.S. Moure, D. Urban& G.A.R.Melo (Eds). Catalogue of Bees (Hymenoptera, Apoidea) in the Neotropical Region. (pp. 214–255). Sociedade Brasileira de Entomologia, Curitiba-PR.

Myers, N., Mittermeier, R.A., Mittermeier, C.G., Fonseca, G.A.B. & Kent, J. (2000). Biodiversity hotspots for conservation priorities. Nature 403, 853-858. doi: 10.1038/35002501

Nemésio, A. & Silveira, F.A. (2006a). Edge effects on the orchid-bee fauna (Hymenoptera: Apidae) at a large remnant of Atlantic Forest in southeastern Brazil. Neotropical Entomology, 35: 313–323. doi: 10.1590/S1519-566X2006000300004

Nemésio, A. & Silveira, F.A. (2006b). Deriving ecological relationships from geographical correlations between host and parasitic species: an example with orchid bees. Journal of Biogeography, 33: 91-97. doi: 10.1111/j.1365-2699.2005.01370.x

Nemésio, A. & Silveira, F.A. (2007). Diversity and distribution of orchid bees (Hymenoptera: Apidae: Euglossina) with a revised checklist of their species. Neotropical Entomology, 36: 874-888. doi: 10.1590/S1519-566X2007000600008

Nemésio, A. & Silveira, F.A. (2010). Forest Fragments with Larger Core Areas Better Sustain Diverse Orchid Bee Faunas (Hymenoptera: Apidae: Euglossina). Neotropical Entomology, 39: 555-561. Doi 10.1590/S1519-566X2010000400014

Nemésio, A. & Vasconcelos, H. L. (2013). Beta diversity of orchid bees in a tropical biodiversity hotspot. Biodiversity and Conservation, 22: 1647-1661. doi: 10.1007/s10531-013-0500-x

Nemésio, A. (2007). The community structure of male orchid bees along the Neotropical Region. Revista Brasileira de Zoologia, 9: 151-158.

Nemésio, A. (2008). Orchid bee community (Hymenoptera, Apidae) at an altitudinal gradient in a large forest fragment in southeastern Brazil. Revista Brasileira de Zoologia, 10: 249-256.

Nemésio, A. (2009). Orchid bees (Hymenoptera: Apidae) of the Brazilian Atlantic Forest. Zootaxa, 2041: 1–242.

Oliveira, M.L. & Campos, L.A.O. (1995). Abundância, riqueza e diversidade de abelhas euglossinae (Hymenoptera, Apidae) em florestas contínuas de terra firme na Amazônia central, Brasil. Revista Brasileira de Zoologia, 12: 547-556. doi: 10.1590/S0101-81751995000300009

Oliveira-Filho, A.T. & Fontes, M.A.L. (2000). Patterns of floristic differentiation among Atlantic Forests in Southeastern Brazil and the influence of climate. Biotropica, 32: 793-810. doi: 10.1111/j.1744-7429.2000.tb00619.x

Oliveira-Filho, A.T., Tameirão-Neto, E., Carvalho, W.A.C.; Werneck, M., Brina, A.E., et al. (2005). Análise florística do compartimento arbóreo de áreas de Floresta Atlântica sensu lato na região das Bacias do Leste (Bahia, Minas Gerais, Espírito Santo e Rio de Janeiro). Rodriguésia: 56: 185-235.

Pereira, T.S, Costa, M.L.M.N., Moraes, L.F.D. & Luchiari, C. (2008). Fenologia de espécies arbóreas em Floresta Atlântica da Reserva Biológica de Poço das Antas, Rio de Janeiro, Brasil. Iheringia Sér. Bot., 63, 329-339.

Peruquetti, R.C., Campos, L.A.O., Coelho, C.D.P., Abrantes, C.V.M. & Lisboa, L.C.O. (1999). Abelhas Euglossini (Apidae) de áreas de Mata Atlântica: abundância, riqueza e aspectos biológicos. Revista Brasileira de Zoologia, 16: 101-118. doi: 10.1590/S0101-81751999000600012

Powell, A.H. & Powell, V.N. (1987). Population dynamics of male euglossine bees in Amazonian forest fragments. Biotropica, 19: 176-179.

RadamBrasil (1983) Levantamento de recursos naturais, v. 32. folha S / F 23 / 24. Rio de Janeiro/ Vitória. Ministério das Minas e Energia, Rio de Janeiro.

Ramalho, A.V., Gaglianone, M.C. & Oliveira, M.L. (2009). Comunidades de abelhas Euglossina (Hymenoptera, Apidae) em fragmentos de Mata Atlântica no Sudeste do Brasil. Revista Brasileira de Entomologia, 53: 95-101. doi: 10.1590/S0085-56262009000100022

Rebêlo, J.M.M. & Garófalo, C.A. (1991). Diversidade e sazonalidade de machos de Euglossini (Hymenoptera, Apidae) e preferências por iscas-odores em um fragmento de floresta no sudoeste do Brasil. Revista Brasileira de Biologia, 51: 787-799.

Rebêlo, J.M.M. & Garófalo, C.A. (1997). Comunidades de machos de Euglossini (Hymenoptera: Apidae) em matas semidecíduas do nordeste do estado de São Paulo. Anais da Sociedade Entomológica do Brasil, 26: 243-255.

Ribeiro, M.C., Metzger, J.P., Martensen, A.C., Ponzoni, F.J. & Hirota, M.M. (2009). The Brazilian Atlantic Forest: How much is left, and how is the remaining forest distributed? Implications for conservation. Biological Conservation, 142: 1141-1153. doi: 10.1016/j.biocon.2009.02.021

RIORURAL-GEF (2007) Marco Zero: Sub-Componente Monitoramento e Avaliação, Microbacia Santa Maria/Cambiocó, São José de Ubá. Relatório Técnico. Rio de Janeiro.

Rosenzweig, M.L. (1995) Species Diversity in Space and Time. Cambridge University Press, New York, NY.

Roubik, D.W. & Hanson, P.E. (2004). Orchids bees of Tropical America: Biology and Field Guide. INBio Press, Heredia, Costa Rica. 370p.

Roubik, D.W. (2001). Ups and downs in pollinator populations: When is there a decline? Conservation Ecology, 5: 2. [online] URL: http: //www. Consecol .org/vol5 /iss1 /art2/. (accessed on 14 March 2013).

Silva, G.C. & Nascimento, M.T. (2001). Fitossociologia de um remanescente de mata sobre tabuleiros no norte do estado do Rio de Janeiro (Mata do Carvão), RJ, Brasil. Revista Brasileira de Botânica, 24: 51-62. doi: 10.1590/S0100-84042001000100006

Silva, O., Rego, M.M.C., Albuquerque, P.M.C. & Ramos, M.C. (2009). Abelhas Euglossina (Hymenoptera: Apidae) em Área de Restinga do Nordeste do Maranhão. Neotropical Entomology, 38: 186-196. doi: 10.1590/S1519-566X2009000200004

Silveira, F.A., Melo, G.A.R. & Almeida, E.A.B. (2002). Abelhas Brasileiras, Sistemática e Identificação. Belo Horizonte, 253p.

Silveira, G.C., Nascimento, A.M., Sofia, S.H. & Augusto, S.C. (2011). Diversity of the euglossine bee community (Hymenoptera, Apidae) of an Atlantic Forest remnant in southeastern Brazil. Revista Brasileira de Entomologia, 55: 109-115.

Smith, A.R., López Quintero, I.J., Moreno Patiño, J.E., Roubik, D.W. & Wcislo, W.T. (2012). Pollen use by *Megalopta* sweat bees in relation to resource availability in a tropical forest. Ecological Entomology, 37: 309–317.

Sofia, S.H. & Suzuki, K.A. (2004). Comunidade de machos de abelhas Euglossina (Hymenoptera: Apidae) em fragmentos florestais no sul do Brasil. Neotropical Entomology, 33: 693-702.

Sofia, S.H., Santos, A.M. & Silva, C.R.M. (2004). Euglossine bees (Hymenoptera, Apidae) in a remnant of Atlantic Forest in Paraná State, Brazil. Iheringia Série Zoologia, 94: 217-222. doi: 10.1590/S0073-47212004000200015

Souza, A.K.P., Hernández, M.I.M. & Martins, C.F. (2005). Riqueza, abundância e diversidade de Euglossina (Hymenoptera, Apidae) em três áreas da Reserva Biológica Guaribas, Paraíba, Brasil. Revista Brasileira de Zoologia, 22: 320-325. doi: 10.1590/S0101-81752005000200004

StatSoft, Inc. (2007). Statistica (data analysis software system), version 8.0. www.statsoft.com.

Storck-Tonon, D.; Morato, E.F. & Oliveira, M.L. (2009). Fauna de Euglossina (Hymenoptera: Apidae) da Amazônia Sul-Ocidental, Acre, Brasil. Acta Amazonica, 39: 693-706.

Sydney, N.V., Goncalves, R.B. & Faria, L.R.R. (2010). Padrões espaciais na distribuição de abelhas Euglossina (Hymenoptera, Apidae) da região Neotropical. Papéis Avulsos de Zoologia, 50: 667-679. doi: 10.1590/S0031-10492010004300001

Talora, D.C. & Morellato, L.P.C. (2000). Fenologia de espécies arbóreas em floresta de planície litorânea do sudeste do Brasil. Revista Brasileira de Botânica, 1: 13-26.

Ter Braak, C.J.F. (1995). Ordination. In: R.H.G. Jongman, C.J.F ter Braak, & O.F.R. van Tongeren (Eds.). Data analysis in community and landscape ecology (pp. 90-212). Cambridge, Cambridge University Press.

Tonhasca, Jr.A., Blackmer, J.L. & Albuquerque, G.S. (2002a). Abundance and diversity of euglossine bees in the fragmented landscape of the Brazilian Atlantic Forest. Biotropica, 34: 416-422.

Tonhasca, Jr.A., Blackmer, J.L. & Albuquerque, G.S. (2002b). Within-hábitat heterogeneity of euglossine bee populations: a re-evaluation of the evidence. Journal of Tropical Ecology, 18: 929-933. doi: 10.1017/S0266467402002602.

Uehara-Prado, M. & Garófalo, C.A. (2006). Small-scale elevational variation in the abundance of *Eufriesea violacea* (Blanchard) (Hymenoptera: Apidae). Neotropical Entomology, 5: 446-451.

Veloso, H.P., Rangel-Filho, A.L.R. & Lima, J.C.A. (1991). Classificação de vegetação brasileira adaptada a um sistema universal. Rio de Janeiro, Instituto Brasileiro de Geografia e Estatística. 123p.

Wcislo, W.T. & Cane, J.H. (1996). Floral resource utilization by solitary bees (Hymenoptera: Apoidea) and exploitation of their stored foods by natural enemies. Annual Review of Entomology, 41: 257-86. doi: 10.1146/annurev.en.41.010196.001353

Whittaker, R.H. (1965). Dominance and diversity in land plant communities. Science, 147: 250-260.

Zahl, S. (1977). Jackknifing an Index of Diversity. Ecology, 58: 907-913. doi: 10.2307/1936227

Zucchi, R., Sakagami, S.F. & Camargo, J.M.F. (1969). Biological observations on a neotropical parasocial bee, *Eulaema nigrita*, with a review on the biology of Euglossinae. Journal of the Faculty of Sciences of Hokkaido Univerdity, 17: 271-380.

7

Trophic Guild Structure of a Canopy Ants Community in a Mexican Tropical Deciduous Forest

GABRIELA CASTAÑO-MENESES[1,2]

1 - Ecología y Sistemática de Microartrópodos, Depto de Ecología y Recursos Naturales, Facultad de Ciencias, Universidad Nacional Autónoma de México, México, DF.

2 - Current Address: Unidad Multidisciplinaria de Docencia e Investigación, Facultad de Ciencias, Universidad Nacional Autónoma de México, Campus Juriquilla, Querétaro, México.

Keywords
Diversity, Chamela, Fogging, Species Richness

Corresponding author
Gabriela Castaño-Meneses
Ecología y Sistemática de Microartrópodos
Deptamiento Ecología y Recursos Naturales, Facultad de Ciencias, UNAM
Ciudad Universitaria, 04510
México, D. F.
E-mail: gabycast99@hotmail.com

ABSTRACT

Seasonality in tropical dry forest shows extreme changes in the physiognomy of forest as well the available resources in each season, thus, the composition and diversity of fauna inhabiting in that ecosystem show seasonal variations in answer to that changes. The ants constitute a very important element in the canopies of tropical forests, and there is few information about their communities in dry forest. In most of ecosystems, the general patterns of ant distribution show increase of their abundance during the wet season, but according with the characteristics of Chamela tropical dry forest in the Pacific Cost of Mexico, the great amount of epiphytes in the area can be an important resource to the ants, and the canopy can be an environment visited for different species of ants during the driest month. In order to study the seasonal variations in species richness, composition and diversity of ant canopy community in a tropical deciduous forest, seven fogging were performed in a watershed of the Chamela Biological Station, Jalisco State, Mexico, including dry and rainy season. A total of 5 563 ant specimens were collected belong to 46 morphospecies from 17 genera. The most species richness genera were *Camponotus* and *Cephalotes*, with 13 and 6 species respectively, and the most abundant ants were species of *Crematogaster, Tapinoma, Cephalotes* and *Camponotus* genera. Nevertheless dominant species were present during all the study, abundance show a great seasonality, with highest values during the dry season. The dominant guild in the canopy was the omnivorous in all study, but differences in guild trophic composition were recorded in each fogging. The ant community in the canopy of Chamela shows important seasonal variations in the composition and trophic guilds dominance, due conditions of this forest, that differences can be result of variations in the exploitation of resources along the year, and vertical migrations of ant species from soil and shrub layer to canopy in the tropical deciduous forest.

Introduction

In tropical regions, ants constitute one of the most abundant and diverse groups in canopies (Tobin, 1995; Davidson et al., 2003; Lach et al., 2010). In Peruvian forest canopy, Wilson (1987) found 135 ant species while in Budongo Forest, Uganda, Schulz and Wagner (2002) recorded 161 species; in the South part of Australia, ants represented about 48% of the total of arboricolous fauna (Andersen & Yen, 1992), while in Borneo constituted near 18% (Stork, 1987) and in New Caledonia close to 7% (Guilbert et al., 1995). Those variations are related with the habitat heterogeneity and the availability of resources, factors that affect the species richness and abundance of the communities, and in the case of tropical ants, they can exploit a great variety of resources that are provided directly or indirectly by several tree species with different phenologies (Ribas et al., 2003; Armbrecht et al., 2004).

The seasonality is a very important component of the ant communities structure in tropical forests (Basset et al., 2003), and in the tropical deciduous forest its importance increases considerably (Dirzo et al., 2011), due to the changes in the availability of resources and the ability of the organisms to use them according to their feeding habits. Thus, species composition of trophic guilds shows changes according to the season, due to trophic guilds that answer differently to environmental conditions modifing the interactions between species and the composition of community (Meyer et al., 2010; Cook et al., 2011).

Into the tropical forest areas, the tropical dry forest ecosystems comprise more than 40% of surface area in the World, but these ecosystems show an important loss of forest area in recent years as a result of accelerated anthropogenic disturbance (Trejo & Dirzo, 2000; Dirzo et al., 2011).

There are few information about the importance of seasonality on the ant activity and structure of communities in the tropical dry forest (Neves et al., 2010), and in Mexico there are only records for the Atlantic Coast (Gove et al., 2005), but there are no studies in the Pacific Coast. The study of ant communities in canopies from Mexico is still unexplored, with a few notes about their abundance (Palacios-Vargas et al., 1999) or their importance as indicators of perturbation (Gove et al., 2005). Thus, in the present work, the temporal variation and the trophic guild distribution in the canopy ant community of Chamela, Jalisco, in the Pacific Cost of Mexico were analyzed in order to study the seasonal pattern shown by ants in this vegetation.

Material and Methods

The study was carried out in the Chamela Biological Station (ChBS) of the Instituto de Biología of Universidad Nacional Autónoma de México (UNAM). This is a natural reserve located at the Pacific Coast of Mexico, in the state of Jalisco (19°83'00"N 105°80'30"W; 150 m elevation). The rainy season, according to Bullock (1986), lasts four months, from July to October, with more than 50% of precipitation during September and October (García-Oliva et al., 2002). Mean annual precipitation and temperature are 788 mm and 24.68°C, respectively (1977–2000; García-Oliva et al., 2002). Details of physical and biological parameters of the reserve have been compiled by Bullock (1988) and Noguera et al. (2002). The flora and vegetation structure of the forest have been described (Lott, 1985; Lott et al., 1987; Balvanera et al., 2002).

A total of seven fumigations were performed in order to sample the canopy. Fogging sessions were made using a Dyna fog machine. The sampling was performed during rainy and dry seasons from 1992 to 1994, in August and September 1992 (rainy season); May (dry season), July (rainy) and November (dry) 1993; and February and May 1994 (dry). Sampling sites were located in the watershed named 4A (Cervantes et al.,

1988), where the tree layer was about 25 m tall. Dominant species in the area are *Guapira macrocarpa* (Miranda), *Celaenodendron mexicanum* Standl., *Lonchocarpus eriocarinalis* Micheli, *Lonchocarpus constrictus* Pittier, *Bursera instabilis* McVaugh & Rzed., *Tabebuia impetiginosa* (Mart.) Standl. and *Caesalpinia eriostachys* Benth., and tree density is about $2,686 \pm 84$ ind. ha^{-1} (Maass et al., 2002a). The average number of trees sampled in each plot was 30 ± 7 ind. In each occasion, a new plot of 100 m^2 was delimited and 50 plastic funnels with 50 cm of diameter were hung randomly in the shrub layer at 50 cm above floor forest at intervals of 50 cm. According to the number and area of the funnels, the biological material retained in each fogging comprises an area of 9.82m^2, and in total were sampled 68.7m^2 in the seven foggings. The average net primary productivity in the forest is 3.2 Mg ha^{-1}y^{-1} (Martínez-Yrízar et al., 1996). The application of insecticide was between 04:00 and 06:00, using a solution of 3% of Resmethrin in kerosene solution. A total of 6L of solution was used in each application. After 5h of the insecticide application, the funnels were washed with 80% ethanol, in order to collect the specimens fell in them. The material was stored in plastic bottles with 1L capacity. The ant specimens obtained were isolated, quantified and identified as morphospecies. The specimens were deposited in Colección de Hormigas del Ecología y Sistemática de Microartrópodos (LESM), Facultad de Ciencias, UNAM. Only workers and soldiers were considered, because they are a better reference to sample canopy habitats (Wilson, 1987). According to their feeding habit preferences, the species were classified as omnivorous, predators, herbivorous, granivorous and nectarivorous (considered the consumption of extrafloral nectaries and hemipteran secretions; Byk & Del-Claro 2010, 2011).

The diversity index per fumigation was estimated by Shannon diversity index. Species richness, Pielou's evenness, and Simpson's dominance indices were also calculated for the study (Ludwing & Reynolds, 1988).

The effect of the season on the ant abundance was evaluated by a nested ANOVA test, nesting month collection within the corresponding season and significant differences were tested by *post hoc* Tukey's test (Zar, 1984). The Spearman correlation between precipitation and temperature and the diversity indices were calculated, as well as the correlation between ant density and the precipitation and temperature. Climatic data were obtained from the meteorological station of Chamela (http://www.ibiologia.unam.mx/ebchamela/www/clima.html). The analyses were performed using *Statistica* version 9.0 software (StatSoft, 2009).

Seasonal variations in the ant community composition were analyzed by ordination analysis through nonmetric dimensional scaling (NMDS), and similarity between groups was tested by a similarity analysis (ANOSIM), using 1000 permutations, following Clarke (1993) and Clarke and Green (1988). Analyses were performed using PAST software (Hammer et al., 2001).

Results

A total of 5,563 specimens belonging to 46 morphospecies belonging to 17 genera of ants were collected during the seven fumigations (Appendix 1). The average density of ants in the canopy was 81 ind/m², while the species ant density was 26 species/m². The Myrmicinae subfamily was the most diverse, represented by 21 morphospecies grouped in nine genera. The most abundant genus in the sampling was *Crematogaster*, with *Crematogaster crinosa* Mayr as the most abundant species, shown an average density of 26 ind/m² in the canopy, representing the 32% of the total. This genus is considered predominantly arboriculous.

The genus *Camponotus* was the most rich in morphospecies number, with 13 species, followed by *Pseudomyrmex* and *Cephalotes* (five each one). All of them are considered predominantly arboricolous.

The calculated diversity indices to the ant community in canopy show values relatively high, compared with other studies, and there are variations between fumigations (Table 1). The highest diversity values were found during the rainy months (August, September), except July, where the diversity and species richness recorded were the lowest during the study. The dominant ants were different in each fumigation, as show by Simpson's dominance index (Table 1; Fig 1). The species better represented in the canopy along the study were *C. crinosa, C. sumichrasti, Tapinoma melanocephalum*

(Fabricius) and *Forelius kieferi* Wheeler (Fig 2), represent 77% of the total. The variation in the species richness, diversity and evenness show that there are high temporal variation in the ant canopy community in Chamela. The NMDS analysis shows the aggregation of September and July, nevertheless the other rainy month, August is located in the same quadrant with November, because both were the months with more species richness. There is an important note that during November, atypical rains fell in Chamela, with a higher amount of precipitation than in rainy months of the same year and in other years (http://www.ibiologia.unam.mx/ebchamela/www/clima.html).

The two samplings from May are grouped, showing similar composition. The ANOSIM test indicated that the observed community in the sampling months was significantly different (global R = 0.64).

The date of collection is an important factor in the structure of the community of ants in the canopy of Chamela. The nested ANOVA test showed a significant effect of the season on the ant density in the canopy (F = 8.45; df = 5, 343; p<0.005), and *post hoc* Tukey's test showed differences between the fumigations performed in July and May in relation to the other months (p<0.05). A possible reason for which July showed a difference with the other rainy months is that it is the month of the begining of the rains, and the conditions can differ regarding August and September, in the middle of the rainy season.

The ant density in the canopy was higher during the dry months' fumigations (May and February). Density average during rainy months it was 73 ind/m², while in the dry months was 92 ind/m². Nevertheless there was not a significant correlation between density of ants and the precipitation recorded during the months of fumigation (r = -0.35, df=5; p>0.05),

Table 1. Parameters of the ant canopy community in Chamela, Jalisco, Mexico. S = Species richness; H'= Shannon diversity index; J´= Pielou's evenness; 1/D = Simpson's dominance index

Fumigation	S	H'	J´	1/D
August-1992	26	2.02	0.62	4.2
September- 1992	17	2.13	0.75	6.3
May-1993	17	1.68	0.59	3.4
July-1993	16	1.16	0.42	1.8
November-1993	29	1.58	0.47	2.8
February-1994	19	1.59	0.54	1.5
May-1994	22	1.34	0.43	2.5
Total	46	1.99	0.52	4.8

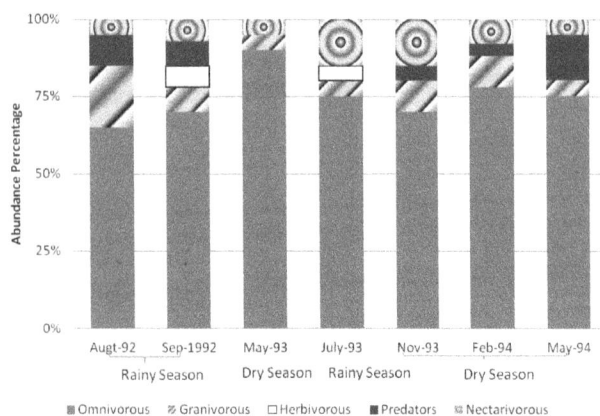

Fig 1 Temporal variation in canopy ant community composition in Chamela, Jalisco. Numbers on the lines indicate the number of species included in Others in each sampling.

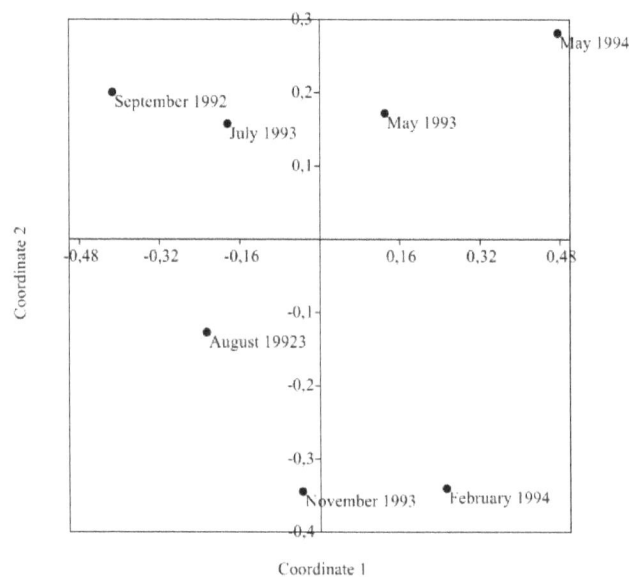

Fig 2 Non-metric multidimensional scaling ordination (NMDS) in two-dimensions of the canopy ant community inhabiting in a tropical dry forests in Chamela, Jalisco, Mexico. Ordination was based on Bray-Curtis dissimilarity index. Stress = 17%.

neither with the temperature (r =-0.17, df = 5; p >0.05; Table 2).

In relation to the trophic guilds of ants found in canopy of Chamela, five guilds were recorded: omnivorous, predators, herbivorous, granivorous and nectarivorous. Omnivorous was the dominant guild in the canopy, representing about 60% of the species founded, followed by granivorous and predators (Fig 3). The trophic guilds distribution was different during the fumigations, and some guilds were found only during the fumigations performed in rainy months, as the case of the herbivorous. Predators increased its abundance during the rainy months while omnivorous increased during the dry months (Fig 3). That pattern produce differences in ant composition along the year, due the variations of feeding habits and the capability of the ant species to use different resources.

Discussion

According to the results of Palacios-Vargas et al. (1999), ants represent the 0.5% of the total arthropods collected by fogging in Chamela, and the pattern of abundance differs of the observed in other groups, as springtails, where the highest abundances were found during rainy season, while ants showed higher abundances in the dry season. The canopy of Chamela showed a particular phenology, because leaves of tree species fall during dry season, but there are many tree species that are in flowering in this season (Bullock & Solís-Magallanes, 1990; Bullock, 2002), and that constitute important resources for many arthropods, including ants. Furthermore, there is a high density and diversity of epipythes in the area (Lott & Atkinson, 2006). These epiphytes can be exploited by ants, and constitute an important refuge during the dry season, due to their capacity to accumulate detritus and water. In epiphytes and branches of trees in Chamela there are important accumulation of organic matter, with amounts higher in the canopy than in the soil (Maass et al., 2002b), and an important phase of the decomposition cycle is developed in the canopy, and many groups of invertebrates can live in that environment, such as Collembola (Palacios-Vargas & Gómez-Anaya, 1993; Palacios-Vargas et al., 1998; Palacios-Vargas & Castaño-Meneses, 2003) that can be potential preys to some groups of ants as *Strumigenys* and *Neivamyrmex* (Brown, 1959; Bolton 1999).

In the present study the most abundant genus in the canopy was *Crematogaster*, an ant genus considered as arboricolous and is frequently found in high populations in rainy forest, different studies show that it represents more than 44% of the total collected arthropods (Basset et al., 1992), and in Thailand is the genus with the highest species richness in the canopy of dominant deciduous tree *Elateriospermum tapos* Blume (Jantarit et al., 2009). The two species of the genus *Crematogaster* found in Chamela (*C. crinosa* and *C. sumichrasti* Mayr) are probably not competitors, nevertheless both are considered as arboricolous, even the first though can be found in the shrub layer and soil in the forest (Castaño-

Table 2. Spearman correlation coefficient between the diversity index with precipitation and temperature monthly average (N = 7) in the canopy of Chamela Biological Station, Chamela, Jalisco, Mexico. H'= Shannon diversity index, S = Species richness, J' = Pielou's evenness index, ns= no significant at α=0.05.

Index	Precipitation	Temperature
H'	0.23 ns	0.38 ns
S	-0.34 ns	0.23 ns
J'	0.27 ns	0.32 ns

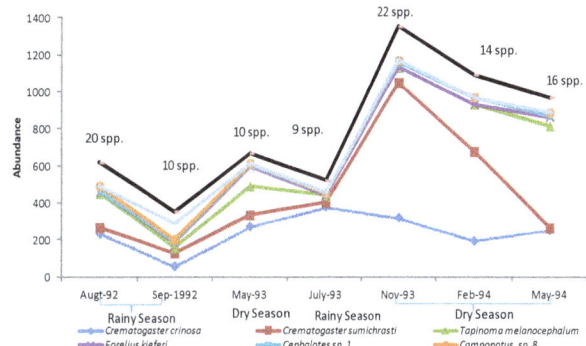

Fig 3 Temporal variation of trophic guilds of ants from the canopy of Chamela, Jalisco, Mexico.
Percentage of ant species from canopy of Chamela included in each trophic guild. N=46. Average percent of each guild: omnivorous: 60%; granivorous: 17%; herviborous: 2%; predators: 14%; nectarivorous: 7%.

Meneses, 2008; Castaño-Meneses et al., 2009). This species has been recorded as very abundant in the upland forest in Peru (Wilson, 1987).

Genera as *Cephalotes*, *Pseudomyrmex* and *Camponotus* were also abundant and species richness in canopy of Chamela. Dominance of that genera has been recorded in different tropical canopies around the World (Wilson, 1987; Guilbert & Casevitz-Weulersse, 1997; Watanasit et al., 2005), thus the composition and dominance in tropical dry forest is similar to the found in tropical rainy forest.

The presence and abundance of *T. melanocephalus* in canopy of Chamela is remarkable, because this is considered as tramp ant species or invasive. In Mexico, *T. melanocephalus* has been reported in canopies of rain forest in Chiapas, associated to orchids (Damon & Pérez-Soriano, 2005), as well in epiphytes from Panama rain forest (Stuntz et al., 2003). This species has one of the widest distribution ranges for any ant species, and its origin has been discussed (Wilson & Taylor, 1967), but, in general. consensus suggests *T. melanocephalum* origin in the Old World tropics, and recent studies indicate that it is most probably originated in the Indo-Pacific (Wetterer, 2009). Nevertheless ,there are some evidence that support the hypothesis of the Neotropical origin of *T. melanocephalum* (Wetterer, 2009), and the presence of alated females in samples of canopy of Chamela, as well the similar seasonal pattern of this species with the other abundant species considered as arboricolous, can be an evidence to support this hypothesis, or well suggest the success of this ant colonizing the forest.

Studies developed in New Caledonia using fogging, recorded 27 species and 14 genera of ants in the canopy (Guilbert & Casevitz-Weulersse, 1997). In the tropical rain forest of Peru, Wilson (1987) recorded 135 species and 40 genera of ants, while in Australia there is a great variation, with records of 37 species in the North region (Majer, 1990), to 102 species in the South (Andersen & Yen, 1992). Species richness recorded in the canopy of Chamela shown values between rainy forest and temperate forest, according with different studied performed in that canopies vegetation, nevertheless different sampling methods has been used (Wilson, 1959; Schonberg et al., 2004; Bos et al., 2007; Jaffe et al., 2007). The genera composition recorded in all canopies is similar in different studies.

The forest canopy can support high populations of organisms which had been recorded in tropical rain forest (Longino & Nadkarni, 1990; Paoletti et al., 1991). It has been proposed that the arboricolous fauna estimation can give a good estimation of the total species in the World (Erwin, 1983; Ødegaard, 2000; Longino et al., 2002). Ants can exploit a great variety of resources due the diversity of their feeding habits. The distribution of trophic guilds was different during the fumigations. That produced differences in ant composition along the year, due the variations of feeding habits and the capability of the ant species to use different resources. The domination of omnivorous is frequent in many ecosystems, including the canopy, and in environments with limited resources (Rojas 2001), due to the specialized feeding habits characteristic of environments with diversity of resources (Lévieux, 1977).

Although no significant correlations were found in the study between abundance, species richness and diversity with precipitation and temperature, these results must be viwed with caution, because the data were collected in atypical climatic conditions recorded during the studied period, as the great precipitation amount in November.

The results show that the ant community in the canopy of Chamela is diverse and have important changes in functional composition along the time. The diversity of trophic guilds showed that the ecosystem present a high productivity supporting different trophic levels in communities of ants and other faunistic groups.

Acknowledgements

This project was supported by IN2078/91 DGAPA-UNAM coordinated by Dr. José G. Palacios-Vargas (Science Faculty, UNAM). The field work was developed with the help of Alfonso Pescador Rubio, José Antonio Gómez Anaya, Alex Cadena Carrión and Alicia Rodríguez Palafox (†). Drs. José G. Palacios-Vargas, Betty Benrey, Alfonso Pescador, Zenón Cano, Francisco Villalobos, Norma García-Calderón, Victor Rico-Gray and Patricia Rojas gave invaluable suggestions on a first draft of manuscript. Two anonymous reviewers and Dr. Jacques Delabie gave invaluable suggestions to improve the manuscript.

References

Andersen, A.N. & Yen, A.Y. (1992). Canopy ant communities in the semi-arid mallee region of Northwestern Victoria. Australian Journal of Zoology, 40: 205-214. doi:10.1071/ZO9920205

Armbrecht, I., Perfecto, I. & Vandermeer, J. (2004). Enigmatic biodiversity correlations: ant diversity responds to diverse resources. Science 304: 284-286. doi: 10.1126/science.1094981

Balvanera, P., Lott, E., Segura, G., Siebe, C. & Islas, A. (2002). Patters of β-diversity in a Mexican tropical dry forest. Journal of Vegetation Science 13:145–158.

Basset, Y., Aberlenc, H.P. & Delvare, B. (1992). Abundance and stratification of foliage arthropods in lowland rainforest of Cameroon. Ecological Entomology, 17: 310-318. doi: 10.1111/j.1365-2311.1992.tb01063.x

Basset, Y., Novotny, V., Miller, S.E., Kitching, R.L. (2003). Arthropods of Tropical Forest. Spatio-temporal dynamics and resource use in the canopy. Cambridge: Cambridge University Press, 474 p.

Bolton, B. (1999). Ant genera of the tribe Dacetonini (Hymenoptera: Formicidae). Journal of Natural History, 33: 1639-1689. doi: 10.1080/002229399299798

Bos, M.M., Steffan-Dewenter, I., Tscharntke, T. (2007). The contribution of cacao agroforest to the conservation of lower canopy ant and beetle diversity in Indonesia. Biodiversity and Conservation, 16: 2429-2444. doi: 10.1007/s10531-007-9196-0

Brown, W.L. (1959). A revision of the dacetine ant genus *Neostruma*. Breviora, 107: 1-13. http://hdl.handle.net/10199/1881

Bullock, S.H. (1986). Climate of Chamela, Jalisco and trends in the South coastal region of Mexico. Archives for Meteorology, Geophysics and Bioclimatology, Series B, 36: 297-316. doi: 10.1007/BF02263135

Bullock, S.H. (1988). Rasgos del ambiente físico y biológico de Chamela, Jalisco, México. Folia Entomológica Mexicana.,77: 5-17.

Bullock, S.H. (2002). La fonología de plantas en Chamela. In Noguera, F.A., Vega-Rivera, J.G. García-Aldrete, A.N. & Quesada-Avendaño, M. (Eds.), Historia Natural de Chamela (pp. 491-498). México: Instituto de Biología, UNAM.

Bullock, S.H. & Solís-Magallanes, J.A. (1990). Phenology of canopy trees of a tropical deciduous forest in Mexico. Biotropica, 22: 22-35.

Byk, J. & Del-Claro, K. (2010). Nectar and pollen-gathering *Cephalotes* ants provide no protection against herbivory: a new manipulative experiment to test ant protective capabilities. Acta Ethologica, 13: 33-38. doi: 10.1007/s10211-010-0071-8

Byk, J. & Del-Claro, K. (2011). Ant-plant interaction in the Neotropical savanna: direct beneficial effects of extrafloral nectar on ant colony fitness. Population Ecology, 53: 327-332. doi: 10.1007/s10144-010-0240-7

Castaño-Meneses, G. (2008). Estructura de la comunidad edáfica de hormigas en la selva baja caducifolia de Chamela, Jalisco, México. In Estrada-Vanegas, E.G. (Ed.), Fauna de Suelo I. Micro, meso y macrofauna (pp. 133-140). Texcoco: Colegio de Postgraduados.

Castaño-Meneses, G., Benrey, B. & Palacios-Vargas, J.G. (2009). Diversity and temporal variation of ants (Hymenoptera: Formicidae) from Malaise traps in a tropical deciduous forest. Sociobiology, 54: 633-645.

Cervantes, L., Domínguez, R. & Maass, M. (1988). Relación lluvia-escurrimiento en un sistema pequeño de cuencas de selva baja caducifolia. IIngeniería hidráulica en México II 1: 30-41.

Clarke, K.R. (1993). Non-parametric multivariate analyses of changes in community structure. Australian Journal of Ecology, 18: 117-143. doi: 10.1111/j.1442-9993.1993.tb00438x

Clarke, K.R. & Green, R.H. (1988). Statistical design and analysis for a "biological effects" study. Marine Ecology Progres Series, 4: 213-226.

Cook, S.C., Eubanks, M.D., Gold, R.E. & Behmer, S.T. (2011). Seasonality directs contrasting food collection behavior and nutrient regulation strategies in ants. PLoS One 6: e25407. doi: 10.1371/journal.pone.0025407. (accessed date: 23 May, 2012).

Damon, A. & Pérez-Soriano, A. (2005). Interaction between ants and orchids in the Soconusco region, Chiapas, Mexico. Entomotropica 20: 59-65.

Davidson, D.W., Cook, S.C., Snelling, R.R. & Chua, T.H. (2003). Explaining the abundance of ants in lowland tropical rainforest canopies. Science 300: 969-972. doi: 10-1126/science.1082074

Dirzo, R., Young, H.S., Mooney, H.A. & Ceballos, G. (2011). Seasonally dry tropical forest. Ecology and Conservation. Washington: Island Press, 400 p.

Erwin, T.L. (1983). Beetles and other insects of tropical forest canopies at Manaus, Brazil, sampled by insecticidal fogging. In Sutton, W.L., Whitmore, T.C. & Chadwick, A.C. (Eds.), Tropical rain forest: ecology and management (pp. 59-79). Oxford: Blackwell Scientific Publications.

García-Oliva, F., Camou, A. & Maass, M. (2002). El clima de la región central de la costa del Pacífico mexicano. In Noguera, F.A., Vega-Rivera, J.H., García-Aldrete, A.N. & Quesada-Avendaño, M. (Eds.), Historia Natural de Chamela (pp. 3-10). Distrito Federal: Instituto de Biología, UNAM.

Gove, A.D., Majer, J.D. & Rico-Gray, V. (2005). Methods for conservation outside of formal reserve systems: the case of ants in the seasonally dry tropics of Veracruz, Mexico. Biological Conservation, 126: 328-338. doi: 10.1016/j.biocon.2005.06.008

Guilbert, E. & Casevitz-Weulersse, J. (1997). Caractérisa-tion de la myrmécofaune de la canopée de forêts primaires de Nouvelle-Calédonie échantillonnée par fogging. In Najt, J. & Matile, L. (Eds.), Zoologia Neocaledonica, Vol. 5. (pp. 357-368). París: Mémoires du Muséum National d'Historie Naturelle.

Guilbert, E., Baylac, M. & Najt, J. (1995). Canopy arthropod diversity in New Caledonian primary forest sampled by fogging. Pan-Pacific Entomology, 71: 3-12.

Hammer, Ø., Harper, D.A.T. & Ryan, P.D. (2001). PAST: Paleontological statistics software package for education and data analysis. Paleontología Electrónica, 4: 9.[7-11-2013]. Available at http://palaeo-electronica.org/2001_1/past/issue1_01.htm.

Jaffe, K., Horchler, P., Verhaagh, M., Gómez, C., Sievert, R., Jaffe, R. & Morawez, W. (2007). Comparing the ant fauna in a tropical and a temperate forest canopy. Ecotropicos, 20: 74-81.

Jantarit, S., Wattanasit, S. & Sotthibandhu, S. (2009). Canopy ants on the briefly deciduous tree (Elateriospermum tapos Blume) in a tropical rainforest, southern Thailand. Songklanakarin Journal of Science and Technology, 31: 21-28.

Lach, L., Parr, C.L. & Abbott, K.L. (2010). Ant ecology. Oxford: Oxford University Press, 402 p.

Lévieux, J. (1977). La nutrition des fourmis tropicales: V. Éléments de synthèse. Les modes d'exploitation de la biocoenose. Insects Sociaux, 24: 235-260. doi: 10.1007/BF02232743

Longino, J.T. & Nadkarni, N.M. (1990). A comparison of ground and canopy leaf litter ants (Hymenoptera: Formicidae) in a neotropical montane forest. Psyche, 97: 81-94. doi: 10.1155/1990/36505

Longino, J.T., Coddington, J. & Colwell, R.K. (2002). The ant fauna of a tropical rain forest: estimating species richness three different ways. Ecology, 83: 689-702. doi: 10.1890/0012-9658(2002)083[0689:TAFOAT]2.0.CO;2

Lott, E.J. (1985). Listados florísticos de México. III La Estación de Biología Chamela, Jalisco. México: Instituto de Biología, UNAM, 47 p.

Lott, E.J., Bullock, S.H. & Solís-Magallanes, J.A. (1987). Floristic diversity and structure of upland and arroyo forests in coastal Jalisco. Biotropica, 19: 228–235.

Lott, E.J. & Atkinson, T.H. (206). 13. Seasonally dry tropical forests: Chamela-Cuixmala, Jalisco, as a focal point for comparison. In Pennington, R.T., Lewis, G.P. & Ratter, J.A. (Eds.), Neotropical savannas and seasonally dry forests: plant diversity, biogeography and conservation (pp. 315-342). Florida: CRC Press.

Ludwing, J.A. & Reynolds, F.F. (1988). Statistical ecology: a primer on methods and computing. New York: Wiley Interscience Publishers, 368 p.

Maass, J.M., Jaramillo, V., Martínez-Yrízar, M., García-Oliva, F., Pérez-Jiménez, A., & Sarukhán, J. 2002. Aspectos funcionales del ecosistema de selva baja caducifolia en Chamela,

Jaslico. In Noguera, F.A., Vega-Rivera, J.H., García-Aldrete, A.N. & Quesada-Avendaño, M. (Eds.), Historia Natural de Chamela (pp. 525-542). México: Instituto de Biología, UNAM.

Maass, J.M., Martínez-Yrízar, A., Patiño, C. & Sarukhán, J. (2002b). Distribution and annual net accumulation of above-ground dead phytomass and its influence on throughfall qua-lity in a Mexican tropical deciduous forest ecosystem. Journal of Tropical Ecology, 18: 821-834. doi: 10.1017/S0266467402002535

Majer, J.D. (1990). The abundance and diversity of arboreal ants in Northern Australia. Biotropica, 22: 191-199.

Martínez-Yrízar, A., Maass, J.M., Pérez-Jiménez, L.A. & Sarukhán, J. (1996). Net primary productivity of a tropical deciduos forest ecosystem in Wester Mexico. Journal of Tropical Ecology, 12: 169-175. doi: http://dx.doi.org/10.1017/S026646740000938X

Meyer, K.M., Schiffers, K., Münkemüller, T., Schädler, M., Calabrese, J.M., Basset, A., Breulmann, M., Duquesne, S., Hidding, B., Huth, A., Schöb, C. & van de Voorde, T.F.J. (2010). Predicting population and community dynamics: the type of aggregation matters. Basic and Applied Ecology, 11: 563-571. doi: 10.1016/jbaae.2010.08.001

Neves, F.S., Braga, R.F., Do Espírito-Santo, M.M., Delabie, J.H.C., Fernandes, G.W. & Sánchez-Azofeifa, G.A. (2010). Diversity of arboreal ants in a Brazilian tropical dry forest: effects of seasonality and successional stage. Sociobiology, 56: 1-18.

Noguera, F.A., Vega-Rivera, J.H., García-Aldrete, A.N. & Quesada-Avendaño, M. (2002). Historia Natural de Chamela. México: Instituto de Biología, UNAM, 568 p.

Ødegaard, F. (2000). How many species of arthropods? Erwin's estimate revised. Biological Journal of the Linnean Society, 71: 583-597. doi: 10.1111/j.1095-8312.2000.tb01279.x

Palacios-Vagas, J.G. & Castaño-Meneses, G. (2003). Seasonality and community composition of springtails in Mexican forest. In Basset, Y., Novotny, V., Miller, S.E. & Kitching, R.L. (Eds.), Arthropods of Tropical forest (pp. 159-169). Cambridge: Cambridge University Press.

Palacios-Vargas, J.G. & Gómez-Anaya, J.A. (1993). Los collembolan (Hexapoda: Apterygota) de Chamela, Jalisco, México. Distribución ecológica y claves. Folia Entomologica Mexicana, 89: 1-34.

Palacios-Vargas, J.G., Castaño-Meneses, G. & Gómez-Anaya, J.A. (1998). Collembola from the canopy of a Mexican tropical deciduous forest. Pan-Pacific Entomologist, 74: 47-54.

Palacios-Vargas, J.G., Castaño-Meneses, G. & Pescador, A. (1999). Phenology of canopy arthropods of a tropical deciduous forest in wetern Mexico. Pan-Pacific Entomologist, 75: 200-211.

Paoletti, M.G., Taylor, R.A.J., Stinner, B.R., Stinner, D.H. & Benzing, D.H. (1991). Diversity of soil fauna in the canopy and forest floor of a Venezuelan cloud forest. Journal of Tropical Ecology, 7: 373-383. doi: 10.1017/S0266467400005654

Ribas, C.R., Schoereder, J.H., Pic, M. & Soares, S.M. (2003). Tree heterogeneity, resource availability, and larger scale processes regulating arboreal ant species richness. Austral Ecology, 28: 305-314. doi: 10.1046/j.1442-9993.2003.01290.x

Rojas, P. (2001). Las hormigas del suelo en México: diversidad, distribución e importancia (Hymenoptera: Formicidae). Acta Zoologica Mexicana, 1: 189-238.

Schonberg, L.A., Longino, J.T., Nadkarni, N.M., Yanoviak, S.P. & Gering, J.C. (2004). Arboreal ant species richness in primary forest, secondary forest, and pasture habitats of a tropical montane landscape. Biotropica, 36: 402-409. doi: 10.1111/j.1744-7429.2004.tb00333.x

Schulz, A. & Wagner, T. (2002). Influence of forest type and tree species on canopy ants (Hymenoptera: Formicidae) in Budongo Forest, Uganda. Oecologia, 133: 224-232. doi: 10.1007/s00442-002-1010-9

StatSoft. (2009). Statistical user guide: complete statistical system StatSoft. Oklahoma.

Stork, N.E. (1987). Guild structure of arthropods from Bornean rain forest trees. Ecological Entomology, 12: 69-80. doi: 10.1111/j.1365-2311.1987.tb00986.x

Stunts, S., Linder, C., Linsenmair, K.E., Simon, U. & Zotz, G. (2003). Do non-myrmocophilic epiphytes influence community structure of arboreal ants? Basic and Applied Ecology, 4: 363-374. doi: 10.1078/1439-1791-00170

Tobin, J.E. (1995). Ecology and diversity of tropical forest canopy ants. In Lowman, M.D. & Nadkarni, N.M. (Eds.), Forest canopies (pp.129-147). San Diego: Academic Press.

Trejo, R.I. & Dirzo, R. (2000). Deforestation of seasonally dry tropical forest towards its northern distribution: a national and local analysis in Mexico. Biological Conservation, 94: 133-142. doi: 10.1016/S0006-3207(99)00188-3

Watanasit, S., Tongjerm, S. & Wiwatwitaya, D. (2005). Composition of canopy ants (Hymenoptera: Formicidae) at Ton Nga Chang Wildlife Sactuary, Songkhla Province, Thailand. Songklanakarin Journal of Science and Technology, 27: 665-673.

Wetterer, J.K. (2009). Worldwide spread of the ghost ant, *Tapinoma melanocephalum* (Hymenoptera: Formicidae). Myrmecological News, 12: 23-33.

Wilson, E.O. (1959). Some ecological characteristics of ants in New Guinea rain forest. Ecology, 40: 437-447.

Wilson, E.O. (1987). The arboreal fauna of Peruvian Amazon Forest: a first assessment. Biotropica 19: 245-251.

Wilson, E.O. & Taylor, R.W. (1967). Ants of Polynesia. Pacific Insects Mongraphs, 14: 1-109.

Zar, J.H. (1984). Biostatistical Analysis. Second Edition. New Jersey: Prentice Hall, 718 p.

Appendix 1 - Monthly and total abundance of ants collected by fogging at the canopy of Biological Station Chamela. 1 = August 1992, 2 = September 1992, 3 = May 1993, 4 = July 1993, 5 = November 1993, 6= February 1994, 7 = May 1994.

Subfamily	Species	1	2	3	4	5	6	7	Total
Amblyoponinae	*Stigmatomma* sp.	1							1
Dolichoderinae	*Forelius keferi* Wheeler, 1934	12	30	160	3			50	255
	Tapinoma melanocephalum (Fabricius, 1793)	20	30	110	40	84	255	550	1071
Ecitoninae	*Neivamyrmex chamelensis* Watkins, 1986		5	4				1	10
Formicinae	*Brachymyrmex* sp. 1					10	15		25
	Brachymyrmex sp. 2						1		1
	Camponotus sp. 1	7					3	13	23
	Camponotus sp. 2	1	2						3
	Camponotus sp. 4	14	10	2	10	20	4	6	64
	Camponotus sp. 5					3	2		5
	Camponotus sp. 6				2	3	4		9
	Camponotus sp. 8	28	10	6	2	10		14	70
	Camponotus sp. 9					1		2	3
	Camponotus sp. 10	4							4
	Camponotus sp. 12	10			2				12
	Camponotus sp. 13					2			2
	Camponotus sp. 14			20	10	13	10		53
	Camponotus sp. 15					13	1		14
	Camponotus sp. 16							3	3
Myrmicinae	*Acromyrmex* sp.		1		1				2
	Carebara sp.		1						1
	Cephalotes sp. 1	15	5	9	8	30	39	1	107
	Cephalotes sp. 2	1		7			5		13
	Cephalotes sp. 3					14			14
	Cephalotes sp. 4	1		7		11		8	27
	Cephalotes sp. 5							1	1
	Cephalotes sp. 7					2			2
	Crematogaster crinosa Mayr, 1862	236	60	280	383	326	200	258	1743
	Crematogaster sumichrasti Mayr, 1870	189	70	60	20	730	481	10	1560
	Temnothorax sp. 2			4			7	5	16
	Temnothorax sp. 3					4			4
	Temnothorax sp. 4	13		1	7	12		5	38
	Temnothorax sp. 5		90	1	2	3		2	98
	Pheidole sp. 1	6	1			1	3		11
	Pheidole sp. 5	5							5
	Pheidole sp. 6	3				2			5
	Pheidole sp. 7	12		1				1	14
	Solenopsis geminata (Fabricius, 1804)	22	30		22	30	49		153
	Strumigenys sp. 2	1				2			3
	Strumigenys sp. 3	14						30	43
Ponerinae	*Pachycondyla* sp.	2	2			13	2	2	21
Pseudomyrmecinae	*Pseudomyrmex* sp. 1	3	3		1	2			9
	Pseudomyrmex sp. 2	4				2	2	2	10
	Pseudomyrmex sp. 3	6		5	2	1	12	6	32
	Pseudomyrmex sp. 4						6	2	8
	Pseudomyrmex sp. 5						2	1	3

8

Comparative Evaluation of Three Chitin Synthesis Inhibitor Termite Baits Using Multiple Bioassay Designs

BK Gautam, G Henderson

Department of Entomology, Louisiana State University, Baton Rouge, LA, USA.

Keywords

Benzoylphenylurea, termite baits, *Coptotermes formosanus*, termite control

Corresponding author

Gregg Henderson
404 Life Sciences Building,
Department of Entomology,
Louisiana State University
Baton Rouge, LA, 70803, USA
E-mail: grhenderson@agcenter.lsu.edu

Abstract

Use of chitin synthesis inhibitors has revolutionized the potential impact of termite baiting systems. Several chitin synthesis inhibitors have been used or tested against subterranean termites. We evaluated the effect of lufenuron on bait matrix consumption and mortality of *Coptotermes formosanus* and compared it with 2 other chitin synthesis inhibitors presently used for termite control: diflubenzuron and noviflumuron. Laboratory no-choice and multi-chamber bioassay designs were employed. At the end of 6 weeks, in both the no-choice and multi-chamber tests, mortality was significantly higher in all the chitin synthesis inhibitor treatments as compared to the controls; however, lufenuron treatment had significantly higher mortality than the other chitin synthesis inhibitors. Multi-chamber tests suggested no sign of feeding deterrence with any of the chitin synthesis inhibitors at the concentrations tested. Consumption of lufenuron cardboard or noviflumuron bait matrix was similar to that of control cardboard in the no-choice tests. We conclude that, based on the overall bait consumption and mortality data, lufenuron was at least as effective as noviflumuron and diflubenzuron.

Introduction

Present termite control methods mainly include application of liquid termiticides and baiting systems. Subterranean termite baiting systems exploit the termites' foraging behavior and their food transfer system (trophallaxis) to reduce or eliminate the colony population from an area (French, 1991; Su et al., 1995). For a typical bait to succeed, foraging termites must find the bait station, recruit more termites, consume sufficient amount of bait matrix containing a toxicant, carry the toxicant back to the colony and transfer a sufficient dose to the rest of the colony members by trophallaxis (Su et al., 1995; Grace et al., 1996; Su & Scheffrahn, 1996; Grace & Su, 2001).

It is suggested that bait treatments have advantages over alternative termite control strategies both by affecting a greater foraging area and by reducing the amount of toxicants placed in the environment (Su & Lees, 2009). The interconnected nests and galleries of a subterranean termite colony often spread over dozens of meters in the field. An excavation conducted by King and Spink (1969) revealed that underground tunneling system of *Coptotermes formosanus* Shiraki extended over 60 m and covered an area of over 0.58 ha. Su and Scheffrahn (1988) estimated 1.4 to 6.8 million foraging individuals in one colony of *C. formosanus*. Due to the huge population size and extensive foraging range of subterranean termite colony, a non-repellent and slow acting control agent has the best potential to impact the majority of individuals in a colony (Su, 1994; Su, 2005).

Chitin synthesis inhibitors (CSI) fall within the benzoylphenylurea group and are slow acting insect growth regulators (IGR) that induce malformation of cuticle and significant reduction of chitin synthesis (van Eck, 1979; Merzendorfer et al., 2012). Inhibition of chitin synthesis disrupts the molting process in insects leading to their death. CSI also prevents normal formation of peritrophic membrane (Zimmermann & Peters, 1987); as a result, the insects become more susceptible to infection by microorganisms such as nuclear polyhedrosis virus (Arakawa et al., 2002). Several chitin synthesis inhibitors such as diflubenzuron, hexaflumuron, noviflumuron, lufenuron, and novaluron have been used or tested with various degree of success in termite baiting systems (Su, 1994;

Cabrera and Thoms, 2006; Vahabzadeh et al., 2007; Lewis & Forschler, 2010; Osbrink et al., 2011).

There are quite a few studies, both laboratory and field, about the noviflumuron based termite bait (though not the durable bait we tested), however, very limited published studies have evaluated the lufenuron and diflubenzuron based termite bait. Lufenuron based termite bait was developed by Syngenta about a decade ago but is still not commercially available for termite control. This could possibly be because of the conflicting reports of the effect of lufenuron on subterranean termites (see for example, Su & Scheffrahn, 1996; Lovelady et al., 2008). The objective of this study was to evaluate the effect of lufenuron on bait matrix consumption and mortality of *C. formosanus* and compare it with 2 other CSIs: diflubenzuron and noviflumuron in the laboratory using two different bioassay designs. We believe that a comparative study in the lab using different bioassay designs would help to determine the efficacy of these chemicals against subterranean termites.

Materials and Methods

Test Materials

Test materials included three chitin synthesis inhibitor bait materials and untreated cardboard as controls. Lufenuron treated cardboard (company code: NB 3401-91; 0.15% lufenuron) and control cardboard were received from FMC Corporation (Philadelphia, PA). They also provided samples of diflubenzuron bait (Advance® Compressed Termite Bait II, BASF Corporation, St. Louis, MO; 0.25% diflubenzuron) and noviflumuron bait (Recruit® HD Termite Bait, Dow AgroSciences LLC, Indianapolis, IN; 0.5% noviflumuron).

Termites

Formosan subterranean termites were collected in June 2012 from Brechtel Park, New Orleans, Louisiana using milkcrate trap methods as described in Gautam and Henderson (2011). In brief, the trap consists of a milkcrate filled with pine 44 wood (*Pinus* sp.) sticks arranged in a lattice structure. The crate is buried in the ground and checked after 1-2 months. When infested with ~10,000 termites or more, the crate is retrieved, brought back to the laboratory and maintained in a trash can by adding water whenever necessary. The termite groups used in the experiment were collected from such laboratory stocks. Termites from two colonies (separated by >500m) were used.

No-choice Tests

No-choice tests were conducted in individual Petri dishes (100 × 15 mm) each containing 50 g of autoclaved sand added with 6 ml of deionized water (12% moisture wt/wt). A pre-weighed cardboard piece or bait material (the lufenuron treated cardboard and untreated control cardboard were cut into pieces with separate scissors, whereas the diflubenzuron and noviflumuron baits were cut with a power operated band saw) was placed on moist sand surface in the dish. The cardboard or bait materials were pressed gently to the moist sand. One hundred termites (90 undifferentiated workers of at least the third instar and 10 soldiers) were released in each dish and the dishes were sealed with Parafilm® to retain moisture. There were 6 replications for each treatment (3/termite colony group) and control. The dishes were then placed in an incubator maintained at 27°C. Moisture content of the dishes was monitored regularly. Termite survival count was made at 6 weeks when the experiment was ended. The cardboard and bait materials were cleaned of debris, dried and weighed to determine the consumption. Consumption of diflubenzuron bait was not recorded as it was difficult to quantify due to its being spread throughout the dish by the termites and mixed with sand.

Multi-chamber Tests

To test how additional food sources may affect bait toxicity, an especially designed 16-chambered test arena with a slight modification from Gautam et al. (2013) was used. The arena consisted of one center chamber (8.5 cm diam × 3.4 cm, Pioneer Plastics©, North Dixon, KY) connected to 5 first peripheral chambers (5.08 cm diam × 3.63 cm, Pioneer Plastics©, North Dixon, KY) with Tygon tubing (5 cm long, 0.64 cm inside diameter, Watts Co., North Andover, MA). The first peripheral containers again connected to 10 second peripheral containers with the same length tubing making two peripheries of the chambers around the center chamber. The side chambers in the first periphery were connected to each other with 10 cm tubing and similarly the side chambers in the second periphery were also connected (Fig. 1). This test arena represented multiple feeding sites as might occur in a typical subterranean termite infested area.

The center chamber contained 60 g of autoclaved sand and the peripheral chambers contained 30 g each. Deionized water was added to each chamber to make the sand moisture content ~15% (wt/wt). The center chamber was the termite release chamber and contained nothing except the moist sand whereas the peripheral chambers contained 3 bait pieces of either lufenuron, diflubenzuron or noviflumuron in 3 randomly selected chambers and the remaining 12 chambers contained pinewood pieces pre-soaked in water for 1 h.

After 1 d of arena set up, 300 termites (90% undifferentiated workers of at least the 3rd instar and 10% soldiers) were introduced in the center chamber. The lids were put back and the arenas were placed on a laboratory bench at 27 ± 1.5°C. There were 4 replications for each treatment and control. Control arenas contained no cardboard or bait, instead wood pieces were placed in all 15 peripheral chambers.

Fig. 1. Image of a laboratory test arena consisting of multiple feeding chambers connected with tubings.

The test arenas were monitored regularly for moisture and water was added when necessary. Live termites were counted at 6 weeks when the experiment ended, and wood and bait material consumption was recorded as in the no-choice tests. In each test arena, consumption of all 12 wood blocks or all 3 bait materials was added to calculate the total consumption. In controls, consumption of all 15 wood blocks was determined. Unlike no-choice tests where consumption of diflubenzuron bait was not quantified, a best effort attempt of consumption was quantified in the multi-chamber tests. Unconsumed diflubenzuron bait was carefully separated from sand and scrapped off the container, then dried and weighed to determine consumption.

Statistical Analysis

Data analysis was done using proc mixed model in SAS 9.3. Data were transformed using either arcsine of the square root method (mortality data) or log transformation method (consumption data) when necessary, especially to improve the normality. Post ANOVA comparisons were made using Tukey's HSD and significance level was determined when $\alpha < 0.05$.

Results and Discussion

No-choice Tests

There was no significant difference in consumption among lufenuron, noviflumuron and control baits ($F = 1.18$; df = 2, 15; $P = 0.334$) (Fig. 2). CSIs had a significant impact on termite mortality ($F = 17.53$; df = 3, 20; $P < 0.0001$). All three CSIs caused significantly higher mortality as compared to the control. When compared among the three CSIs, lufenuron caused significantly higher mortality (87%) than either diflubenzuron (55%) or noviflumuron (65%) at 6 weeks (Fig. 3).

Multi-chamber Tests

Among the 3 CSIs, bait matrix consumption was significantly different ($F = 44.54$; df = 2, 9; $P < 0.0001$). Consumption of diflubenzuron bait was significantly higher than noviflumuron bait or lufenuron bait and the consumption of noviflumuron bait was significantly higher than the lufenuron bait (Fig. 4). Similarly, wood consumption was significantly impacted by the CSI baits ($F = 247.19$; df = 3, 10; $P < 0.0001$). Wood consumption was significantly lower in all the CSI tests at 6 weeks (range: 120-333 mg) compared to the controls (4190 mg). Consumption of wood was not significantly different among the CSI baits (Fig. 5).

As in no-choice tests, termite mortality was significantly impacted by the CSI baits in multi-chamber tests ($F = 21.25$; df = 3, 10; $P < 0.0001$) and the mortality varied depending on the type of CSI. Approximately 100% mortality was observed in lufenuron tests at 6 weeks, which was significantly higher than the mortality in diflubenzuron or noviflumuron tests. There was no significant difference in mortality between diflubenzuron and noviflumuron (Fig. 6).

The results showed that the 3 CSIs tested induced significant mortality of *C. formosanus* foragers with lufenuron causing the highest percentage mortality in 6-week test. It is interesting to note that the relatively low concentration of

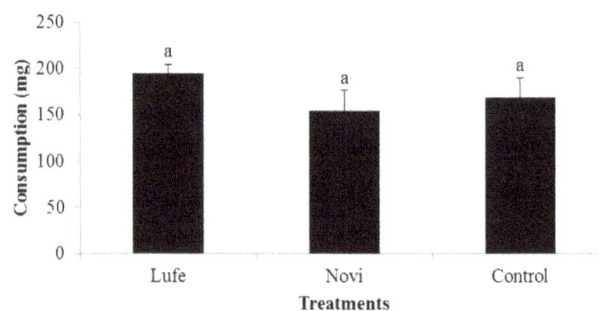

Fig. 2. Consumption (mean ± SEM) of 2 chitin synthesis inhibitor baits and control cardboard in no-choice tests at 6 weeks after exposure. Same letters above the bar indicate not significantly different from each other. Lufe = lufenuron, Novi = noviflumuron.

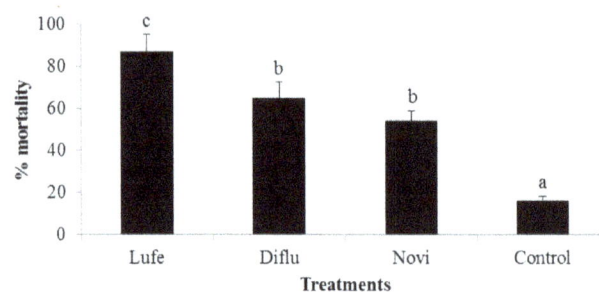

Fig. 3. Percentage mortality (mean ± SEM) of *C. formosanus* exposed to 3 different baits and control in no-choice tests at 6 weeks after exposure. Different letters above the bars indicate significantly different from each other. Lufe = lufenuron, Diflu = diflubenzuron, Novi = noviflumuron.

lufenuron (1500 ppm) inflicted a significantly higher mortality than the higher concentrations of either diflubenzuron (2500 ppm) or noviflumuron (5000 ppm) in 6 weeks. These results are consistent with the findings by Vahabzadeh et al. (2007) who reported that lufenuron caused significantly higher mortality than diflubenzuron in 6 weeks for *Reticulitermes flavipes* (Kollar). Rojas and Ramos (2004) found significantly lower queen fecundity in the lufenuron treatments but not in the diflubenzuron or hexaflumuron treatments when compared with the control. In 6 weeks, only lufenuron treatment caused ~100% mortality of *C. formosanus* individuals in multi-chamber tests suggesting a faster knock-down of lufenuron. The multi-chamber test, where only 3 of 16 chambers were provisioned with bait matrix, was particularly designed to dilute the baiting effect which may occur in the field. We observed that there was no noticeable impact on termite mortality due to the presence of other food sources. There are successful field results of lufenuron termite bait developed by Syngenta (Greensboro, NC), using a cardboard matrix containing 1500 ppm lufenuron (Lovelady et al. 2008, Haverty et al. 2010). Lovelady et al. (2008), in their review paper, reported that all of the 4 baited *R. flavipes* colony sites in Columbus, OH were eliminated of termites within 3.5-10.5 months of lufenuron bait placement. The authors further reported that the lufenuron bait successfully eliminated subterranean termites from 78 out of ~100 infested structures in a large multi-site study covering all of the major termite regions across the US (22 States).   Likewise, Haverty et al. (2010) reported the cessation of termite activity within 70 days of the lufenuron bait placement in all the 6 baited colony sites containing 21 colonies of *R. hesperus* Banks in Placerville, CA. Successful elimination of an aerial colony of *R. flavipes* from a six-story apartment building has also been reported using 1500 ppm lufenuron bait (Bowen & Kard 2012). Wang et al. (2013) reported that a combination of a lower concentration (1000 ppm) of lufenuron and opportunistic pathogens such as *Pseudomonas aeruginosa* (Schroeter) Migula caused higher percentage mortality of *C. formosanus* in a relatively short period. Su and Scheffrahn (1996), however, reported only ~60% mortality of *C. formosanus* individuals in the laboratory with

4000 ppm or 8000 ppm lufenuron in 6 weeks after exposure.

Contrary to the reports that lufenuron treatment elicited feeding deterrence at >1000 ppm for *C. formosanus* and >50 ppm for *R. flavipes* (Su & Scheffrahn, 1996), the present multi-chamber tests indicate that lufenuron does not cause noticeable feeding deterrence to *C. formosanus* at 1500 ppm. Feeding deterrence was not reported by Vahabzadeh et al. (2007) at any concentration (0.1-1000 ppm) tested. In fact, the authors mentioned that lufenuron was highly acceptable to *R. flavipes* and was the most palatable of the 4 chemicals evaluated: lufenuron, diflubenzuron, hexaflumuron and triflumuron. Similarly, Lovelady et al. (2008) and Haverty et al. (2010) stated that *R. flavipes* and *R. hesperus* readily consumed the cardboard bait matrix loaded at a rate of 1500 ppm lufenuron.

Noviflumuron, a CSI which replaced hexaflumuron used in Recruit III termite bait (Sentricon System, Dow AgroSciences), is a more extensively studied chemical. Dow AgroScience scientists reported that this chemical is non-deterrent at up to 10,000 ppm on filter paper to *R. flavipes* (Karr et al., 2004) and successful results in suppressing the populations of subterranean termites by others has also been reported (Su 2005; Cabrera & Thoms, 2006; Thoms et al., 2009).

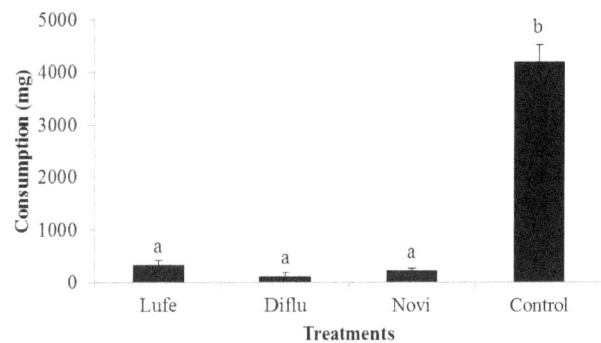

Fig. 5. Consumption (mean ± SEM) of wood in treated and control arenas in multi-chamber tests at 6 weeks after exposure. Different letters above the bars indicate significantly different from each other. Lufe = lufenuron, Diflu = diflubenzuron, Novi = noviflumuron.

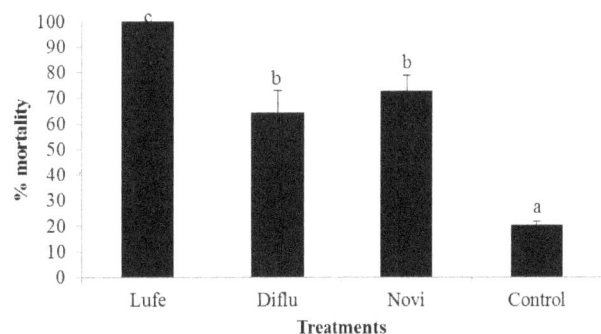

Fig. 6. Percentage mortality (mean ± SEM) of *C. formosanus* exposed to 3 different baits and control in multi-chamber tests at 6 weeks after exposure. Different letters above the bars indicate significantly different from each other. Lufe = lufenuron, Diflu = diflubenzuron, Novi = noviflumuron.

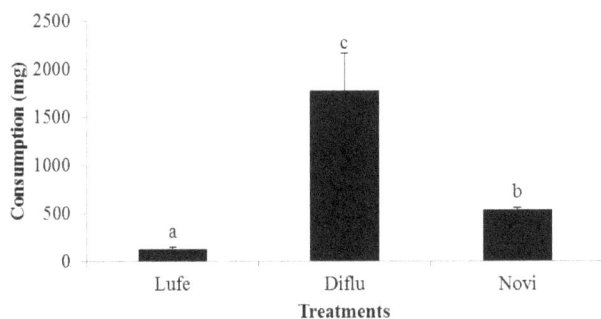

Fig. 4. Consumption (mean ± SEM) of 3 chitin synthesis inhibitor baits in multi-chamber tests at 6 weeks after exposure. Different letters above bars indicate significantly different from each other. Lufe = lufenuron, Diflu = diflubenzuron, Novi = noviflumuron.

Relatively low mortality caused by diflubenzuron in both the no-choice and multi-chamber tests in 6 weeks indicated a slow action of this chemical similar to that of noviflumuron. Osbrink et al. (2011) reported that field tests of bait matrix containing 1000 ppm diflubenzuron had no noticeable impact on *C. formosanus* or *R. flavipes* populations when checked regularly for 3 years after bait placement. We are not sure, however, if the lower concentration of diflubenzuron used in their study was the reason for not achieving a significant mortality. However, other studies have also reported not so promising results with diflubenzuron (Green et al., 2008; Su & Scheffrahn; 1993). Although Rojas and Ramos (2001) reported that diflubenzuron caused 100% mortality of *C. formosanus* in 9 weeks after exposure in the laboratory tests, their field testing was less impressive (Rojas et al., 2008). Nevertheless, the present results demonstrated that lufenuron is an effective CSI against *C. formosanus* and is relatively fast acting.

Acknowledgements

We thank Louisiana Department of Agriculture and Forestry and FMC Corporation for partial funding and material support in this study. We thank two other anonymous referees for their comments and criticisms on the manuscript. This manuscript is approved for publication by the Director of the Louisiana Agricultural Experiment Station as manuscript number 2013-234-9630.

References

Arakawa, T., Furuta, Y., Miyazawa, M. & Kato, M. (2002). Flufenoxuron, an insect growth regulator, promotes peroral infection by nucleopolyhedrovirus (BmNPV) budded particles in the silkworm, *Bombyx mori* L. Journal of Virological Methods, 100: 141–147.

Bowen, CJ. & Kard, B. (2012). Termite aerial colony elimination using lufenuron bait (Isoptera: Rhinotermitidae). Journal of Kansas Entomological Society, 85: 273-284.

Cabrera, BJ. & Thoms, EM. (2006). Versatility of baits containing noviflumuron for control of structural infestations of Formosan subterranean termites (Isoptera: Rhinotermitidae). Florida Entomologist, 89: 20-31.

French, JRJ. (1991). Baits and foraging behavior of Australian species of *Coptotermes*. Sociobiology, 19: 171-186.

Gautam, BK. & Henderson, G. (2011). Effects of sand moisture level on food consumption and distribution of Formosan subterranean termites (Isoptera: Rhinotermitidae) with different soldier proportions. Journal of Entomological Science, 46: 1-13.

Gautam, BK., Henderson, G. & Wang, C. (2014). Localized treatments using commercial dust and liquid formulations of fipronil against *Coptotermes formosanus* Shiraki (Isoptera: Rhinotermitidae) in the laboratory. Insect Science, 21: 174-180.

Grace, JK. & Su N-Y. (2001). Evidence supporting the use of termite baiting systems for long-term structural protection (Isoptera). Sociobiology, 37: 301-310.

Grace, JK., Tome, CHM., Shelton, TG., Ohsiro, RJ. & Yates, JR. (1996). Baiting studies and considerations with *Coptotermes formosanus* (Isoptera: Rhinotermitidae) in Hawaii. Sociobiology, 28: 511-520.

Green III, F., Arango, R. & Esenther, G. (2008). Community- wide suppression of *R. flavipes* from Endeavor, Wisconsin-search for the holy grail, pp. 1-10. *In* The International Research Group on Wood Protection, IRG/WP 08-10674. IRG Secretariat Stockholm, Sweden.

Haverty, MI., Tabuchi, RL., Vargo, EL., Cox, DL., Nelson, LJ. & Lewis, VR. (2010). Response of *Reticulitermes hesperus* (Isoptera: Rhinotermitidae) colonies to baiting with lufenuron in northern California. Journal of Economic Entomology, 103:770–780.

Jones, SC. (1989). Field evaluation of fenoxycarb as a bait toxicant for subterranean termite control. Sociobiology, 15: 33-41.

Karr, LL., Sheets, JJ., King, JE. & Dripps, J E. (2004). Laboratory performance and pharmacokinetics of the benzoylphenylurea noviflumuron in eastern subterranean termites (Isoptera: Rhinotermitidae). Journal of Economic Entomology, 97: 593-600.

King, EGJ. & Spink, WS. (1969). Foraging galleries of the Formosan subterranean termite, *Coptotermes formosanus,* in Louisiana. Annals of the Entomological Society of America, 62: 536-542.

Lewis, JL. & Forschler, BT. (2010). Impact of five commercial baits containing chitin synthesis inhibitors on the protist community in *Reticulitermes flavipes* (Isoptera: Rhinotermitidae). Environmental Entomology, 39: 98-104.

Lovelady, C., Cox, D., Zajac, M. & Cartwright, B. (2006). Lufenuron termite bait. pp. 68-69. *In* S. C. Jones (ed.), Proceedings of the 2008 national conference on urban Entomology, Tulsa, OK.

Merzendorfer, H. (2013). Chitin synthesis inhibitors: old molecules and new developments. Insect Science, 20: 121–138.

Merzendorfer, H., Kim, HS., Chaudhari, SS., Kumari, M., Specht, CA., Butcher, S., Brown, SJ., Robert Manak, J., Beeman, RW., Kramer, KJ. & Muthukrishnan, S. (2012). Genomic and proteomic studies on the effects of the insect growth regulator diflubenzuron in the model beetle species *Tribolium castaneum*. Insect Biochemistry and Molecular Biology, 42: 264–276.

Osbrink, WLA., Cornelius, ML. & Lax, AR. (2011). Area-wide field study on effect of three chitin synthesis inhibitor baits on populations of *Coptotermes formosanus* and *Reticu-*

litermes flavipes (Isoptera: Rhinotermitidae) Journal of Economic Entomology, 104: 1009-1017.

Robertson, AS. & Su, N-Y. (1995). Discovery of an effective slow-acting insect growth regulator for controlling subterranean termites. Down to Earth, 50: 1-7.

Rojas, MG. & Morales-Ramos, JA. (2001). Bait matrix for delivery of chitin synthesis inhibitors to the Formosan subterranean termite (Isoptera: Rhinotermitidae). Journal of Economic Entomology, 94: 506-510.

Rojas, MG. & Morales-Ramos, JA. (2004). Disruption of reproductive activity of *Coptotermes formosanus* primary reproductives by three chitin synthesis inhibitors. Journal of Economic Entomology, 97:2015–2020.

Rojas, MG., Morales-Ramos, J., Lockwood, M., Etheridge, L., Carroll, J., Coker, C. & Knight, P. (2008). Area-wide management of subterranean termites in south Mississippi using baits. USDA-ARS Tech. Bull. 1917.

Su, N.-Y. (1994). Field evaluation of a hexaflumuron bait for population suppression of subterranean termites (Isoptera: Rhinotermitidae). Journal of Economic Entomology, 87: 389–397.

Su, N.-Y. (2005). Response of the Formosan subterranean termites (Isoptera: Rhinotermitidae) to baits or nonrepellent termiticides in extended foraging arenas. Journal of Economic Entomology, 98: 2143-2152.

Su, N-Y. & Scheffrahn, RH. (1988). Foraging population and territory of the Formosan subterranean termite (Isoptera: Rhinotermitidae) in an urban environment. Sociobiology, 14: 353-359.

Su, N-Y. & Scheffrahn, RH. (1993). Laboratory evaluation of two chitin synthesis inhibitors, hexaflumuron and diflubenzuron, as bait toxicants against Formosan and eastern subterranean termites (Isoptera: Rhinotermitidae). Journal of Economic Entomology, 86: 1453-1457.

Su, N-Y. & Scheffrahn, RH. (1996). Comparative effects of two chitin synthesis inhibitors, hexaflumuron and lufenuron, in a bait matrix against subterranean termites (Isoptera: Rhinotermitidae). Journal of Economic Entomology, 89: 1156-1160.

Su, N-Y., Thomas, EM., Ban, PM. & Scheffrahn, RH. (1995). Monitoring baiting stations to detect and eliminate foraging populations of subterranean termites (Isoptera, Rhinotermitidae) near structures. Journal of Economic Entomology, 88: 932-936.

Tamashiro, M., Yates, JR., Yamamoto, RT. & Ebesu, RH. (1991). Tunneling behavior of the Formosan subterranean termite and basalt barriers. Sociobiology, 19: 163-170.

Thoms, EM., Eger, JE., Messenger, MT., Vargo, E., Cabrera, B., Riegel, C., Murphree, S., Mauldin, J. & Scherer, P. (2009). Bugs, baits, and bureaucracy: completing the first termite bait efficacy trials (quarterly replenishment of novifumuron) initiated after adoption of Florida Rule, Chapter 5E-2.0311. American Entomologist, 55: 29-39.

Vahabzadeh, RD., Gold, RE. & Austin, JW. (2007). Effects of four chitin synthesis inhibitors on feeding and mortality of the Eastern subterranean termite, *Reticulitermes flavipes* Kollar (Isoptera: Rhinotermitidae). Sociobiology, 50: 833-859.

van Eck, WH. (1979). Mode of action of two bezoylphenyl ureas as inhibitors of chitin synthesis in insects. Insect Biochemistry, 9: 295–300.

Wang C., Henderson, G. & Gautam, BK. (2013). Lufenuron suppresses the resistance of Formosan subterranean termites (Isoptera: Rhinotermitidae) to entomopathogenic bacteria. Journal of Economic Entomology, 160: 1812-1818.

Zimmermann, D. & Peters, W. (1987). Fine structure and permeability of peritrophic membranes of *Calliphora erythrocephala* (Meigen) (Insecta: Diptera) after inhibition of chitin and protein synthesis. Comparative Biochemistry and Physiology B, 86: 353–360.

9

Trap-nesting solitary wasps (Hymenoptera: Aculeata) in an insular landscape: Mortality rates for immature wasps, parasitism, and sex ratios

AL Oliveira Nascimento, CA Garófalo

Universidade de São Paulo (FFCLRP-USP), Ribeirão Preto, SP – Brazil

Keywords
Atlantic forest, colonization, natural enemies, phenology, wasp abundance

Corresponding author
Ana L. Oliveira Nascimento
Faculdade de Filosofia, Ciências e
Letras de Ribeirão Preto
Universidade de São Paulo
Av. Bandeirantes, 3900
Ribeirão Preto, São Paulo, Brazil.
14040-901
E-mail: analuizanascimento@uol.com.br

Abstract

The aim of this study was to examine the species composition and the abundance of solitary wasps that nest in preexisting cavities in the Ilha Anchieta State Park, Brazil. Sampling was made during two years utilizing trap-nests. Of the 254 nests obtained, 142 nests were built by 14 species belonging to four genera and four families. In the remaining 112 nests all immatures were dead by unknown causes or had been parasitized by natural enemies. The occupation of trap-nests occurred almost throughout the study period and the wasps nested more frequently during the super-humid season. *Trypoxylon lactitarse, Pachodynerus nasidens, Trypoxylon* sp.2 aff. *nitidum* and *Podium denticulatum* were the most abundant species. The sex ratios of *T. lactitarse* and *Trypoxylon* sp.2 aff. *nitidum* were significantly male-biased, whereas those of *Trypoxylon* sp.5 aff. *nitidum* and *P. nasidens* were significantly female-biased. Sex ratios of *P. denticulatum* and *P. brevithorax* were not significantly different from 1:1. Natural enemies emerging from the nests were identified as belonging to the families Chrysididae, Ichneumonidae and Chalcididae (Hymenoptera), the genus *Melittobia* (Hymenoptera, Eulophidae), and the species *Amobia floridensis* (Townsend, 1892) (Diptera: Sarcophagidae). The number of cells with dead immatures from unknown factors was significantly higher than the number of cells parasitized by insects.

Introduction

There are more than 34,000 described species of aculeate wasps in the world, of which about 14,700 species belong to the families Pompilidae, Crabronidae, Sphecidae, and Vespidae (subfamily Eumeninae) (Morato et al., 2008). For the neotropical region, 1,858 species of Crabronidae and Sphecidae have been recorded (Amarante, 2002; 2005), of which 645 occur in Brazil (Morato et al., 2008). Ninety per cent of the species known have solitary behaviour and diversified nesting habits (O'Neill, 2001). The female captures several insects or spiders to provision the cells, exerting a very significant predation pressure on ecosystems (Krombein, 1967). Around 5% of all species of solitary wasps have the habit of nesting in pre-existing cavities (Krombein, 1967). This characteristic has facilitated the study of these solitary species, because females are attracted to nest in human-made trap-nests.

In studies made with solitary wasp species that nest in pre-existing cavities, the use of trap-nests has provided information not only on the occurrence of species in a given habitat, but also on the wasp community composition and their natural enemies, on the nesting biology of the species, on food sources used to rear the immatures (e.g., Krombein, 1967; Gathmann et al., 1994; Camillo et al., 1995; Camillo & Brescovit, 1999; Zanette et al., 2004; Tylianakis et al., 2006; Buschini et al., 2006; Asís et al., 2007; Buschini, 2007; Santoni & Del Lama, 2007; Buschini & Woiski, 2008; Ribeiro & Garófalo, 2010; Musicante & Salvo, 2010; Loyola & Martins, 2011; Polidori et al., 2011). Habitat quality, the effects of habitat fragmentation and of landscape complexity on community composition and predatory-prey interactions (Tscharntke et al., 1998; Morato, 2001; Steffan-Dewenter, 2002; Kruess & Tscharntke, 2002; Tylianakis et al., 2007; Loyola & Martins, 2008; González et al.,2009; Holzschuh et al., 2009; Schüepp et al., 2011), and how urban environments can support such insects (Zanette et al., 2005) have also been assessed with the use of trap nests.

Given the important role of solitary wasps in terrestrial ecosystems (LaSalle & Gauld, 1993), any effort to understand them and unsure their survival is fully justified. In this context, biological inventories are the basic tools for the initial survey of biological diversity, as well as for monitoring changes in different components of biodiversity in response to the impacts of both natural processes and human activities (Lewinsohn et al., 2001). The present study reports data on the occupation of trap-nests by solitary wasps from a survey carried out in the Ilha Anchieta State Park, a reserve within the Atlantic Forest domain. Among tropical biomes, the Atlantic Forest is the one that has suffered the most fragmentation and degradation by human intervention, which threatens the high species diversity and the high degree of endemism of this biome (Myers et al., 2000).

Material and Methods

Study Area

The study was conducted within the Ilha Anchieta State Park (45° 02' – 45° 05' W and 23° 31' – 23° 34' S). The park occupies the entire island (828 ha) and has only one perennial stream, which is located in an area of coastal forest (restinga). The topography is rugged and mountainous, with slopes typically greater than 24°. Low-lying level areas (with slopes under 6°) are found at two beaches ('Grande' and 'Presídio'). Areas of intermediate slope are located in the valley bottoms and on the flat hilltops of the island (Rocha-Filho & Garófalo, 2013). The vegetation found on Ilha Anchieta has been described by Guillaumon et al. (1989), following Rizzini (1977), and includes anthropogenic fields, rocky coast, Atlantic Forest, Gleichenial, mangrove, and restinga.

The climate of Ilha Anchieta is tropical rainy (Koppen, 1948). It has two distinct seasons a year: one super humid, from October to April, with monthly rainfall above 100 mm, and the other less humid, from May to September, when the monthly rainfall is below 100 mm. During the study, from September 2007 to August 2009, the average monthly temperatures ranged from 22.5 °C to 25.8 °C in the super-humid periods and from 18.0 °C to 21.4 °C in the less humid periods. The monthly rainfall totals ranged from 190.3 mm to 488.3 mm in the super-humid periods and from 0 to 220.9 mm in the less humid periods. Temperature and rainfall data of the Ubatuba region were obtained from the Centro Integrado de Informações Agrometereológicas - (http://www.ciiagro.sp.gov.br). The super-humid and less-humid terminology follows Talora and Morellato (2000).

Methods

The design of the trap-nests used in the present study followed Camilo et al. (1995): they consisted of tubes made of black cardboard, with one end closed with the same material. These tubes, with dimensions of 5.8 cm long × 0.6 cm diameter (small tube, ST) and 8.5 cm long × 0.7 cm diameter (large tube, LT) were inserted into holes drilled into wooden plates (length: 30 cm, height: 12 cm, thickness: 5.0 cm). Each wooden plate had a total of 55 holes available, 2.0 cm apart from each other and distributed in five rows. We also used another type of trap-nest made of bamboo canes (BC) which were cut so that a nodal septum closed one end of the cane. The bamboo canes had variable length and their internal diameter ranged from 0.4 to 2.9 cm although all sizes were not equally represented. In order to protect the bamboo canes from the sun and the rain, they were inserted, in bundles of 10-15 units, into a PVC tube, 40 cm long × 12 cm diameter. The plates containing the cardboard tubes and the PVC tubes were arranged on iron stands and placed at the sampling sites. Three sampling sites (site 1: 45° 04' 06.5" W and 23° 32' 22.0" S; site 2: 45° 03' 55.7" W and 23° 32' 24.3" S, and site 3: 45° 03' 52.1" W and 23° 32' 23.6" S) were established on the 'Presídio' beach. Distances between sites ranged from 100 m, between sites 2 and 3, to 308 m, between sites 1 to 3. Sites 1 and 2 were separated by a distance of 220 m. The iron stands were installed behind or next to existing buildings in the area. The geographical coordinates of all three sampling points were recorded with a GPS receiver. The iron stands installed at each sampling site carried 55 large tubes, 55 small tubes, and 90 bamboo canes distributed in three PVC tubes.

The trap-nests were inspected once a month from September 2007 to August 2009. Each inspection was made using a penlight. When traps contained completed nests, they were collected and replaced with empty ones. These nests were taken to the laboratory in Ribeirão Preto, SP, and each trap-nest was placed in a transparent glass or plastic tube, 4.0-5.0 cm longer than the trap and with an internal diameter of 0.9 cm. The nests were kept at room temperature and observed daily until the adults emerged. As the adults emerged into the glass or plastic tube, the trap was removed and the individuals were collected. A few days after the last emergence, the nest was opened and its contents analysed. Cells and nests from which no adult emerged were also opened, and the cause and stage of mortality were recorded. The analysis of the food present in brood cells allowed identification of the genus (or subfamily in the case of Eumeninae) of the breeding species. Voucher specimens of wasps were deposited in the Collection of Bees and Solitary Wasps (Coleção de Abelhas e Vespas Solitárias - CAVES) of the Department of Biology of the Faculty of Philosophy, Science and Letters at Ribeirão Preto –USP.

Statistical Analysis

Statistical tests followed Zar (1999) and were performed using the statistical package SigmaStat for Windows, Version 3.10 (2004 – Systat Software, Inc.). Kruskal-Wallis (H) test was used to verify differences among the diameters of bamboo canes occupied by different species. To isolate the species that differs from the others it was used a multiple comparison procedure (Dunn's Method). Chi-square tests were used to compare mortality rates for immature wasps, parasitism rates, and sex ratios. Pearson correlation analyses were performed to test for the strength of associations between nesting frequency and climatic conditions (average monthly temperature and rainfall). Throughout the text, all means (X) are reported with their associated standard error (SE) as X ± SE.

Results

Occupation of trap-nests by solitary wasps

We collected 150 nests in the first year (September 2007 - August 2008) and 104 nests in the second year (September 2008 - August 2009). Of the nests established in the first year, 78 nests were built by 12 species and of those established in the second year, 64 nests were made by six species. These species belong to four genera (*Trypoxylon, Pachodynerus, Podium,* and *Auplopus*) of four families (Crabonidae, Vespidae, Pompilidae, and Sphecidae). Crabronidae was the family with the highest number of species (nine), followed by Vespidae and Pompilidae, each with two species. Sphecidae was represented by only one species (Table 1). In the 112 remaining nests (72 nests of the first year and 40 of the second one) all immatures were dead from unknown causes or had been parasitized by insects, so that only the genus of the nesting species could be identified.

Most species (n = 10) nested exclusively in the bamboo cane traps. *Pa. nasidens, Pa. brevithorax,* and *Trypoxy-*

lon sp.2 aff. *nitidum* were the only species that used all three types of trap-nest. *Pachodynerus nasidens* nested more frequently in the small and large tubes whereas nests of *Trypoxylon* (n = 75) were found mainly in bamboo canes (n = 69). *Podium denticulatum* occupied only the large tubes and bamboo canes, but most nests were built in bamboo canes (Table 1).

The analysis of the diameters of bamboo canes occupied by *T. lactitarse, Trypoxylon* sp.2 aff. *nitidum, Pa. nasidens* and *Po. denticulatum* revealed significant differences among them (Kruskal-Wallis H = 56.7; df = 3; P < 0.001). The cavities used by *T. lactitarse* were larger than those occupied by *Trypoxylon* sp.2 aff. *nitidum* and *Pa. nasidens,* and those used by *Po. denticulatum* were larger than those used by *Trypoxylon* sp.2 aff. *nitidum.* On the other hand, no significant difference was found among the diameters of bamboo canes used by *Po. denticulatum, T. lactitarse* and *Pa. nasidens* as well as between the canes occupied by *Pa. nasidens* and *Trypoxylon* sp.2 aff. *nitidum* (Table 2).

Phenology of nesting

Trypoxylon lactitarse, Trypoxylon sp.2 aff. *nitidum, Pa. nasidens,* and *Auplopus* sp.1 nested in both years, *Pa. brevithorax* and *Auplopus pratens* nested only in the second year, and the remaining eight species occupied the traps only in the first year (Table 1). The occupation of trap-nests occurred throughout the study period, with the exception of December 2007, when no nests were built. In both years, the wasps nested more frequently during the super-humid season (109 nests in the first year and 76 in the second year) than

Table 1. Species of solitary wasps that nested in the trap-nests in the Ilha Anchieta State Park, Ubatuba, state of São Paulo, from September 2007 to August 2009, number of nests built and type of trap-nest (ST = small tube; LT = large tube; BC = bamboo cane) occupied.

Species	Number of nests		Frequency of utilization of each type of trap-nest		
	1ST year	2ND year	ST	LT	BC
Trypoxylon lactitarse (Saussure, 1867)	13	21			34
Trypoxylon aurifrons (Shuckard, 1837)	4				4
Trypoxylon punctivertex (Richards, 1934)	2				2
Trypoxylon albitarse (Fabricius, 1804)	1				1
Trypoxylon sp.1 aff. *nitidum*	3				3
Trypoxylon sp.2 aff. *nitidum*	17	11	5	1	22
Trypoxylon sp.5 aff. *nitidum*	5				5
Trypoxylon (*Trypargilum*) sp.1	3				3
Trypoxylon (*Trypargilum*) sp.2	1				1
Pachodynerus nasidens (Latreille, 1812)	7	24	13	11	7
Pachodynerus brevithorax (Saussure, 1852)		5	3	1	1
Podium denticulatum (Smith, 1856)	20			2	18
Auplopus pratens (Dreisbach, 1963)		2			2
Auplopus sp. 1	2	1			3
Total	78	64	21	15	106

Table 2. Diameter of bamboo canes occupied by solitary wasps in the Ilha Anchieta State Park, Ubatuba, state of São Paulo, from September 2007 to August 2009.

Species	Trap-Nest Diameter (cm)	
	Range	X ± SE (N)
Trypoxylon lactitarse	0.8-1.1	0.96 ± 0.02 (34)
Trypoxylon aurifrons	0.5-0.7	0.6 ± 0.05 (4)
Trypoxylon punctivertex	0.5-0.6	0.55 ±0.07 (2)
Trypoxylon albitarse	1.1	1.1 (1)
Trypoxylon sp.1 aff. *nitidum*	0.6	0.6 (3)
Trypoxylon sp. 2 aff. *nitidum*	0.4 – 0.7	0.5± 0.02 (22)
Trypoxylon sp. 5 aff. *nitidum*	0.6-0.7	0.6 ±0.02 (5)
Trypoxylon (Trypargilum) sp.1	0.6	0.6 (3)
Trypoxylon (Trypargilum) sp.2	0.9	0.9 (1)
Pachodynerus nasidens	0.6-0.9	0.67± 0.04 (7)
Pachodynerus brevithorax	1.0	1.0 (1)
Podium denticulatum	0.8-1.1	0.93 ±0.02 (18)
Auplopus pratens	1.2-1.5	1.35 ±0.14 (2)
Auplopus sp.1	1.1 -1.5	1.33 ±0.12 (3)

during the less-humid season (31 nests in the first year and 28 in the second year). In both years, the highest frequency of nesting during the hot/super-humid season (average temperature = 23.9 ± 1.1°C) occurred in March. During the cool/less-humid season (average temperature = 19.9 ± 1.06°C), July of the first year and August of the second year were the months with the highest numbers of nests built (Fig 1).

Trypoxylon lactitarse, Pa. nasidens, Trypoxylon sp.2 aff. *nitidum* and *Po. denticulatum* were the most abundant

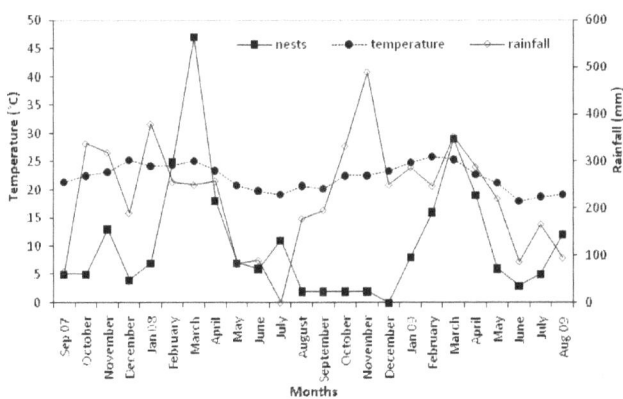

Fig 1. Climatic conditions (average temperature and rainfall) and number of nests established in trap-nests by solitary wasps in the Ilha Anchieta, state of São Paulo, from September 2007 to August 2009.

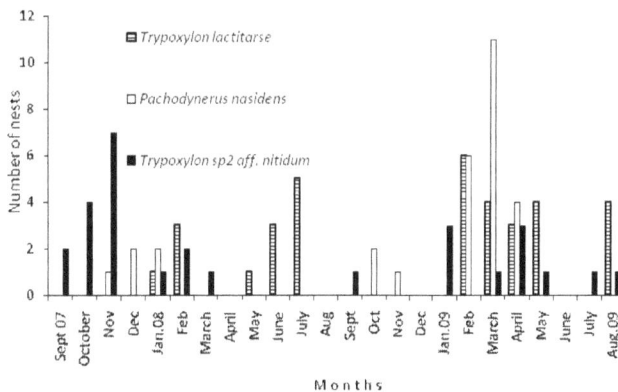

Fig 2. Number of nests built by *Trypoxylon lactitarse, Pachodynerus nasidens* and *Trypoxylon* sp.2 aff. *nitidum* in trap-nests in the Ilha Anchieta, Ubatuba, state of São Paulo, from September 2007 to August 2009.

species in terms of number of nests. *Trypoxylon lactitarse* showed an overall higher frequency of nesting in the second year (Table 1). More traps were occupied in the cool/less-humid season in the first year (n= 9 nests), and more traps in the hot/super-humid season in the second year (n = 13 nests). In both years, the females were active for only five months (Fig 2).

Pachodynerus nasidens was the second most abundant species. This species built 7 nests in the first year and 24 in the second one (Table 1). Except for two nests constructed in July of the first year, the nests were constructed in the hot/super-humid season (Fig 2). *Trypoxylon* sp.2 aff. *nitidum* occupied more traps in the first year (17 nests) than in the second year (11 nests) (Table 1). The nesting peak occurred in November of the first year, whereas January and April were the months with the highest number of nests in the second year. Regardless of the total number of nests built the females of *Trypoxylon* sp.2 aff. *nitidum* were active for a longer period than the females of other species (Fig 2). Females of *Po. denticulatum*, the fourth most abundant species, occupied trap-nests only in February, March, and April of the first year. Analysis of the temporal distribution of nesting of the four most abundant species in relation to average monthly temperature and rainfall only revealed a significant correlation between the number of nests and temperature for *Pa. nasidens* (r = 0.45; P = 0.02; df = 24).

Period of development and sex ratio

The total duration from oviposition to adult emergence was not determined. However, the maximum interval between nest collection and adult emergence may provide an estimate. Excepting two males of *T. punctivertex* and one female and six males of *Po. denticulatum*, which emerged from 165 to 192 days after the nests had been removed from the field, other individuals had the maximum interval ranged from 59 to 66 days (Table 3). These figures indicate an egg-to-emergence period about of two months. On the other hand, the results obtained for *T. punctivertex* and *P. denticulatum* indicate the occurrence of diapause in nests established in February (*Po. denticulatum*), March (*T. punctivertex* and *Po. denticulatum*), and April (*Po. denticulatum*).

The sex ratios of *T. lactitarse* and *Trypoxylon* sp.2 aff. *nitidum* were significantly male-biased, whereas those of *Trypoxylon* sp.5 aff. *nitidum* and *Pa. nasidens* were significantly female-biased. Sex ratios of *Po. denticulatum* and *Pa. brevithorax* were not significantly different from 1:1 (Table 4). Females

and males of the remaining species were produced in similar numbers from the nests of *A. pratens* and *T. punctivertex*, whereas from the nests of *T. albitarse* and *Trypoxylon* sp.2 only males emerged (Table 4).

Mortality of immatures

Of 768 brood cells constructed in 254 nests, 240 (31%) contained dead immatures from unknown causes, and 199 (26%) had been parasitized by insects. The number of cells with dead

Table 3. Period (in days) between the collection of the nests of solitary wasps that occupied the trap-nests in the Ilha Anchieta State Park, Ubatuba, state of São Paulo, from September 2007 to August 2009, and emergence of individuals produced.

Species	Period (in days) between collection date and emergence			
	Males		Females	
	Range	X ± SE(N)	Range	X ± SE(N)
Trypoxylon lactitarse	9-61	39.1±1.30(65)	20-56	37±1.35(30)
Trypoxylon aurifrons	17-20	17.4±0.39(8)	17-20	18±0.98(3)
Trypoxylon punctivertex	177	177(2)	09-12	10.5±1.49(2)
Trypoxylon albitarse	66	66(1)		
Trypoxylon sp. 1 aff. *nitidum*	15-30	24.3±3.6(4)	22	22(1)
Trypoxylon sp. 2 aff *nitidum*	9-35	23.6±0.97(50)	9-37	22.9±1.36(22)
Trypoxylon sp. 5 aff. *nitidum*	26	26(1)	11-26	16.1±1.52(14)
Trypoxylon (*Trypargilum*) sp. 1	28-42	32.2±2.59(5)	30	30(1)
Trypoxylon (*Trypargilum)* sp. 2	24	24(1)		
Pachodynerus nasidens	11-58	22±3.19(28)	5-59	21.5± 1.77(50)
Pachodynerus brevithorax	11-19	17±1.22(6)	11-21	17.3±2.15(4)
Podium denticulatum	11-177	91.1±20.0(14)	10-192	68.7±31.89(7)
Auplopus pratens	17-18	17.5±0.50(2)	21-22	21.5±0.50(2)
Auplopus sp.1	2/20	6.8±3.33(5)	21	21(1)

Table 4. Number of males and females emerged and sex ratio for each species that nested in the Ilha Anchieta State Park, Ubatuba, state of São Paulo, from September 2007 to August 2009. (Values of χ^2 in bold indicate a sex ratio significantly different at $P < 0.05$ and df = 1, from 1:1).

Species	Number of males	Number of females	Sex ratio (no. ♂ : no. ♀)
Trypoxylon lactitarse	65	30	2.2♂: 1 ♀ (χ^2= **12.9**)
Trypoxylon aurifrons	8	3	2.6♂: 1 ♀
Trypoxylon punctivertex	2	2	1♂: 1 ♀
Trypoxylon albitarse	1		
Trypoxylon sp.1 aff. *nitidum*	4	1	4♂: 1 ♀
Trypoxylon sp.2 aff. *nitidum*	50	22	2.3♂: 1♀ (χ^2= **10.9**)
Trypoxylon sp.5 aff. *nitidum*	1	14	0.1♂: 1 ♀ (χ^2= **11.3**)
Trypoxylon (*Trypargilum*) sp.1	5	1	5♂: 1 ♀
Trypoxylon (*Trypargilum*) sp.2	1		
Pachodynerus nasidens	28	50	(0.56♂:1♀) (χ^2= **6.2**)
Pachodynerus brevithorax	6	4	1.5♂: 1 ♀ (χ^2= 0.4)
Podium denticulatum	14	7	(2♂:1♀) (χ^2= 2.33)
Auplopus pratens	2	2	1♂: 1 ♀
Auplopus sp.1	5	1	5♂: 1 ♀

immatures was significantly higher than the number of cells parasitized (χ^2=4.10; P<0.05; df=1) only in nests of *Trypoxylon* whose species could not be determined (Table 5). For *Trypoxylon* (*Trypargilum*) sp.2 and *A. pratens*, the offspring had a survival rate of 100%. In the case of *T. albitarse*, the only cause of non-emergence was death of the immature from unknown causes, whereas for *Trypoxylon* (*Trypargilum*) sp1 the one cell from which a wasp did not emerge had been parasitized. Although in relatively small numbers all others species had immature mortality due to unknown causes and parasite attack (Table 5).

Natural enemies and hosts

The natural enemies were identified as belonging to the families Chrysididae (Hymenoptera), Ichneumonidae (Hymenoptera), and Chalcididae (Hymenoptera), the genus *Melittobia* (Hymenoptera: Eulophidae), and the species *Amobia floridendis* (Townsend, 1892) (Diptera: Sarcophagidae).

Trypoxylon sp.2 aff. *nitidum* was the species with the highest number of nests attacked (15) and cells parasitized (21), followed by *T. lactitarse, Pa. nasidens*, and *Pa. brevithorax* with 17, 5 and 4 cells parasitized, respectively (Table 6). Among the nests that could not be identified to the level of species due to total mortality of the brood cells from unknown causes or parasitism, the largest number (61 nests)

belonged to the genus *Trypoxylon*, followed by *Podium* (10 nests), Eumeninae (3 nests), and *Auplopus* (2 nests)(Table 6). No natural enemy was found in the nests of *A. pratens, T. albitarse*, or *Trypoxylon* (*Trypargilum*) sp.2.

Among the natural enemies associated with the nests, *Amobia floridensis* (Diptera: Sarcophagidae) was the main enemy, attacking the nests of six species and responsible for 58.3% of all parasitized cells. *Melittobia* sp. (Hymenoptera: Eulophidae) caused 23.6% of the combined mortalities of *T. punctivertex, Trypoxylon* sp.2 aff. *nitidum, Pa. nasidens*, and *Po. denticulatum*. Individuals of the family Chrysididae were reared exclusively in the nests of *Trypoxylon*, parasitizing 22 cells (11.1%) in 17 nests. Individuals of Ichneumonidae attacked more often nests of *T. lactitarse*, parasitizing 6 cells in 6 nests. In addition to this host, nests (n = 2) of two other species of *Trypoxylon* also had cells parasitized (one in each nest), and a cell of *Auplopus* sp. was also attacked by this enemy. Individuals of Chalcididae attacked exclusively nests of *Auplopus*, and parasitized two cells in two nests (Table 6).

Discussion

Although there are tendencies to make a comparative analysis between the number of species occupying trap-nests and/or their abundances obtained in inventories carried out

Table 5. Numbers of brood cells, individuals produced, cells with dead immatures and parasitized cells in nests of solitary wasps that occupied trap-nests in the Ilha Anchieta State Park, Ubatuba, state of São Paulo, from September 2007 to August 2009. (* Nests of unidentified species) **(Value in bold indicates a statistically significant, at P <0.05 and df = 1, difference between the number of immature wasps mortalities due to unknown causes compared to parasite attack).

Species	No. of brood cells	No. individuals produced	No. of cells with dead immatures	No. of parasitized cells	χ^2**
Trypoxylon lactitarse	133	95	21	17	$\chi^2 = 0.42$
Trypoxylon aurifrons	16	11	2	3	
Trypoxylon punctivertex	9	4	2	3	
Trypoxylon albitarse	3	1	2		
Trypoxylon sp. 1 aff. *nitidum*	10	5	2	3	
Trypoxylon sp. 2 aff. *nitidum*	109	72	16	21	$\chi^2 = 0.67$
Trypoxylon sp. 5 aff. *nitidum*	21	15	5	1	
Trypoxylon (*Trypargilum*) sp. 1	7	6		1	
Trypoxylon (*Trypargilum*) sp. 2	1	1			
Pachodynerus nasidens	91	78	8	5	
Pachodynerus brevithorax	17	10	3	4	
Podium denticulatum	27	21	4	2	
Auplopus pratens	4	4			
Auplopus sp.1	9	6	2	1	
Subtotal	457	329	67	61	
*Trypoxylon**	265	0	149	116	$\chi^2 = \textbf{4.10}$
*Podium**	29	0	16	13	$\chi^2 = 0.31$
Eumeninae*	11	0	5	6	
*Auplopus**	6	0	3	3	
Subtotal	311	0	173	138	
Total	768(100%)	329 (43%)	240 (31%)	199 (26%)	

Table 6. Host species, number of nests and brood cells attacked by natural enemies of trap-nesting solitary wasps in the Ilha Anchieta State Park, Ubatuba, state of São Paulo, from September 2007 to August 2009. (* Nests of unidentified species).

Host Species	Natural Enemies				
	Number of nests attacked and (number of brood cells attacked = c)				
	Amobia floridensis	*Melittobia sp.*	Chrysididae	Ichneumonidae	Chalcididae
Trypoxylon lactitarse	7(11c)			6(6 c)	
Trypoxylon aurifrons	1(1c)		1(2c)		
Trypoxylon punctivertex		1(2c)	1(1c)		
Trypoxylon sp. 1 aff. *nitidum*			2(3c)		
Trypoxylon sp. 2 aff. *nitidum*	3(5c)	4(7c)	7(8c)	1(1c)	
Trypoxylon sp. 5 aff. *nitidum*			1(1c)		
Trypoxylon (*Trypargylum*) sp. 1				1(1c)	
Pachodynerus nasidens	3(3c)	1(2c)			
Pachodynerus brevithorax	2(4c)				
Podium denticulatum	1(1c)	1(1c)			
Auplopus sp.1					1(1c)
Subtotal	17(25c)	7(12c)	12 (15 c)	8(8c)	1(1c)
*Trypoxylon**	42(79c)	11(27c)	5 (7c)	3(3 c)	
Podium *	9(12c)	1(1c)			
Eumeninae*		2(6c)			
*Auplopus**		1(1c)		1(1c)	1(1c)
Subtotal	51(91c)	15(35c)	5(7c)	4(4c)	1(1c)
Total	68(116c)	22(47c)	17(22c)	12(12c)	2(2c)

at different geographical locations, the results of these analyses should be interpreted with caution. Differences in sampling methods, the type and arrangement of the trap-nests in the study area, the arrangement of natural cavities, and the sampling periods may hinder the formulation of reliable comparisons (Aguiar et al., 2005). Among the 14 species collected from the trap-nests set up on Ilha Anchieta, seven: *T. lactitarse, T. albitarse* (Coville, 1981), *T. aurifrons* (Amarante, 2002), *T. punctivertex, Po. denticulatum* (Bohart & Menke, 1976), *Pa. brevithorax* and *Pa. nasidens* (Willink & Roig-Alsina, 1998), have in common a broad geographic distribution. Within Brazil these species have been reported for three farmland habitats (Santa Carlota Farm, Cajuru, SP, Camillo et al., 1995), a riparian forest area surrounded by crops and pastures (Ituiutaba, MG, Assis & Camillo, 1997), an urban forest (Belo Horizonte, MG, Loyola & Martins, 2006), three habitats in a municipal park (Araucárias Municipal Park, Gurarapuava, PR, Buschini & Woiski, 2008), and tropical savanna and riparian forest areas (Ingai, MG, Pires et al., 2012). They have also been observed in a region of temperate deciduous forest in (Ontario, Canada, Taki et al., 2008) and forest fragments covered by Chaco Serrano vegetation (Córdoba, Argentina, Musicante & Salvo, 2010). Of the seven remaining species, *A. pratens* has been recorded only in Brazil (Fernandez, 2000), *Trypoxylon* sp.1 aff. *nitidum*

and *Trypoxylon* sp.2 aff. *nitidum* had occurrences reported by Assis and Camillo (1997), and for the other four species, there is no information available.

Trypoxylon lactitarse, the species that occupied the highest number of trap-nests on Ilha Anchieta, was also the dominant species in the studies by Camillo et al. (1995), Assis and Camillo (1997), Buschini and Woiski (2008), and Musicante and Salvo (2010), which evidences a great plasticity to adapt to different environments. When compared to these studies, *T. lactitarse* on Ilha Anchieta showed significantly elevated reproductive activity, occupying a large number of trap-nests. The only exception was the study by Musicante and Salvo (2010), carried out in forest fragments dispersed in an agricultural matrix strongly dominated by wheat in winter and soya beans or corn in summer, in which the number of *T. lactitarse* nests was smaller than observed on Ilha Anchieta. *Trypoxylon* sp.2 aff. *nitidum* built a larger number of nests on Ilha Anchieta than reported for the habitats studied by Assis and Camillo (1997). Likewise, *Po. denticulatum* nested more frequently on Ilha Anchieta than in the areas studied by Camillo et al. (1996) and Assis and Camillo (1997), who used sampling periods similar to those used in the present study. Although most nests on Ilha Anchieta were built during the hottest and rainiest period of the year, individual analyses for the four most abundant species showed an association between temperature and nesting activity only for *Pa. nasidens*. The relatively small number of occupied trap-nests hinders a more robust analysis of the action of the climatic factors on the phenology of these species.

Krombein (1967) and Fricke (1991) have suggested that the diameter of the cavity used by a given species is not only related to the size of the female, but also to its prey. These parameters may explain the lack of overlap in the diameters of the bamboo canes used by *T. lactitarse* and other species of the same genus. The exceptions were *T. albitarse* and *T. (Tripargilum)* sp.2, which presented marginal overlap with *T. lactitarse*, due to the larger size of the individuals of the latter species compared to the other species (Assis & Camillo, 1997) and/or the larger size of the spiders used to provision the cells. *Auplopus* species occupied the bamboo canes with the largest diameters. This is certainly related to the fact that cells are built with material (clay) taken into the cavity, inside which the female needs space to work on the construction. It's interesting to observe that among the most abundant species *T. lactitarse, Po. denticulatum* and *Pa. nasidens* nested in cavities with similar diameters. This could suggest the occurrence of competition among those species depending up of the availability of cavities with those diameters. Considering all the variables that may contribute to the selection of a cavity for nesting, any project aimed at the preservation of species with the characteristic of nesting in pre-existing cavities should provide a large number of options in terms of diameters of the cavities offered.

The method used in the present study does not directly yield data about the duration of the development period of immature wasps (from oviposition to emergence). However, by combining the time intervals between nest collection and the emergence of individuals with information on the distribution of nesting activities throughout the year, it is possible to affirm that species such as *T. lactitarse, T.* sp.2 aff. *nitidum, Pa. nasidens*, and *Po. denticulatum*, have more than one generation a year. For species such as *T. lactitarse* and *Pa. nasidens*, the maximum periods observed between nest collection and emergence reflect mainly the influence of a lower temperature resulting in a longer development period. The exceedingly long periods observed for *Po. denticulatum* indicate the occurrence of diapause in some immature wasps of this species, as also reported by Ribeiro and Garófalo (2010), based on studies carried out on the campus of the University of São Paulo at Ribeirão Preto, state of São Paulo. Only two nests of *T. punctivertex* were collected, both in March 2008; from one nest a female rapidly emerged followed much later by two males that had evidently undergone diapause, from the other nest a female rapidly emerged. Considering that diapause acts as a protection mechanism against unfavourable seasons, thereby reducing the extinction risk (Martins et al., 2001), the prepupal diapause observed in *Po. denticulatum* and *T. punctivertex* nests could be a survival strategy against adverse conditions. The occurrence of nests from which some wasps emerge with and others without diapause, as observed with *T. punctivertex*, shows that some adults should pass through an adverse season by sheltering somewhere.

Trap-nest diameter (Buschini, 2007) and food availability (Polidori et al., 2011) are two factors that may affect the sex ratio of the offspring of species that nest in pre-existing cavities. Considering the four most abundant species sampled on Ilha Anchieta, *T. lactitarse* and *T.* sp.2 aff. *nitidum* showed a male-biased sex ratio, *Pa. nasidens* a female-biased sex ratio, whereas the sex ratio of *Po. denticulatum* was not significantly different from 1:1. Analogous results have been reported by Borges and Blochtein (2001) for *T. lactitarse*, whereas Buschini (2007) reported similar proportions for the two sexes for the two generations of *T. lactitarse* collected during a study in the state of Paraná. Ribeiro and Garófalo (2010) found a sex ratio of 1:1 for *Po. denticulatum*, as observed in the present study. It is important to highlight that there are multiple factors that may affect the sex ratio of a population. A first step towards evaluating these factors will be the acquisition of more extensive data sets, based on larger sample of nests and to assess the food resources availability in the study area.

Several studies carried out with solitary wasps occupying trap-nests have shown wide variations in the mortality rates of immatures. The loss rate for immatures reported by some authors is typically close to 50% (Krombein, 1967; Freeman & Jayasingh, 1975; Jayasingh & Freeman, 1980; Camillo et al., 1996; Buschini & Wolf, 2006), but some species have reported much lower and others much higher mortality rates (Danks, 1971; Camillo et al., 1993; Gathmann et al., 1994; Assis & Camillo, 1997; Tormos et al., 2005). In the present study, the brood mortality rate based on the total set of collected nests was around 57%, but analysing separately each of the seven species that produced the largest number of brood cells (*T. aurifrons*, 16 brood cells to *T. lactitarse*, 133 brood cells), the loss rate for immatures ranged from 14.3% (*Pa. nasidens*) to 41.2% (*P. brevithorax*). Curiously, the two extremes were represented by Eumeninae species. Except for *Pa. brevithorax*, the other six species that utilized more frequently the trap-nests showed loss rates for immatures that are much lower than the values reported by other authors.

Unknown factors and natural enemies caused brood mortalities in similar proportions in nests built by identified species. Among the natural enemies that attacked the trap-nests, parasitization by individuals of the family Chrysididae was restricted to species of *Trypoxylon* as the host, whereas individuals of the family Chalcididae only attacked the nests of *Auplopus*. Considering the other identified enemies associated with the nests collected on Ilha Anchieta, *Amobia floridensis* was the agent that caused the highest brood mortality, attacking the nests of *Trypoxylon, Pachodynerus*, and *Podium*. With a distribution from the United States to Brazil (Pape et al., 2004), the associations between species of Eumeninae and *Trypoxylon* are well known through reports by Krombein (1967), Freeman and Jayasingh (1975), and but the association between *A. floridensis* and *Podium* is described here for the first time. According to Krombein (1967), the *Amobia* female follows the host female back to its nest, lands, and remains close to the nest entrance. When the host female leaves its nest to forage in the field, the fly enters the nest and larviposits. *Amobia floridensis* larvae compete with the host species larvae for caterpillars and spiders placed in the cells of Eumeninae and *Trypoxylon*, respectively, and according to the results obtained in the present study, they also compete for the cockroaches stored in the cells of *Podium*.

The dispersal ability of flying arthropods is usually determined by body size, as larger species have a better flight capacity (Gathmann et al., 1994; Steffan-Dewenter & Tscharntke, 1999). There is no information available on the flight capacity of the wasp species sampled on Ilha Anchieta, but it is reasonable to propose that the colonization of the island may have occurred through the transportation of adult individuals on the vessels that daily cross the 600 m channel that separates the island from the mainland. Another means by which cavity-nesting wasp species may cross water barriers is within pieces of floating wood, a process known to occur for solitary bees (Michener, 2000). Regardless of the colonization process for Ilha Anchieta, the number of nests collected in the present study may be considered relatively small, given the area of the island and the number of wasp species sampled. However, the number of trap-nests occupied by females may be related to the availability of natural cavities within the study area. One of the tourist attractions on Ilha Anchieta are the ruins of a prison with several buildings constructed in masonry, whose current dilapidated state affords many nesting opportunities for wasps. This fact together with the decline of 30.9% in the number of trap-nests occupied and 50% of species nesting in the second year compared to the first suggests that the populations of solitary wasps on Ilha Anchieta may be dependent on the mainland populations for the permanent occupation of the island.

Acknowledgments

We are grateful to J.C. Serrano for wasp identifications, C.A. Mello-Patiu for *Amobia floridensis* identification, E.S.R. Silva for technical help, the Instituto Florestal de São Paulo for permission to work in the Ilha Anchieta State Park, the staff at Ilha Achieta for their support, two anonymous reviewers for providing valuable comments on the manuscript, and Research Center on Biodiversity and Computing (BioComp). A.L.O. Nascimento received grants from FAPESP (08/06023-4) and CAPES. Research supported by FAPESP (04/15801-0) and CNPq (305274/2007-0).

References

Aguiar, C.M.L., Garófalo, C.A. & Almeida, G.F. (2005). Trap-nesting bees (Hymenoptera, Apoidea) in areas of dry semideciduous forest and caatinga, Bahia, Brazil. Revista Brasileira de Zoologia, 22: 1030-1038.

Amarante, S.T.P. (2002). A synonymic catalog of the Neotropical Crabronidae and Sphecidae (Hymenoptera: Apoidea). Arquivos de Zoologia, 37: 1-139.

Amarante, S.T.P. (2005). Addendum and corrections to a synonymic catalog of Neotropical Crabronidae and Sphecidae. Papéis Avulsos de Zoologia, 45: 1-18.

Asís, J. D., Benéitez, A., Tormos, J., Gayubo & Tomé, A.M. (2007). The significance of the vestibular cell in trap nesting wasps (Hymenoptera:Crabronidae): Does its present reduce mortality?. Journal of Insect Behavior, 20: 289-303.

Assis, J.M.F. & Camillo, E. (1997). Diversidade, Sazonalidade e Aspectos Biológicos de Vespas Solitárias (Hymenoptera: Sphecidae: Vespidae) em Ninhos Armadilhas na Região de Ituiutaba, MG. Anais da Sociedade Entomológica do Brasil, 26: 335-347.

Bohart, R.M. & Menke, A.S. (1976). Sphecidae wasps of the world. A generic revision. Berkeley: University of California Press, 695 p.

Borges, F.B. & Blochtein, B. (2001). Aspectos da biologia de cinco espécies de *Trypoxylon* (Hymenoptera, Sphecidae, Trypoxylini) no Rio Grande do Sul. Biociências, 9: 51-62.

Buschini, M.L.T. & Wolff, L.L. (2006). Notes on the biology of *Trypoxylon* (*Trypargilum*) *opacum* Brèthes (Hymenoptera, Crabronidae) in southern Brazil. Brazilian Journal of Biology, 66: 907-917.

Buschini, M.L.T. (2007). Life-history and sex allocation in *Trypoxylon* (syn. *Trypargilum*) *lactitarse* (Hymenoptera; Crabronidae). Journal of Zoological Systematics and Evolutionary Research, 45: 206-213.

Buschini, M.L.T. & Woiski, T. D. (2008). Alpha-beta diversity in trap nesting wasps (Hymenoptera, Aculeata) in southern Brazil. Acta Zoológica, 89: 351-358.

Camillo, E. & Brescovit, A.D. (1999). Aspectos biológicos de *Trypoxylon* (*Trypargilum*) *lactitarse* Saussure e *Trypoxylon* (*Trypargilum*) *rogenhoferi* Khol (Hymenoptera, Sphecidae) em ninhos-armadilha, com especial referencia a suas presas. Anais da Sociedade Entomologica do Brasil, 28: 251-162.

Camillo, E., Garófalo, C.A., Serrano, J.C. & Muccillo, G.(1995). Diversidade e abundância sazonal de abelhas e vespas solitárias em ninhos-armadilha (Hymenoptera, Apocrita, Aculeata). Revista Brasileira de Entomologia, 39: 459-70.

Camillo, E., Garófalo, C.A., Assis, J.M.F. & Serrano, J.C. (1996). Biologia de *Podium denticulatum* Smith em ninhos armadilhas (Hymenoptera, Sphecidae, Sphecinae). Anais da Sociedade Entomológica do Brasil, 25: 439-450

Coville, R.E. (1981). Biological observations on three *Trypoxylon* wasps in the subgenus *Trypargilum* from Costa Rica: *T. nitidum schultessi, T. saussurei* and *T. lactitarse* (Hymenoptera: Sphecidae). Pan-Pacific Entomology, 57: 332-340.

Danks, H.V. (1971). Populations and nesting-sites of some aculeate Hymenoptera nesting in Rubus. Journal of Animal Ecology, 40: 63-70.

Fernandez, F. (2000). Avispas cazadoras de arañas (Hymenoptera: Pompilidae) de la Région Neotropical. Biota Colombiana, 1: 3-24.

Freeman, B.E.& Jayasingh, D.B. (1975). Population dynamics of *Pachodynerus nasidens* (Hymenoptera) in Jamaica. Oikos, 2: 86-91.

Fricke, J.M. (1991). Trap-nest bore diameter preferences among sympatric *Passaloecus* spp. (Hymenoptera: Sphecidae). Great Lakes Entomology, 24: 123-125.

Gathmann, A.; Greiler, H.J.& Tscharntke, T. (1994). Trap-nesting bees and wasps colonizing set-aside fields: succession and body size, management by cutting and sowing. Oecologia, 98: 8-14.

González, J.A., Gayubo, S.F., Asís, J.D.& Tormos, J. (2009). Diversity and biogeographical significance of solitary wasps (Chrysididae, Eumeninae, and Spheciforme) at the Arribes del Duero Natural Park, Spain: Their importance for insect diversity conservation in the Mediterranean Region. Environmental Entomology, 38: 608-626.

Guillaumon, J.R., Marcondes, M.A.P., Negreiros, O.C., Mota, I.S.; Emmerich, W., Barbosa, A.F., Branco, I.H.D.C., Camara, J.J.C., Ostini, S., Pereira, R.T.L., Scorvo Filho, J.D., Shimomichi, P.Y., Silva, S.A. & Melo Neto, J.E. (1989). Plano de Manejo do Parque Estadual da Ilha Anchieta. I.F. – Série Registros. São Paulo: Instituto Florestal, 103p.

Holzschuh, A., Steffan-Dewenter, I. & Tscharntke, T. (2009). Grass strip corridors in agricultural landscape enhance nest-site colonization by solitary wasps. Ecological Applications, 19: 123-132.

Jayasingh, D.B. & Freeman, B.E. (1980). The comparative population dynamics of eight solitary bees and wasps (Aculeata, Apocrita, Hymenoptera) trap-nested in Jamaica. Biotropica, 12: 214-219.

Köppen, W. (1948). Climatologia: con un estudio de los climas de la tierra. México: Fondo de Cultura Econômica, 479p

Krombein, K.V. (1967). Trap-nesting wasps and bees: Life histories, nests and associates. Washington: Smithsonian Press, 569p.

Kruess, A. & Tscharntke, T. (2002). Grazing intensity and diversity of grasshoppers, butterflies, and trap-nesting bees and wasps. Conservation Biology, 16: 1570-1580.

LaSalle, J. & Gauld, I.D. (1993). Hymenoptera: their diversity, and their impact on the diversity of other organism. In: LaSalle, J. & Gauld, I.D. (eds.), Hymenoptera and Biodiversity (pp. 1-26). Wallinford (UK): CAB International.

Lewinsohn, T.M., Prado, P.I.K.L. & Almeida, A.M. (2001). Inventários bióticos centrados em recursos: insetos fitófagos e plantas hospedeiras, p. 174-189. In: Garay, I. & Dias, B. (orgs.). Conservação da Biodiversidade em Ecossistemas Tropicais. Editora Vozes.

Loyola, R.D. & Martins, R.P. (2006). Trap-occupation by solitary wasps and bees (Hymenoptera: Aculeata) in a forest urban remnant. Neotropical Entomology, 35: 41-48.

Loyola, R. D. & Martins, R. P. (2008). Habitat structure components are effective predictors of trap-nesting Hymenoptera diversity. Basic and Applied Ecology, 9: 735-742.

Loyola, R. D. & Martins, R. P. (2011). Small-scale area effect on species richness and nesting occupancy of cavity-nesting bees and wasps. Revista Brasileira de Entomologia, 55: 69-74.

Martins, R.P., Guerra, S.T.M. & Barbeitos, M.S. (2001).Variability in egg-to-adult development time in the bee *Ptilothrix plumata* and its parasitoids. Ecological Entomology, 26: 609-616.

Morato, E.F. (2001). Efeitos da fragmentação florestal sobre abelhas e vespas solitárias na Amazônia Central. II. Estratificação vertical. Revista Brasileira de Zoologia, 18: 737-747.

Morato, E. F., Amarante, S.T. & Silveira, O.T. (2008). Avaliação ecológica rápida da fauna de vespas (Hymenoptera: Aculeata) do Parque Nacional da Serra do Divisor, Acre, Brasil. Acta Amazonica, 38: 789-798.

Michener, C. D. (2000). The Bees of the World. Baltimore: Johns Hopkins, 913p.

Musicante, M. L. & Salvo, A. (2010).Nesting biology of four species of *Trypoxylon* (*Trypargilum*) (Hymenoptera: Crabronidae) in Chaco Serrano woodland, Central Argentina. Revista de Biologia Tropical, 58: 1177-1188.

Myers, N., Mittermeier, R.A., Mittermeier, C.G., Fonseca, G.A.B. & Kent, J. (2000). Biodiversity hotspots for conservation priorities. Nature, 403: 853-858.

O'Neill, K.M. (2001). Solitary wasps: behaviour and natural history. New York: Comstock Publishing Associates, 406p.

Pape, T., Wolff, M. & Amat, E.C. (2004). The blow flies, bot flies, woodlouse flies and flesh flies (Diptera: Calliphoridae, Oestridae, Rhinophoridae, Sarcophagidae) of Colombia. Biota Colombiana, 5: 201-208.

Pires, E.P., Pompeu, D.C.& Souza-Silva, M. (2012). Nidificação de vespas e abelhas solitárias (Hymenoptera: Aculeata) na Reserva Biológica Boqueirão, Ingá, Minas Gerais. Bioscience Journal, 28: 302-311.

Polidori, C., Boesi, R. & Borsato, W. (2011). Few, small, and male: Multiple effects of reduced nest space on the offspring of the solitary wasp, *Euodynerus* (*Pareuodynerus*) *posticus* (Hymenoptera: Vespidae). Comptes Rendus Biologies, 334: 50-60.

Ribeiro, F. & Garófalo, C.A. 2010. Nesting behavior of *Podium denticulatum* Smith (Hymenoptera: Sphecidae). Neotropical Entomology, 39: 885-891.

Rizzini, C. (1997). Tratado de Fitogeografia do Brasil. Rio de Janeiro: Âmbito Cultural, 747p.

Rocha-Filho, L.C., Garófalo, C.A. (2013). Community ecology of euglossine bees in the coastal atlantic forest of São Paulo state, Brazil. Journal of Insect Science,13: 23.

Santoni, M.M. & Del Lama, M.A. (2007). Nesting biology of the trap-nesting Neotropical wasp *Trypoxylon* (*Trypargilum*) *aurifrons* Shuckard (Hymenoptera, Crabronidae). Revista Brasileira de Entomologia, 51: 369-376.

Schüepp, C., Herrmann, J.D., Herzog, F. & Schmidt-Entling, M.H. (2011). Differential effects of habitat isolation and landscape composition on wasps, bees, and their enemies. Oecologia, 165: 713-721.

Steffan-Dewenter, I. (2002).Landscape context affects trap-nesting bees, wasps, and their natural enemies. Ecological Entomology, 27: 631-637.

Steffan-Dewenter, I & Tscharntke, T. (1999). Effects of Habitat Isolation on Pollinator Communities and Seed Set. Oecologia, 121: 432-440.

Taki, H., Viana, B.F., Kevan, P.G., Silva, F.O. & Buck, M. (2008). Does Forest loss affect the communities of trap-nesting wasps (Hymenoptera: Aculeata) in forests? Landscape vs. local habitat conditions. Journal of Insect Conservation, 12: 15-21.

Talora, D.C. & Morellato, P.C. (2000). Fenologia de espécies arbóreas em floresta de planície litorânea do sudeste do Brasil. Revista Brasileira de Botânica, 23: 13-26.

Tormos, J., Asís, J.D., Gayubo, S.F., Calvo, J. & Martins, M.A. (2005). Ecology of crabronid wasps found in trap nests from Spain (Hymenoptera: Spheciformes). Florida Entomologist, 88: 278-284.

Tscharntke, T., Gathmann, A. & Steffan-Dewenter, I. (1998). Bioindication using trap-nesting bees and wasps and their natural enemies: community structure and interactions. Journal of Applied Ecology, 35; 708-719.

Tylianikis, J. M., Klein, A.M., Lozada, T. & Tscharntke, T. (2006). Spatial scale of observation affects alpha, beta and gamma diversity of cavity-nesting bees and wasps across a tropical land-use gradient. Journal of Biogeography, 33: 1295–1304. DOI: 10.1111/j.13652699.2006.01493.x

Tylianakis, J.M., Tscharntke, T. & Lewis, O.T. (2007). Habitat modification alters the structure of tropical host-parasitoid food webs. Nature: 445: 202-205. DOI: 10.1038/nature05429.

Willink, A., Roig-Alsina, A. (1998). Revision del gênero *Pachodynerus* Saussure (Hymenoptera: Vespidae, Eumeninae). Contributions of the American Entomological Institute, 30: 1-117.

Zanette, L.R.S., Martins, R.P. & Ribeiro, S.P. (2005). Effects of urbanization on Neotropical wasp and bee assemblages in a Brazilian metropolis. Landscape and Urban Planning., 71: 105-121. DOI: 10.1016/j.landurbplan.2004.02.003

Zanette, L. R. S., Soares, L. A., Pimenta, H. C., Gonçalves, A. M. & Martins, R. P. (2004). Nesting biology and sex ratios of *Auplopus militaris* (Lynch-Arribalzaga 1873) (Hymenoptera Pompilidae). Tropical Zoology, 17: 145-154.

Zar, J.H. (1999). Biostatistical Analysis. New Jersey: Prentice Hall, 663p.

10

The Social Wasp Fauna of a Riparian Forest in Southeastern Brazil (Hymenoptera, Vespidae)

GA Locher[1], OC Togni[1], OT Silveira[2], E Giannotti[1]
Abstract

1 - *Instituto de Biociências, Universidade Estadual Paulista, Rio Claro, SP, Brazil.*
2 - *Museu Paraense Emílio Goeldi, Belém, PA, Brazil.*

Keywords
Richness, abundance, Polistinae, attractive traps, active sampling

Corresponding author
Gabriela de Almeida Locher
Departamento de Zoologia
Instituto de Biociências
Universidade Estadual Paulista
13506-900   Rio Claro, SP, Brazil
E-Mail: gabriela.locher@gmail.com

An inventory of social wasps was carried out on a monthly basis, from March 2010 to March 2011 in a section of riparian forest along the Passa-Cinco River in Ipeúna, São Paulo state, Brazil. Two active collecting methods (active collecting and point sampling using a liquid bait) and one passive method (baited PET bottle trap) were used. The results increased the data on the diversity of social wasps of the State and were used to compare richness, equitability and diversity obtained with several collecting methods which have been employed to social wasps. Thirty-one species belonging to eight genera were recorded; the most abundant were *Agelaia vicina* (de Saussure) and *Agelaia pallipes* (Olivier). Both species belong to the tribe Epiponini, which was dominant in the sample. Regarding sampling methods, the active collecting ones sampled the greatest richness value and the highest Shannon-Wiener diversity index and also presented the largest number of exclusive species. However, the other methods have also obtained exclusive species.

Introduction

Native forests have been exploited for years in a degrading manner, resulting in a number of environmental problems such as species extinctions, local climate change, soil erosion, siltation and eutrophication of watercourses (Ferreira & Dias, 2004). In order to monitor the effects of changes in the environment, to develop conservation strategies on a local scale, information on biodiversity of an area is essential. The concept of diversity can be divided into *alpha*, *beta* and *gamma*, namely the biological diversity in natural and changed communities, the biodiversity exchange rate between different communities, and the species richness of all the communities which integrate a landscape, respectively (Moreno, 2001). An inventory of a local fauna which generates data on alpha diversity is the first step in conservation biology.

Brazil hosts a great diversity of social wasps, reaching a total of 319 species, about 33% of the social wasp species described worldwide (Prezoto et al., 2007). Due to this high diversity and the importance that has been assigned to this group of insects, researches related to diversity of social wasps in this country are increasing in number and have covered different regions, environments and biomes (Richards, 1978; Rodrigues & Machado, 1982; Marques, 1989; Marques et al., 1993; Diniz & Kitayama, 1994; Mechi, 1996; Santos, 1996; Lima et al., 2000; Raw, 2003; Marques et al., 2005; Mechi, 2005; Melo et al., 2005; Hermes & Köhler, 2006; Santos et al., 2006; Silva-Pereira & Santos, 2006; Souza & Prezoto, 2006; Elpino-Campos et al., 2007; Santos et al., 2007; Morato et al., 2008; Ribeiro Junior, 2008; Silveira et al., 2008; Souza et al., 2008; Gomes & Noll, 2009; Santos et al., 2009a; Santos et al., 2009b; Alvarenga et al., 2010; Auad et al., 2010; Lima et al., 2010; Prezoto & Clemente, 2010; Santos & Presley, 2010; Souza et al., 2010; Pereira & Antonialli-Junior, 2011; Silva et al. 2011; Souza et al., 2011; Tanaka Junior & Noll, 2011; Jacques et al., 2012; Silveira et al., 2012; Simões et al., 2012 ; Auko & Silvestre, 2013; Grandinete & Noll, 2013; Togni et al. 2014). In the State of São Paulo, studies on diversity of social wasps are still scarce, mainly focusing on the Northwestern region, with studies in localities such as Paulo de

Faria, Pindorama and Neves Paulista (Gomes & Noll, 2009), Patrocínio Paulista (Lima et al., 2010), Magda, Bebedouro, Matão and Barretos (Tanaka Junior & Noll, 2011); on the Central Eastern region, with researches in Rio Claro (Rodrigues & Machado, 1982), Luíz Antônio, Corumbataí (Mechi, 1996), and Santa Rita do Passa Quatro (Mechi, 2005); and finally at the Littoral in Ubatuba (Togni et al., 2014).

According to the above information, this study aimed to conduct an inventory of the diversity of social wasps in an area of Riparian Forest in the Central Eastern region of São Paulo State, increasing the data on the diversity of social wasps of the State and to compare richness, equitability and diversity obtained through the several methods used.

Materials and Methods

Social wasps (Vespidae, Polistinae) were sampled monthly from March 2010 to March 2011 along the western edge of a riparian forest fragment (semideciduous alluvial forest, according to Rodrigues, 1999) by the margins of the Passa-Cinco River, in a landscape with the surrounding matrix composed by sugarcane plantations, in the rural municipality of Ipeúna - State of São Paulo (22 ° 24'52 .55 "S, 47 ° 43'35 .55" W). The Passa-Cinco is one of the main rivers of the Corumbataí River Basin, in a region that presents a subtropical climate (Köppen Cwa type) characterized by dry winter and rainy summer, with an average temperature in the warmest month of up to 22˚C (Palma-Silva, 1999). Originally, the watershed vegetation was mainly composed of semideciduous forests, and "cerrado" vegetation spots in smaller areas (Koffler, 1993 as cited in Valente & Vettorazzi, 2002). According to Valente and Vettorazzi (2002) the sub-basin of Passa-Cinco River had in year 2000 51.72% of its surface occupied by pasture areas, 14.3% by sugarcane, 15.67% native forest, 10.5% planted forest, 0.74% Cerrado, 0.5% urban areas and 6.47% by others.

Three different methods were used for sampling the wasps: baited PET bottle trap, active collecting, and point sampling using a liquid bait. The baited PET bottle trap method consisted in 10 attractive traps made with 2 L PET type bottles (Polyethylene Terephthalate Bottles) installed along a 1000 meter trail bordered by riparian vegetation and sugarcane plantation. The traps were set on the edge of the vegetation at a height of approximately 1.5 meters (Ribeiro Junior, 2008), with a 100 meters interval between traps (Togni et al., 2014). Each bottle had four circular orifices at the middle, and contained 200 ml of attractive liquid (modified from Souza & Prezoto, 2006) consisting of natural industrialized guava juice and sugar solution (Togni et al., 2014). After a week, traps were removed, the attractive liquid sifted and all social wasps encountered were fixed in 70% ethanol.

Active collecting (without the use of any attractive) consisted in active searching for individuals along the same 1000 meter trail mentioned before collected with an entomo-logical net, at the time of greatest foraging activity between 10:00 am to 3:00 pm (see Prezoto et al., 2008). Rounds of active collecting were made at two different days each month by four independent collectors on the same transect where the bottle traps were placed. The collected individuals were associated with the nearest trap point for further analysis and comparisons.

The method of point sampling using a liquid bait was made with a sucrose solution (1: 4, commercial sugar: water) with 2 cm³ of salt for each half liter of solution which was sprayed over the vegetation at 10 marked points along the sampling trail (with a distance of 100 meters between points). At least 30 minutes after spraying, active searching was performed by two independent collectors on the vegetation for 3 minutes with the aid of an entomological net (modified from Liow et al., 2001; Lima et al., 2010; Tanaka Junior & Noll, 2011).

The species relative abundances were computed by dividing the number of collected individuals of each species by the total abundance found (expressed as percentage) and the richness value is the total number of species encountered.

The richness value obtained by each method was divided by the total richness sampled to evaluate the efficiency of each of them. We calculated the Shannon-Wiener index and Equitability of samples obtained by each method according to the formulas from Krebs (1998) in order to compare the diversity sampled by each of them.

The rarefaction curve based on samples was obtained using the Mao Tau estimator (Colwell et al., 2004) calculated with EstimateS- version 8.2 (Colwell, 2009) and used to compare the three methods based on the monthly richness sampling effort. The species accumulation curve for months of collection was calculated through the addition of new species found every month.

To analyze the equality of samples obtained from each sampling method, the Kruskal-Wallis test (BioEstat 5.0 version 5.0 (Ayres et al., 2007)) was used. For the test, the raw abundance data of each species, and the relative abundances (percentage), were used.

Results and Discussion

Social wasps community in Ipeúna - SP

After 13 months of sampling activity in the rural area in Ipeúna, São Paulo state, 31 species of eight genera of social wasps were found. A total of 954 individuals was collected and 86.58% belonged to the Epiponini tribe, while only 9.43% belonged to the tribe Polistini and 3.98% to the tribe Mischocyttarini (Table 1). Other studies have also noted the predominance of Epiponini in the total abundance of social wasps found. Representatives of this tribe constituted more than 70% in many studies conducted in Brazil (Santos, 1996; Mechi, 2005; Hermes & Köhler, 2006; Santos et al., 2006;

Elpino-Campos et al., 2007; Morato et al., 2008; Gomes & Noll, 2009; Santos et al., 2009b; Silva & Silveira, 2009; Auad et al., 2010; Alvarenga et al., 2010; Lima et al., 2010; Pereira & Antonialli-Junior, 2011; Tanaka Junior & Noll, 2011; Simões et al., 2012; Grandinete & Noll, 2013; Togni et al., 2014). However, when only nests are the targets of surveys, the abundance of Epiponini is often smaller, as observed in Juiz de Fora, Minas Gerais (Lima et al., 2000), with only 11.03% of the colonies belonging to this tribe and in Biritinga, Bahia, were 67.43% of the total abundance found was of Epiponini, yet only 25.71% of the nests found belonged to this tribe (Santos & Presley, 2010). This is probably due to greater depredation of Epiponini nests by humans because of their large size, to destruction by rain and also by the greater facility of finding Mischocittarini and Polistini nests next to human constructions. Differently from the other polistine tribes in the Neotropics, Epiponini is composed of swarming species that may construct large nests and consequently have greater abundance of individuals (Jeanne, 1991; Carpenter & Marques, 2001), a factor that certainly concurs to the high occurrence of specimens from this group in active and passive collections, highlighting the importance of this tribe to the ecosystem studied.

The species with greater abundances (Table 1) were *Agelaia vicina* (404 individuals, relative abundance = 42.35%) and *A. pallipes* (138 individuals, relative abundance = 14.47%). *Agelaia vicina* is known as the polistine species with the largest nests and colonies, and Zucchi et al. (1995) reported the existence of a colony with more than one million adult individuals and nest with over 7.5 million cells. Oliveira et al. (2010) added that the high growth rate, large population and high number of queens provide nests with such proportions. The magnitude of these colonies certainly explains the larger abundance of individuals of this species in the present work, which might in fact be produced by just a few colonies in the area. Oliveira et al. (2010) suggest that *A. vicina* should be considered in many environments as a keystone species, i.e. as populations which determine the stability (integrity and persistence over time) of a community through their activities and abundances (Paine, 1969). In addition to having very large colonies and nests, *A. vicina* presents elevated offspring production rate, requiring therefore a large amount of prey, consisting of a large diversity of arthropods, and severely impacting their populations locally (Oliveira et al., 2010). Among the species belonging to the tribe Mischocyttarini, that with the greatest abundance was *Mischocyttarus drewseni* with 26 individuals collected, i.e. 2.73% of the total. *M. drewseni*, which is widely distributed in Brazil (Richards, 1978), rarely presents colonies with more than 30 individuals (Jeanne, 1972), and was diagnosed by Souza et al. (2010) as an indicator of degraded areas and strong human pressure in riparian forest of Rio das Mortes, in Barroso, Minas Gerais. The species with the greatest abundance belonging to the Polistini tribe was *Polistes versicolor* with 37 specimens (3.88%). *Plt. versicolor*

is another species with a wide distribution in Brazil (Richards, 1978), which has been suggested as a biological control agent in certain crops, such as for control of the eucalyptus defoliator (Elisei et al., 2010). A colony of *Plt. versicolor* can collect around 4015 preys during a year, consisting in a majority (but not exclusively) of lepidopteran larvae (Prezoto et al., 2006).

Among the species found in the area, only *Apoica pallens* had not been recorded by Richards (1978) in the State of São Paulo, but it has been sampled in other inventories (Rodrigues & Machado, 1982; Mechi, 1996; Togni et al., 2014). While recorded by Richards (1978) for São Paulo, *Polistes ferreri* and *Mischocyttarus mattogrossoensis* were not found by other authors. These species may be rather easily misidentified due to close similarity with other more abundant species. The wasp *M. mattogrossoensis* is apparently restricted to cerrado areas (Raw, 2003) as shown by the recorded occurrences in Brazil: Mato Grosso (Richards, 1978; Diniz & Kitayama, 1994; 1998), Goiás (Raw, 2003) and Mato Grosso do Sul (Grandinete & Noll, 2013). However this species was also recorded in an urban environment in Cruz das Almas, Bahia (Marques et al., 1993).

Comparing the studies on diversity of social wasps carried out in the State of São Paulo (Rodrigues & Machado, 1982; Mechi, 1996; Mechi, 2005; Gomes & Noll, 2009; Lima et al., 2010; Tanaka Junior & Noll, 2011; Togni et al., 2014) a survey accomplished at Rio Claro (Rodrigues & Machado, 1982; in an environment comprised of *Eucalyptus* plantations with secondary regrowth of semideciduous seasonal forest, approximately 25 kilometers away from Ipeúna) showed the greatest richness (33 species), with the present study coming next with 31 species. The survey at Rio Claro, although using a quite distinct sampling method (active search for colonies during 13 years) and being performed in a area with a much larger area with 2230 hectares, showed 21 species in common with that of Ipeúna. Nine species (*Polybia bifasciata, Pseudopolybia vespiceps, Agelaia vicina, Mischocyttarus mattogrossoensis, Polistes acteon, Plt. ferreri, Plt. geminatus, Plt. lanio* and *Plt. subsericeus*) were exclusive to Ipeúna, whereas 12 species (*Polybia platycephala, Protonectarina sylveirae, Protopolybia exigua, P. sedula, Parachartergus pseudapicalis, Apoica flavissima, Mischocyttarus araujoi, M. cerberus, M. labiatus, M. latior, Polistes canadensis* and *Plt. consobrinus*) were exclusive to Rio Claro. Despite these differences, it is important to emphasize that 63.63% of the species sampled more than 30 years ago in Rio Claro were found in Ipeúna (even considering the distinctly smaller collection effort), thus showing a noticeable similarity between those social wasps communities.

It is truly remarkable that even a small fragment of riparian vegetation with very strong anthropic influences due to nearby large plantations, has also obtained a very impressive species richness. It is noteworthy that the idea of high diversity in this degraded area, may in fact be indicative of

Table 1. Abundance of wasp species collected with the three methods (AC - Active collection, PS - Point Sample using a liquid bait and BT - attractive PET bottle traps), total abundance and relative abundance (%) of species collected in Passa Cinco River's riparian forest in Ipeúna. (AC - Active collection, PS - Point Sample, BT - Bottle Traps).

Species	AC	PS	BT	Total	Relative Abundance
Agelaia multipicta (Haliday, 1836)	1	2	0	3	0.31
Agelaia pallipes (Olivier, 1791)	20	32	86	138	14.47
Agelaia vicina (de Saussure, 1854)	159	93	152	404	42.35
Apoica pallens (Fabricius, 1804)	0	0	1	1	0.1
Brachygastra augusti (de Saussure, 1854)	1	1	0	2	0.21
Brachygastra lecheguana (Latreille, 1824)	3	1	0	4	0.42
Polybia chrysothorax (Lichtenstein, 1796)	4	0	10	14	1.47
Polybia dimidiata (Olivier, 1791)	33	5	17	55	5.77
Polybia fastidiosuscula de Saussure, 1854	0	9	32	41	4.3
Polybia ignobilis (Haliday, 1836)	19	4	23	46	4.82
Polybia jurinei de Saussure,1854	0	0	3	3	0.31
Polybia gr. *occidentalis* sp 1 (Olivier, 1791)	12	16	2	30	3.14
Polybia gr. *occidentalis* sp 2 (Olivier, 1791)	7	12	7	26	2.73
Polybia paulista H. von Ihering, 1896	10	10	3	23	2.41
Polybia sericea (Olivier, 1791)	12	5	3	20	2.1
Polybia bifasciata de Saussure, 1854	0	1	0	1	0.1
Pseudopolybia vespiceps (de Saussure, 1864)	7	4	0	11	1.15
Synoeca cyanea (Fabricius, 1775)	4	0	0	4	0.42
Epiponini's total	292	195	339	826	86.58
Mischocyttarus cassununga (von Ihering, 1903)	5	0	0	5	0.52
Mischocyttarus drewseni de Saussure, 1857	22	4	0	26	2.73
Mischocyttarus mattogrossoensis Zikán, 1935	3	0	0	3	0.31
Mischocyttarus rotundicollis (Cameron, 1912)	3	0	1	4	0.42
Mischocyttarini's total	33	4	1	38	3.98
Polistes actaeon Haliday, 1836	2	0	0	2	0.21
Polistes billardieri Fabricius, 1804	3	2	1	4	0.42
Polistes cinerascens de Saussure, 1854	6	0	0	8	0.84
Polistes lanio (Fabricius, 1775)	0	0	2	2	0.21
Polistes ferreri de Saussure, 1853	4	1	3	8	0.84
Polistes simillimus Zikán, 1951	7	5	3	15	1.57
Polistes geminatus Fox, 1898	4	0	0	4	0.42
Polistes subsericeus de Saussure, 1854	6	4	0	10	1.05
Polistes versicolor (Olivier, 1791)	6	14	17	37	3.88
Polistini's total	38	26	26	90	9.43
Total abundance	363	225	366	954	100
Total richness	26	20	18	31	
Exclusive richness	6	1	3		

an impoverishment of the social wasps fauna of the state of São Paulo that may be happening for a long time, making thirty-one species appear to be a high diversity when compared with other inventories carried out in the state.

Comparison of sampling methods

With regard to the methods used for sampling social wasps (active collecting, point sample using liquid bait, and baited bottle traps), a remarkable aspect to note was their large complementarities, with only 38.71% of the species being collected by all three methods (Table 1).

By using the active collection method, 363 individuals belonging to 26 species (seven genera) were obtained. Six species were captured exclusively by this method: *Synoeca cyanea, Mischocyttarus cassununga, M. mattogrossoensis, Polistes actaeon, Plt. cinerascens* and *Plt. geminatus*. Among these species, *Plt. actaeon* and *M. cassununga* have already been collected by attractive PET bottle traps by Souza et. al. (2011) e Jacques et al. (2012) respectively. *Synoeca cyanea*, as well as *M. cassununga* were also previously sampled by passive trapping, actually with Malaise trap by Auad et al. (2010). The other species were captured in other studies with different active collecting methods, as active collecting with liquid bait (*Plt. geminatus* (Tanaka Junior & Noll, 2011), *S. cyanea* and *M. cassununga* (Lima et al., 2010), *M. mattogrossoensis* (Grandinete & Noll, 2013)) and active collection in quadrants and point sampling (*M. cassununga* and *Plt. cinerascens*, (Souza & Prezoto, 2006)).

The method of point sampling using a liquid bait was responsible for the capture of 225 individuals belonging to 20 species (six genera), with only *Polybia bifasciata* being exclusive. This species was previously captured by Souza and Prezoto (2006) with the method of point sampling, while Togni (2009), Jacques et al. (2012) and Simões et al. (2012) obtained specimens using attractive traps.

The method of baited PET bottle traps collected 366 specimens belonging to 18 species (five genera), of which *Apoica pallens, Polybia jurinei* and *Polistes lanio* were exclusives to this technique. These species, however, had already been sampled by different methods in other studies: *Apoica pallens*, though being a species with nocturne foraging habit (Hunt et al., 1995), which would explain absence in the active collections during daytime, was otherwise captured actively during the day by Mechi (1996), Souza and Prezoto (2006), Elpino-Campos et al. (2007), Silveira et al.(2008), Clemente (2009), Silva and Silveira (2009) and Souza et al. (2011). *Polybia jurinei* was found by active collection by Souza and Prezoto (2006), Elpino-Campos et al. (2007), Silva and Silveira (2009) and Jacques et al. (2012); by active collection in flowering plants by Mechi (1996; 2005), and by active collecting with the use of liquid bait by Noll and Gomes (2009), Lima et al. (2010) and Grandinete and Noll (2013). *Polistes lanio* was collected by active collection by Simões *et al.* (2012) and by

active collection while scanning flowering plants by Mechi (1996; 2005).

The most efficient method, regarding the number of species collected, was the active collection, responsible for capturing 83.87% of the species, followed by point sampling using a liquid bait (64.52%) and baited PET bottle traps (58.06%). When adding the samples obtained through active collection and point sample using liquid bait, quite different from the method that uses attractive PET bottle traps which is characteristically passive, 28 species were sampled (90.32% of total richness) with 13 being exclusive of those methods.

The diversity index of Shannon-Wiener (H') for the total sample was 2.248 and the evenness (J') was 0.655. Comparing the methods employed regarding these indexes, it was noted that the highest Shannon-Wiener index was obtained by active collection (H' = 2.297), which is even higher than that obtained for the entire sample, whereas the lowest index was obtained by the attractive PET bottle traps method (H'= 1.854). The equitability obtained in the area was very similar between active collection (J' = 0.705) and point sampling using a liquid bait (J' = 0.717) and was lower for samples collected using attractive PET bottle traps (J '= 0.641). In this way, we note that the attractive PET bottle traps were responsible for less homogeneous samples regarding the species abundance, with a less equity in these samples and therefore a lower diversity index.

When the methods are analyzed as a sole group regarding the difference of abundance sampled, there is no significant difference between them (H' = 5.202, df = 2, (p) Kruskal-Wallis = 0.0742) and the same occurs when using the species relative abundance of each method in the analysis (H' = 4.348, df = 2, (p) Kruskal-Wallis = 0.114).

By comparison of the curves of species accumulation with the rarefaction curves (Figure 1) it is possible to see that the curve of richness accumulation obtained through active collection is that closest to an asymptote, seeming to stabilize in the last four collection days. It is worth noting that the curve obtained for the rarefaction of baited bottle traps tends to an asymptote in a similar way to the curve of active collection, although with a smaller number of species than that obtained by the other techniques, indicating a false idea of good sampling.

Therefore, the method shown to be more efficient was the active collection, since it was responsible for the largest number of species obtained exclusively by one method, for the highest Shannon-Wiener index and a rarefaction curve compatible with the curve of expected species. This method had already been reported as the most efficient in respect of richness by other researches (Souza & Prezoto, 2006; Elpino-Campos et al., 2007; Silva & Silveira, 2009; Souza et al., 2011; Jacques et al., 2012; Simões et al., 2012). The low equitabilities and richness obtained for attractive traps, indicate the need for additional sampling with other methods. However, the occurrence of exclusive species demonstrates the importance

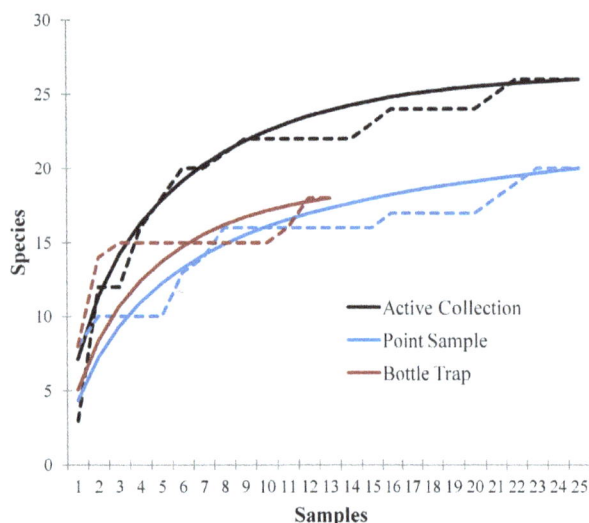

Fig 1. Species accumulation curves (dashed line) and rarefaction curves (continuous line) through samples (collecting days), for each methodology employed in the surveying of the Ipeúna social wasp fauna: Active collecting (Active Collection), Point sampling using a liquid bait (Point Sample), and attractive PET bottle trap (Bottle Trap).

of this technique. Therefore the use of more than one collection method is indicated in conducting inventories of social wasps, as proven in other studies (Silveira, 2002; Souza & Prezoto, 2006; Ribeiro Junior, 2008; Clemente, 2009; Togni, 2009; Souza et al., 2011; Jacques et al., 2012; Simões et al., 2012).

Acknowledgments

We thank the Brazilian National Council for Scientific and Technological Development (CNPq) for financial support and the Brazilian Institute for the Environment and Renewable Natural Resources (SISBIO / IBAMA) for the authorization for collection and transport of specimens, number 22250-1.

References

Alvarenga, R.D., de Castro, M.M., Santos-Prezoto, H.H. & Prezoto, F. (2010). Nesting of social wasps (Hymenoptera, Vespidae) in urban gardens in Southeastern Brazil. Sociobiology, 55 (2): 445-452.

Auad, A.M., Carvalho, C.A., Clemente, M.A. & Prezoto, F. (2010). Diversity of social wasps (Hymenoptera) in a silvipastoral system. Sociobiology, 55 (2): 627-636.

Auko, T.H. & Silvestre, R. (2013). Faunal composition of wasps (Hymenoptera: Vespoidea) in a seasonal forest from Serra da Bodoquena National Park, Brazil. Biota Neotropica, 13: 292-299. doi: 10.1590/S1676-06032013000100028

Ayres, M., Ayres Junior, M., Ayres, D.L. & Santos, A.A. (2007). BioEstat – Aplicações estatísticas nas áreas das ciên-

cias bio-médicas. Belém, PA: Ong Mamiraua.

Carpenter, J.M. & Marques, O.M. (2001). Contribuição ao estudo dos vespídeos do Brasil (Insecta, Hymenoptera, Vespoidae). Cruz das Almas: Universidade Federal da Bahia, 147 p.

Clemente, M.A. (2009) Vespas sociais (Hymenoptera, Vespidae) do Parque Estadual do Ibitipoca-MG: estrutura, composição e visitação floral. Mestrado em Ciências Biológicas (Universidade Federal de Juiz de Fora, Juiz de Fora).

Colwell, R.K. (2009). EstimateS: Statistical estimation of species richness and shared species from samples. User's Guide and application published at: <http://purl.oclc.org/estimates>.

Colwell, R.K., Mao, C.X. & Chang, J. (2004). Interpolating, extrapolating, and comparing incidence-based species accumulation curves. Ecology, 85: 2717-2727. doi: 10.1890/03-0557

Diniz, I.R. & Kitayama, K. (1994). Colony densities and preferences for nest habitats of some social wasps in Mato Grosso State, Brazil (Hymenoptera, Vespidae). Journal of Hymenoptera Research, 3: 133-143.

Diniz, I.R. & Kitayama, K. (1998). Seasonality of vespid species (Hymenoptera : Vespidae) in a central Brazilian cerrado. Revista de Biologia Tropical, 46: 109-114.

Elisei, T., Nunes, J.V., Ribeiro Junior, C., Fernandes Junior, A.J. & Prezoto, F. (2010). Uso da vespa social *Polistes versicolor* no controle de desfolhadores de eucalipto. Pesquisa Agropecuária Brasileira, 45: 958-964. doi: 10.1590/S0100-204X2010000900004

Elpino-Campos, Á., Del-Claro, K. & Prezoto, F. (2007). Diversity of social wasps (Hymenoptera: Vespidae) in Cerrado fragments of Uberlândia, Minas Gerais State, Brazil. Neotropical Entomology, 36: 685-692. doi: 10.1590/S1519-566X2007000500008

Ferreira, D.A.C. & Dias, H.C.T. (2004). Situação atual da mata ciliar do Ribeirão São Bartolomeu em Viçosa, MG. Revista Árvore, 28: 617-623.

Gomes, B. & Noll, F.B. (2009). Diversity of social wasps (Hymenoptera, Vespidae, Polistinae) in three fragments of semideciduous seasonal forest in the northwest of São Paulo State, Brazil. Revista Brasileira de Entomologia, 53: 428-431. doi: 10.1590/S0085-56262009000300018

Grandinete, Y.C. & Noll, F.B. (2013). Checklist of Social (Polistinae) and Solitary (Eumeninae) Wasps from a Fragment of Cerrado "Campo Sujo" in the State of Mato Grosso do Sul. Sociobiology, 60 (1): 101-106. doi: 10.13102/sociobiology.v60i1.101-106

Hermes, M.G. & Köhler, A. (2006). The flower-visiting social wasps (Hymenoptera, Vespidae, Polistinae) in two areas of Rio Grande do Sul State, southern Brazil. Revista Brasileira de Entomologia, 50: 268-274. doi: 10.1590/S0085-56262006000200008

Hunt, J.H., Jeanne, R.L. & Keeping, M.G. (1995). Observations on *Apoica pallens*, a nocturnal neotropical social wasp (Hymenoptera: Vespidae, Polistinae, Epiponini). Insect Sociaux, 42: 223-236. doi: 10.1007/BF01240417

Jacques, G.C., Castro, A.A., Souza, G.K., Silva-Filho, R., Souza, M.M. & Zanuncio, J.C. (2012). Diversity of Social Wasps in the Campus of the "Universidade Federal de Viçosa" in Viçosa, Minas Gerais State, Brazil. Sociobiology, 59: 1053-1063.

Jeanne, R.L. (1972). Social biology of the Neotropical wasp *Mischocyttarus drewseni*. Bulletin of the Museum of Comparative Zoology, 144 (3): 63-150.

Jeanne, R.L. (1991). The swarm-founding Polistinae. In Ross, K.G. & R.W.Matthews (Eds.), The social biology of wasps, (pp. 191-231). New York: Cornell University.

Krebs, C.J. (1998). Ecological methodology. New York: Addison Wesley Longman, 620 p.

Lima, A.C.O., Castilho-Noll, M.S.M., Gomes, B. & Noll, F.B. (2010). Social wasp diversity (Vespidae, Polistinae) in a forest fragment in the northeast of São Paulo state sampled with different methodologies. Sociobiology, 55: 613-623.

Lima, M.A.P., Lima, J.R.d. & Prezoto, F. (2000). Levantamento dos gêneros, flutuação das colônias e hábitos de nidificação de vespas sociais (Hymenoptera, Vespidae) no Campus da UFJF, Juiz de Fora, MG. Revista Brasileira de Zoociências, 2: 69-80.

Liow, L.H., Sodhi, N.S. & Elmqvist, T. (2001). Bee diversity along a disturbance gradient in tropical lowland forests of south-east Asia. Journal of Applied Ecology, 38: 180-192.

Marques, O.M. (1989) Vespas sociais (Hymenoptera, Vespidae) em Cruz das Almas - Bahia: Identificação taxonômica, hábitos alimentares e de nidificação. Mestrado em Agronomia (Universidade Federal da Bahia, Cruz das Almas).

Marques, O.M., Carvalho, C.A.L. & Costa, J.M. (1993). Levantamento das espécies de vespas sociais (Hymenoptera, Vespidae) no município de Cruz das Almas - Estado da Bahia. Insecta, 2: 1-9.

Marques, O.M., Santos, P.A., Vinhas, A.F., Souza, A.L.V., Carvalho, C.A.L. & Meira, J.L. (2005). Social wasps (Hymenoptera: Vespidae) visitors of nectaries of *Vigna unguiculata* (L.) Walp. in the region of Recôncavo of Bahia. Magistra, 17: 64-68.

Mechi, M.R. (1996) Levantamento da fauna de vespas aculeata na vegetação de duas áreas de cerrado. Doutorado em Ciências, área de concentração em Ecologia (Universidade Federal de São Carlos, São Carlos).

Mechi, M.R. (2005). Comunidade de vespas Aculeata (Hymenoptera) e suas fontes florais. In Pivello, V.R. & Varanda, E.M. (Eds.), O cerrado Pé-de-Gigante: Ecologia e conservação - Parque Estadual de Vassununga (pp. 256-266). São Paulo: SMA.

Melo, A.C., Santos, G.M.d.M., Cruz, J.D.d. & Marques, O.M. (2005). Vespas Sociais (Vespidae). In Juncá, F.A., Funch, L. & Rocha, W. (Eds.), Biodiversidade e conservação da Chapada Diamantina (pp. 244-257). Brasília: Ministério do Meio Ambiente.

Morato, E.F., Amarante, S.T. & Silveira, O.T. (2008). Avaliação ecológica rápida da fauna de vespas (Hymenoptera, Aculeata) do Parque Nacional da Serra do Divisor, Acre, Brasil. Acta Amazonica, 38: 789-798. doi: 10.1590/S0044-59672008000400025

Moreno, C.E. (2001). Métodos para medir la biodiversidad. Zaragoza: ORCYT/UNESCO & SEA, 84p.

Noll, F.B. & Gomes, B. (2009). An improved bait method for collecting hymenoptera, especially social wasps (Vespidae: Polistinae). Neotropical Entomology, 38: 477-481. doi: 10.1590/S1519-566X2009000400006

Oliveira, O.A.L., Noll, F.B. & Wenzel, J.W. (2010). Foraging behavior and colony cycle of *Agelaia vicina* (Hymenoptera: Vespidae; Epiponini). Journal of Hymenoptera Research, 19: 4-11.

Paine, R.T. (1969). A note on trophic complexity and community stability. The American Naturalist, 103 (929): 91-93.

Palma-Silva, G.M. (1999) Diagnóstico ambiental, qualidade de água e índice de depuração do Rio Corumbataí (SP). Dissertação (Mestrado em Gestão Integrada de Recursos) (Universidade Estadual Paulista Julio de Mesquita Filho, Rio Claro).

Pereira, M.G.C. & Antonialli-Junior, W.F. (2011). Social wasps in riparian forest in Batayporã, Mato Grosso do Sul State, Brazil. Sociobiology, 57: 153-163.

Prezoto, F. & Clemente, M.A. (2010). Vespas sociais do Parque Estadual do Ibitipoca, Minas Gerais, Brasil. MG Biota, 3 (4): 22-32. Retrived from: http://www.ief.mg.gov.br/images/stories/MGBIOTA/mgbiotaV3n4/mgbiotav.3.n.4.pdf

Prezoto, F., Santos-Prezoto, H.H., Machado, V.L.L. & Zanuncio, J.C. (2006). Prey captured and used in *Polistes versicolor* (Olivier) (Hymenoptera: Vespidae) nourishment. Neotropical Entomology, 35: 707-709. doi: 10.1590/S1519-566X2006000500021

Prezoto, F., Ribeiro Júnior, C., Cortes, S.A.O. & Elisei, T. (2007). Manejo de vespas e marimbondos em ambiente urbano. In Pinto, A.D.S., Rossi, M.M. & Salmeron, E. (Eds.), Manejo de pragas urbanas (pp. 123-126). Piracicaba: Editora CP2.

Prezoto, F., Ribeiro Júnior, C., Guimarães, D.L. & Elisei, T. (2008). Vespas sociais e o controle biológico de pragas: atividade forrageadora e manejo das colônias. In Vilela, E.F., Santos, I.a.D., Schoereder, J.H., Serrão, J.E., Campos, L.a.D.O. & Lino-Neto, J. (Eds.), Insetos sociais: da biologia à aplicação (pp. 413-427). Viçosa: Editora UFV.

Raw, A. (2003). The social wasps of the Federal District of

Brazil.4 p. Retrived from: http://www.uesc.br/anthonyraw/DF%20social%20wasps%20www.pdf.

Ribeiro Junior, C. (2008) Levantamento de vespas sociais (Hymenoptera: Vespidae) em uma eucaliptocultura. Mestrado em Ciências Biológicas (Universidade de Juiz de Fora, Juiz de Fora).

Richards, O.W. (1978). The social wasps of the Americas excluding the Vespinae. London: British Museum (Natural History), 580 p.

Rodrigues, R.R. (1999). A vegetação de Piracicaba e municípios do entorno. Circular Técnica IPEF, (189): 1-17 Retrived from: http://www.ipef.br/publicacoes/ctecnica/nr189.pdf.

Rodrigues, V.M. & Machado, V.L.L. (1982). Vespídeos sociais: espécies do Horto Florestal "Navarro de Andrade" de Rio Claro, SP. Naturalia, 7: 173-175.

Santos, B.B. (1996). Ocorrência de vespídeos sociais (Hymenoptera, Vespidae) em pomar em Goiânia, Goiás, Brasil. Revista do Setor Ciências Agrárias, 15: 43-46.

Santos, G.M.M. & Presley, S.J. (2010). Niche overlap and temporal activity patterns of social wasps (Hymenoptera: Vespidae) in a brazilian cashew orchard. Sociobiology, 56: 121-131.

Santos, G.M.M., Aguiar, C.M.L. & Gobbi, N. (2006). Characterization of the social wasp guild (Hymenoptera: Vespidae) visiting flowers in the caatinga (Itatim, Bahia, Brazil). Sociobiology, 47: 1-12.

Santos, G.M.M., Bispo, P.C. & Aguiar, C.M.L. (2009a). Fluctuations in richness and abundance of social wasps during the dry and wet seasons in three phyto-physiognomies at the tropical dry forest of Brazil. Environmental Entomology, 38: 1613-1617.

Santos, G.M.M., Cruz, J.D.d., Marques, O.M. & Gobbi, N. (2009b). Diversidade de vespas sociais (Hymenoptera: Vespidae) em áreas de cerrado na Bahia. Neotropical Entomology, 38: 317-320. doi: 10.1590/S1519-566X2009000300003

Santos, G.M.M., Bichara Filho, C.C., Resende, J.J., Cruz, J.D.d. & Marques, O.M. (2007). Diversity and community structure of social wasps (Hymenoptera: Vespidae) in three ecosystems in Itaparica island, Bahia State, Brazil. Neotropical Entomology, 36: 180-185. doi: 10.1590/S1519-566X2007000200002

Silva-Pereira, V. & Santos, G.M.M. (2006). Diversity in bee (Hymenoptera: Apoidea) and social wasp (Hymenoptera: Vespidae, Polistinae) community in "Campos Rupestres", Bahia, Brazil. Neotropical Entomology, 35: 165-174. doi: 10.1590/S1519-566X2006000200003

Silva, S.S. & Silveira, O.T. (2009). Vespas sociais (Hymenoptera, Vespidae, Polistinae) de floresta pluvial Amazônica de terra firme em Caxiuanã, Melgaço, Pará. Iheringia, Série Zoologia, 99: 317-323.

Silva, S.S., Azevedo, G.G. & Silveira, O.T. (2011). Social wasps of two Cerrado localities in the northeast of Maranhão state, Brazil (Hymenoptera, Vespidae, Polistinae). Revista Brasileira de Entomologia, 55: 597-602. doi: 10.1590/S0085-56262011000400017

Silveira, O.T. (2002). Surveying Neotropical social wasps: an evaluation of methods in the "Ferreira Penna" Research Station (ECFPn), in Caxiuanã, PA, Brazil (Hymenoptera, Vespidae, Polistinae). Papéis Avulsos de Zoologia, 42: 299-323. doi: 10.1590/S0031-10492002001200001

Silveira, O.T., Costa Neto, S.V. & Silveira, O.F.M. (2008). Social wasps of two wetland ecosystems in brazilian Amazonia (Hymenoptera, Vespidae, Polistinae). Acta Amazonica, 38 (2): 333-344. doi: 10.1590/S0044-59672008000200018

Silveira, O.T., Silva, S.S., Pereira, J.L.G. & Tavares, I.S. (2012). Local-scale spatial variation in diversity of social wasps in an Amazonian rain forest in Caxiuanã, Pará, Brazil (Hymenoptera, Vespidae, Polistinae). Revista Brasileira de Entomologia, 56: 329-346. doi: 10.1590/S0085-56262012005000053

Simões, M.H., Cuozzo, M.D. & Frieiro-Costa, F.A. (2012). Diversity of social wasps (Hymenoptera, Vespidae) in Cerrado biome of the southern of the state of Minas Gerais, Brazil. Iheringia, Série Zoologia, 102: 292-297. doi: 10.1590/S0073-47212012000300007

Souza, A.R.D., Venâncio, D.F.A., Zanuncio, J.C. & Prezoto, F. (2011). Sampling methods for assessing social wasps species diversity in a eucalyptus plantation. Journal of Economic Entomology, 104: 1120-1123. doi: 10.1603/EC11060

Souza, M.M. & Prezoto, F. (2006). Diversity of social wasps (Hymenoptera: Vespidae) in Semideciduous forest and cerrado (savanna) regions in Brazil. Sociobiology, 47: 135-147.

Souza, M.M., Silva, M.J., Silva, M.A. & Assis, N.R.G. (2008). A capital dos marimbondos - vespas sociais Hymenoptera, Vespidae do município de Barroso, Minas Gerais. MG Biota, 1 (3): 24-38.

Souza, M.M., Louzada, J., Serrão, J.E. & Zanuncio, J.C. (2010). Social wasps (Hymenoptera: Vespidae) as indicators of conservation degree of riparian forests in southeast Brazil. Sociobiology, 56: 387-396.

Tanaka Junior, G.M. & Noll, F.B. (2011). Diversity of social wasps on Semideciduous Seasonal Forest fragments with different surrounding matrix in Brazil. Psyche, 2011. doi: 10.1155/2011/861747

Togni, O.C. (2009) Diversidade de vespas sociais (Hymenoptera, Vespidae) na mata atlântica do litoral norte do estado de São Paulo. Mestrado em Ciências Biológicas - Zoologia (Universidade Estadual Paulista Julio de Mesquita Filho, Rio Claro).

Togni, O.C., Locher, G.A., Giannotti, E. & Silveira, O.T.

(2014). The social wasp community (Hymenoptera, Vespidae) in an area of Atlantic Forest, Ubatuba, Brazil. Check List - Journal of Species List and Distribution, 10: 10-17.

Valente, R.O.A. & Vettorazzi, C.A. (2002). Análise da estrutura da paisagem na Bacia do Rio Corumbataí, SP. Scientia Forestalis, 62: 114-129.

Zucchi, R., Sakagami, S.F., Noll, F.B., Mechi, M.R., Mateus, S., Baio, M.V. & Shima, S.N. (1995). *Agelaia vicina*, a swarm-founding Polistine with the largest colony size among wasps and bees (Hymenoptera: Vespidae). Journal of the New York Entomological Society, 103: 129-137.

11

Molecular Phylogeny of the Ant Subfamily Formicinae (Hymenoptera, Formicidae) from China Based on Mitochondrial Genes

ZL CHEN[1], SY ZHOU[1], DD YE[1], Y CHEN[1], CW LU[1]

1 - College of Life Sciences, Guangxi Normal University, Guilin, China

Keywords
Ant phylogeny; Formicidae; *Cyt b, COI, COII*

Corresponding author:
Shan-Yi Zhou
College of Life Sciences
Guangxi Normal University
Guilin, 541004, China.
E-Mail: syzhou5612@yahoo.com.cn

Abstract

To resolve long-standing discrepancies in the relationships among genera within the ant subfamily Formicinae, a phylogenetic study of Chinese Formicine ants based on three mitochondria genes (*Cyt b, COI, COII*) was conducted. Phylogenetic trees obtained in the current study are consistent with several previously reported trees based on morphology, and specifically confirm and reinforce the classifications made by Bolton (1994). The tribes Lasiini, Formicini, Plagiolepidini and Camponotini are strongly supported, while Oecophyllini has moderate support despite being consistent across all analyses. We have also established that the genus *Camponotus* and *Polyrhachis* are indeed not monophyletic. Additionally, we found strong evidence for *Polyrhachis paracamponota,* as described by Wu and Wang in 1991, to be corrected as *Camponotus* based on molecular, morphological and behavioral data.

Introduction

Ants are one of the most successful groups of eusocial insects. They act as an important part of the animal biomass in tropical rainforests and occupy key positions in many terrestrial environments (Wilson & Hölldobler 2005). Resolving the phylogeny of major ant lineages is vital for understanding the factors contributing to their success. Previous studies based on morphological (Baroni Urbani *et al.* 1992, Bolton 2003), fossil-based (Grimaldi *et al.* 1997, Dlussky 1999, Ward & Brady 2003, Bolton 2003), and molecular (Astruc *et al.* 2004, Saux *et al.* 2004, Ward & Brady 2003, Ward & Downie 2005, Ward *et al.* 2005, Brady *et al.* 2006, Moreau *et al.* 2006, Ouellette *et al.* 2006) data provided useful framework for understanding the relationships among ant subfamilies. However, relationships among genera within the subfamilies are not well understood. In addition, the genus-level phylogeny and classification of ant subfamilies remain controversial in many respects.

Formicinae is one of the most abundant ant subfamilies

in the Holarctic (Wilson 1955). According to Bolton (2012), Formicinae includes 49 extant genera and over 3700 species and subspecies in the world. Although the subfamily includes a large number of abundant and ecologically important species that are often subjected to ecological and sociobiological studies, little is known about their phylogeny. Although there are several classifications based on a variety of morphological characteristics, such as sexual traits and larval morphology (Wheeler 1922, Emery 1925, Wheeler & Wheeler 1985, Agosti 1991, Bolton 1994, 2003), the tribes or genus-groups represent artificial assemblages and are used inconsistently by different myrmecologists or even by the same myrmecologist at different times. In particular, some aspects of worker morphology show a strong tendency towards convergence, making it challenging to infer phylogenetic relationships from morphological characteristics alone (Ward 2007). Indeed, Bolton has acknowledged that some tribes in his tribal arrangements would likely need to be re-evaluated (Bolton 2003).

No molecular phylogenetic study has been performed on the subfamily Formicinae in China to date. This study aimed to establish molecular relationships among Formicinae members relative to previously established frameworks and to take a deeper look into species level relationships within more ambiguous assemblages. This was done by obtaining sequences of the mitochondrial genes cytochrome b (*Cyt b*), cytochrome oxidase subunit 1 (*COI*) and cytochrome oxidase subunit 2 (*COII*) and comparing them using Bayesian Inference (BI) (Nylander 2004), Maximum Parsimony (MP) and Neighbour Joining (NJ) (Swofford 2002).

Materials and Methods

Taxon sampling

In this study, a total of 47 species representing 14 genera from five tribes were selected to test the groups suggested by the tribal structure and dendrograms of Wheeler (1922), Emery (1925), Wheeler and Wheeler (1985), Agosti (1991), and Bolton (1994, 2003). *Cerapachys sulcinodis* from the subfamily Cerapachyinae and *Radoszkowskius oculata* from the family Mutillidae were added as outgroups. Apart from *R. oculata*, all other vouchers of Formicinae and *C. sulcinodis*, consisting of nestmate specimens from the same collection event have been deposited in the collection of Guangxi Normal University. Detailed information of the species studied is listed in Appendix 1.

DNA extraction, PCR, and sequencing alignment

Total genomic DNA was extracted from ground whole workers, of which the gasters were removed to minimize contamination from gut bacteria, using standard CTAB methods (slightly modified from Navarro *et al.* 1999). DNA sequence data from three protein-coding mitochondrial genes, namely *Cyt b*, *COI*, and *COII*, were obtained using conventional PCR methods (Villesen *et al.* 2004, Ward & Downie 2005). The sequences and positions on the mitochondrial DNA of the primers used for PCR and sequencing are shown in Table 1.

The primers J2791 and H3665 were used to amplify fragments of mitochondrial DNA that correspond to the 3' end of *COI*, *ITS*, and tRNA-leucine and the 5' end of *COII*. Fragments were sequenced in both directions, and the result-

ing chronograms were assembled and edited using DNAStar (Bioinformatics Pioneer DNAStar, Inc., WI). Sequence for each gene fragment was aligned using CLUSTALX v.1.83 (Thompson *et al.* 1997). Sites from the intergenic spacer (*ITS*) and tRNA-leucine were not used in the analyses. All new DNA sequences generated in this study were submitted to the NCBI GenBank database. Sequence data of the outgroup *R. oculata* was obtained via GenBank direct submission by Wei, S.J. and Chen, X.X. All GenBank accession numbers related to this study are listed in Appendix 1.

Phylogenetic analyses

Reconstruction of phylogenetic relationships among taxa was conducted using NJ, MP, and BI methods. NJ analysis was performed using PAUP* Version 4.0b10 (PPC) (Swofford 2002). Estimates of nodal support on distance trees were obtained using bootstrap analyses (1000 replications). MP analysis was also unweighted and performed using PAUP* Version 4.0b10 (PPC) (Swofford 2002). It involved the use of a heuristic search with random sequence addition (10 replicates each) and the TBR branch-swapping algorithm. Bayesian phylogenetics was used to estimate tree topology using MRBAYES v.3.1.2 (Ronquist & Huelsenbeck 2003). Data were partitioned by gene to yield a total of three data partitions, and the best-fitting model for each partition was selected using MRMODELTEST v. 2.2 (Nylander 2004) under Akaike information criteria (Posada & Buckley 2004).

Results

DNA sequence composition

Table 2 shows the nucleotide content and substitution of three fragment sequences. The final data matrix contained 1830 characters (1049 variable sites, 897 parsimony-informative sites, 152 singleton sites) from the following gene fragments: *Cyt* b-447 characters (270 variable sites, 232 parsimony-informative sites, 38 singleton sites), COI-825 aligned characters (433 variable sites, 379 parsimony-informative sites, 54 singleton sites), and COII-558 characters (341 variable sites, 289 parsimony-informative sites, 52 singleton

Table 1. Sequences of primmer used in this study. Position refers to coordinates in the *Solenopsis invicta* mitochondrion complete genome, GenBank accession numbers: HQ215540. Primer combinations are as follows, with the forward primer listed Wrst for each pair: CB-11400–CB-11884, LCO1490–HCO2198, J2791–H3665, J2791–*COI*-R, CO-F–H3665.

Designation	Sequence (5'–3')	Position	Reference
CB-11400	TATGTACTACCHTGAGGDCAAATATC	9381-9406	Modified from Folmer *et al.* 1994
CB-11884	ATTACACCNCCTAATTTATTAGGRAT	9840-9865	Modified from Folmer *et al.* 1994
LCO1490	GGTCAACAAATCATAAAGATATTGG	117-141	Modified from Folmer *et al.* 1994
HCO2198	TAAACTTCAGGGTGACCAAAAAATCA	700-726	Modified from Folmer *et al.* 1994
J2791	ATACCHCGDCGATAYTCAGA	1300-1319	Modified from Chiotis *et al.* 2000
CO-R	TCRTGRAAGAAGATTATTA	1650-1668	This study
CO-F	CTTTTATTAAAAATHAACAC	1586-1605	This study
H3665	CCACARATTTCWGAACATTG	2177-2196	Modified from Chiotis *et al.* 2000

sites). The base composition of these three fragments varied among the studied species. On average, the base composition was: T 40.8%, C 17.8%, A 31.9%, and G 9.5%, with a strong AT bias (72.7%) as is commonly found in other insect mitochondrial genomes (Vogler & Pearson 1996). The A+T contents of the third, second and first codon position from the three fragments were 84.2%, 66.2%, and 67.4%, respectively. The transitions of nucleotide substitution were more common than transversion with a transition. Numerically, the transversion between A and T was the highest among the four types of nucleotide transversions, whereas the transition between C and T was the highest of the two types of nucleotide transitions.

Amino acid composition and substitution saturation

The complete 1830 nucleotide sequence encoded 610 amino acids of 20 different types. Leucine (Leu) was the most frequent (13.53%) followed by isoleucine (Ile) (13.30%). Cysteine (Cys) was the least frequent, with a constant content of 0.29%. All three protein-coding genes were tested for saturation. These were achieved by plotting the numbers of

observed substitutions versus the uncorrected p-distance estimates. The scattergrams (Fig. 1) show that TV increased along the uncorrected p-distance and TS reached saturation between certain pairs of taxa.

Phylogenetic trees

Phylogenetic analyses (Figs. 2 to 4) showed that the outgroups *C. sulcinodis* and *R. oculata* were well-resolved from the Formicinae taxa at the base of the trees with high confidence values (0.94 Bayesian posterior probability (PP), 100% NJ bootstrap, 99% MP bootstrap). As shown in Figure 5E (this Figure was synthesized from Figs. 2 to 4), all consensus trees strongly indicated that the 14 genera of Formicinae could be divided into five lineages, which we labeled as clades I-V, and consisted of genera from the tribes Lasiini, Formicini, Oecophyllini, Plagiolepidini and Camponotini, respectively. Our findings are consistent with morphological classifications of Bolton (1994) (Figs. 5E and 5F).

Clade I included four genera: *Lasius*, *Nylanderia*, *Prenolepis*, *Pseudolasius* (1.0 PP, 84% NJ bootstrap, 54% MP bootstrap). *Pseudolasius* appeared to be a sister group of

Table 2. The content and substation of nucleotide sequences. Cs, conserved sites; V, variable sites; Pi, parsimony-informative sites; S, Singleton sites; ii, identical pairs; si, transitional pairs; sv, transversional pairs; R, Ts/Tv.

genes	Cs	V	Pi	S	Nuleotide content (%)					Nuleotide substitution			
					T	C	A	G	A+T	ii	si	sv	R
COI (825)	392	433	379	54	40.8	18.3	30.1	10.7	70.9	664	78	83	0.94
COII (558)	217	341	289	52	40.7	16.6	35.2	7.4	75.9	440	52	66	0.79
Cyt b (447)	177	270	232	38	40.7	18.4	31.2	9.7	71.9	349	45	52	0.88
Total (1830)	781	1049	879	152	40.8	17.8	31.9	9.5	72.7	1454	175	200	0.87

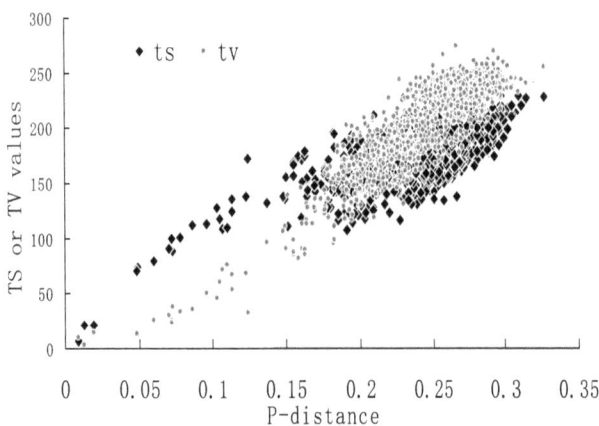

Fig.1. Scatterplots showing the number of substitutions (y-axes; TS, transitions; TV, transversions) versus uncorrected p-distance (x-axes) at each codon position.

(*Lasius* + (*Nylanderia* + *Prenolepis*)) in all three trees. These analyses showed that *Nylanderia* is a sister genus of *Prenolepis* with very strong support (1.0 PP, 90% NJ bootstrap, 89% MP bootstrap). A supported clade of ((*Formica* + *Polyergus*) + (*Proformica* + *Cataglyphis*)) (1.0 PP, 73% NJ bootstrap, 73% MP bootstrap) forms Clade II. Our analyses showed *Formica* as a sister genus of *Polyergus* (1.0 PP, 97% NJ bootstrap, 97% MP bootstrap), and *Proformica* as a sister genus of *Cataglyphis* with very strong support (1.0 PP, 97% NJ bootstrap, 91% MP bootstrap) in all trees. Clade III included only one species (*Oecophylla smaragdina*) and was placed as a sister group to Clade II. Although this species was not supported by strong bootstrap values (0.58 PP, 54% NJ bootstrap, 16% MP bootstrap), it was a consistent feature in all reconstructions. Clade IV comprised of three genera: *Anoplolepis*, a sister group to (*Plagiolepis* + *Lepisiota*). The genus *Plagiolepis* and *Lepisiota* also formed a sister group with good support in all trees. Clade V included *Camponotus* and *Polyrhachis* with very strong support (1.0 PP, 100% NJ bootstrap, 87% MP bootstrap). However the species-level phylogeny of the genera remains unresolved except for the distinct subclade of

(*C. mitis* + (*C. vanispinus* + (*C. jianghuaensis* + *C. albosparsus*))). *C. singularis* is a sister species of other species of the genus *Camponotus* (including *Polyrhachis paracamponota*, excluding *C. yiningensis*) with very strong support (98% NJ bootstrap) in the NJ tree (Fig. 3) and modest support (67% MP bootstrap) in the MP tree (Fig. 2). However, in the BI tree (Fig. 4), *C. parius* first clustered with *C. wasmanni* with strong support (1.0 PP) and then as a sister group of *C. sin-*

gularis plus the rest of the species of *Camponotus* (including *P. paracamponota*, excluding *C. yiningensis*). *C. yiningensis* was tightly associated with *Polyrhachis* with very strong support (1.0 PP, 100% NJ bootstrap, 87% MP bootstrap), and further studies on its status are needed. The species *P. paracamponota* clustered with *Camponotus*, and was distinct from *Polyrhachis*.

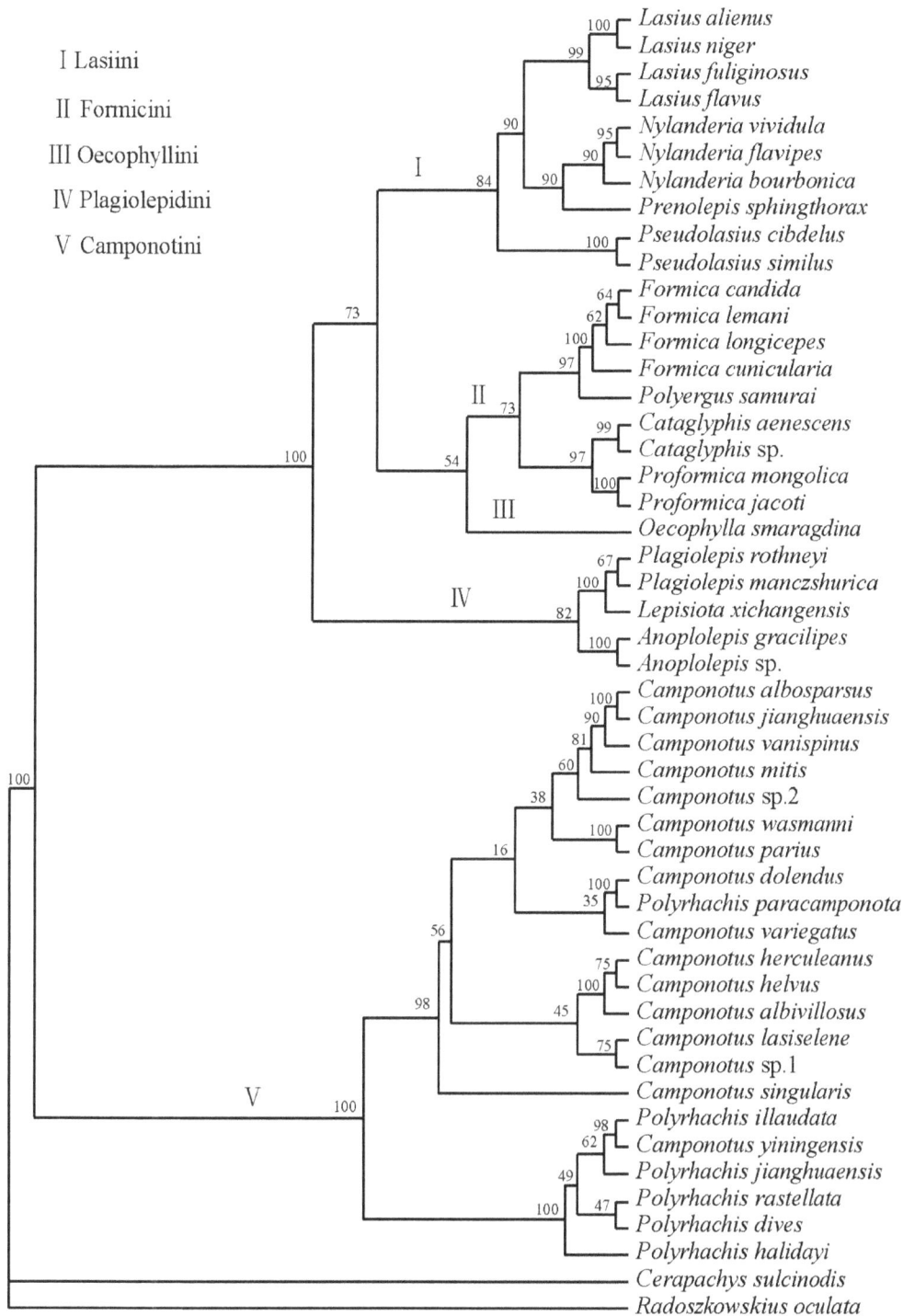

Fig. 2. Maximum-parsimony (MP) consensus tree from 1000 bootstrap replicates, obtained from 48 species of the concatenated sequences of the *Cytb* gene (447 bp), *COI* gene (825 bp) and *COII* gene (558 bp), with *Cerapachy sulcinodis* and *Radoszkowskius oculata* as the outgroups.

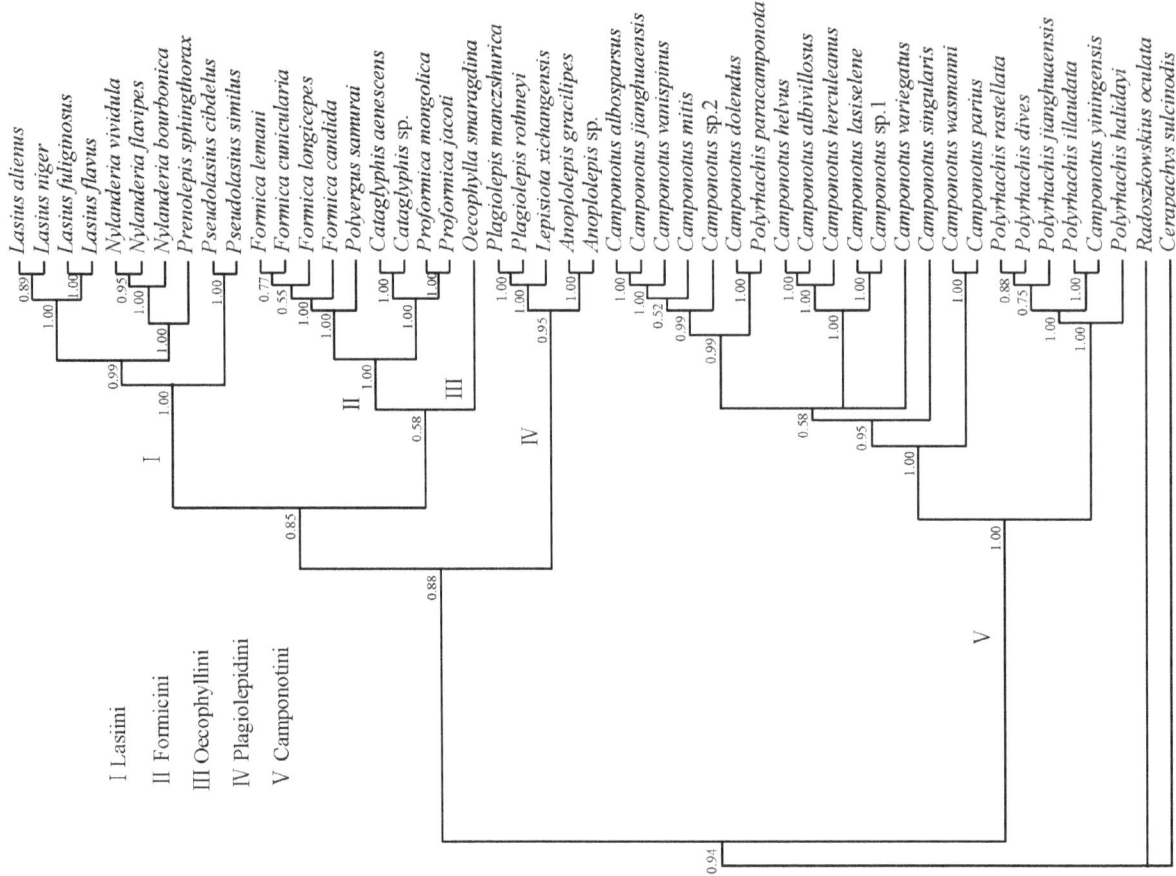

Fig. 3. Neighbor-joining (NJ) consensus tree from 1000 bootstrap replicates, obtained from 48 species of the concatenated sequences of the *Cyt b* gene (447 bp), *COI* gene (825 bp) and *COII* gene (558 bp), with *Cerapachy sulcinodis* and *Radoszkowskius oculata* as the outgroups.

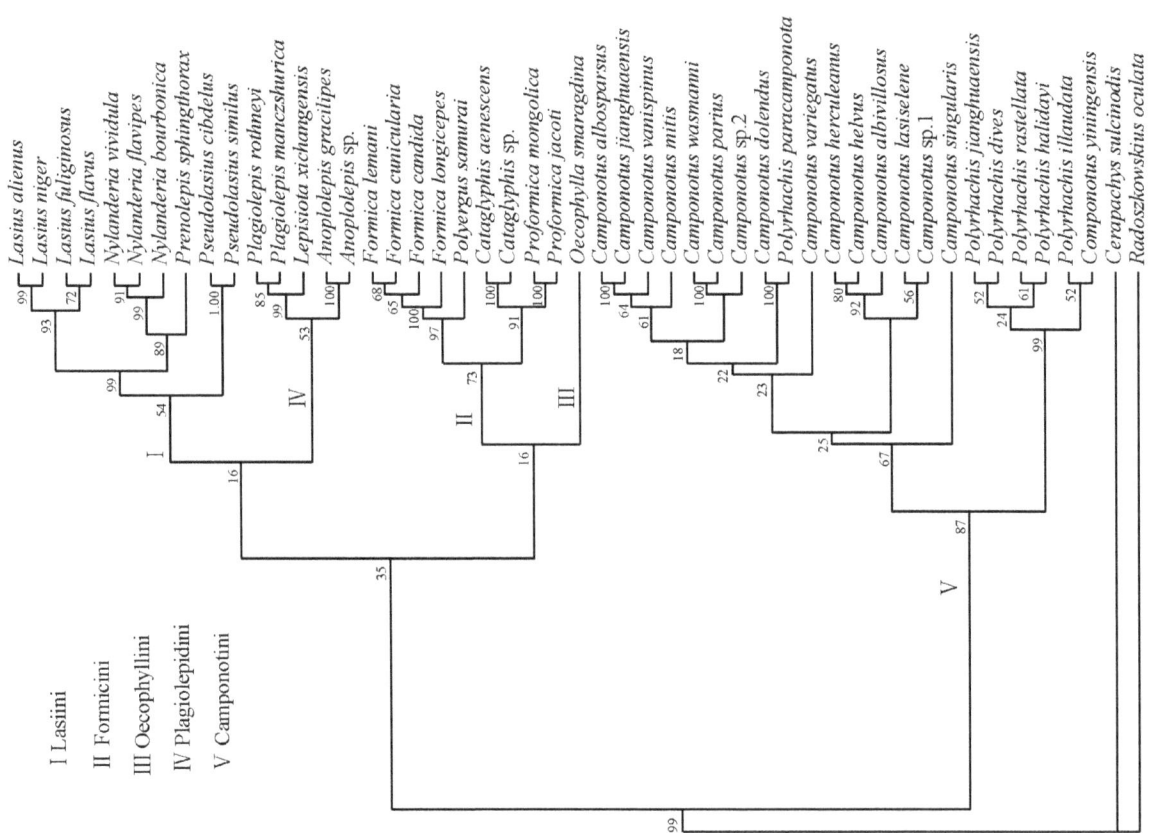

Fig. 4. Bayesian (BI) majority-rule consensus tree, obtained from 48 species of the concatenated sequences of the *Cyt b* gene (447 bp), *COI* gene (825 bp) and *COII* gene (558 bp) three partitions all under the same best-fit model (GTR+I+G) selecting by AIC in Modeltest, with *Cerapachy sulcinodis* and *Radoszkowskius oculata* as the outgroups.

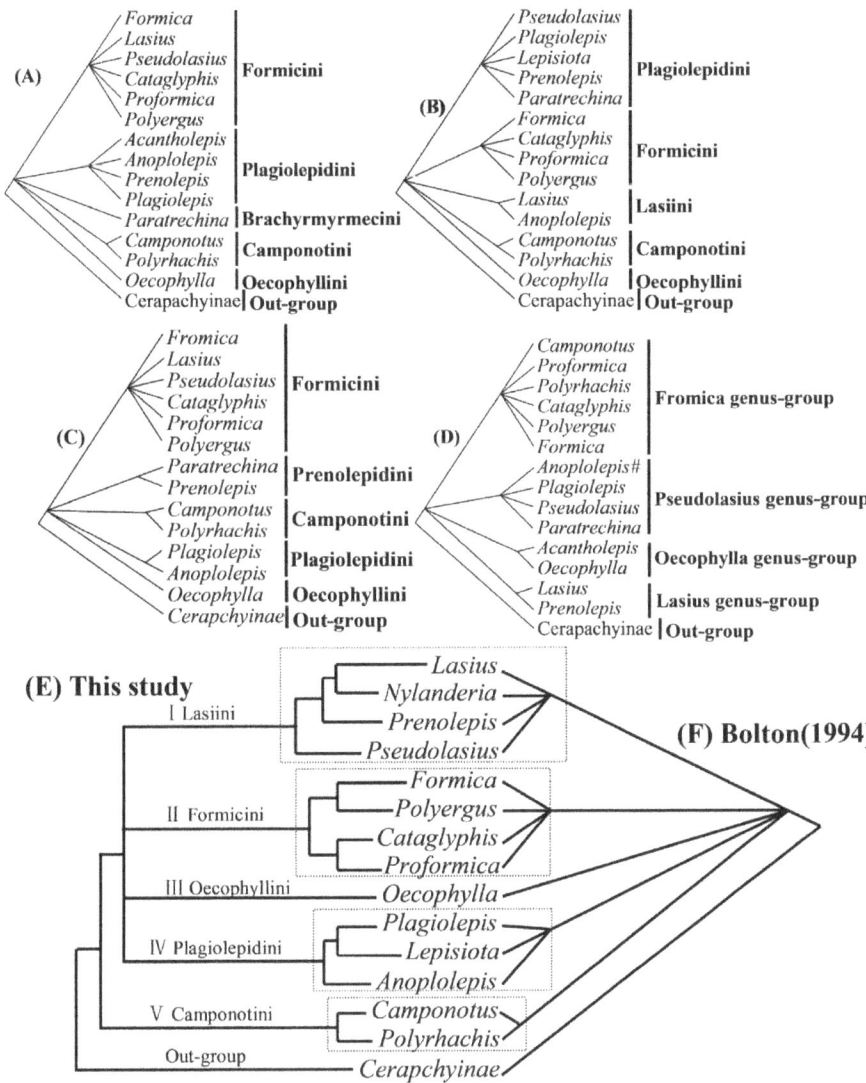

Fig. 5 Classifications of Formicine genera based on the schemes of: (A) Wheeler WM 1922; (B) Bolton 2003; (C) Wheeler, WM *et al.* 1985; (D) Agosti 1991; (E) This study; (F) Bolton 1994. {NB: only positions for species of interest in this phylogeny are noted; there are changes in classifications of other genera which are not being used in this study }.

Discussion

Results of the phylogenetic relationships of Formicinae in this study (Figs. 2 to 4, 5E) showed both similarities and differences compared with those of previous studies (Fig. 5A-5D, 5F). Surprisingly, results of our molecular phylogenetic trees have better fit with the morphological cladogram of Bolton (1994), with which they are congruent, than with that of Bolton (2003).

Clade I is best characterized morphologically with the worker alitrunk not conspicuously constricted or otherwise specialized and the mesonotum typically convex in profile view. The workers of *Lasius*, *Nylanderia* and *Prenolepis* shared the following morphological characters (Bolton 1994): mandibles roughly triangular with four to seven teeth, antennae 12-segmented, the torula close to but not touching the posterior clypeal margin. A propodeal spiracle present at or near the declivity of the propodeum, and the petiolar node in profile usually inclined forward, with a short anterior face and much longer posterior face. These data support the earlier hypothesis proposed by Bolton in 1994, into which *Pseudo-*

lasius, *Prenolepis*, *Nylanderia* and *Lasius* were placed and formed the tribe Lasiini, but disagrees with that of Bolton (2003), in which the genera *Plagiolepis* and *Lepisiota* were added to form the tribe Plagiolepidini. In addition, these four genera formed a strongly supported group in all trees, especially in the case of the sister genus relationship between *Nylanderia* and *Prenolepis* (1.0PP, 90% NJ bootstrap, 99% MP bootstrap). These results are consistent with those of previous morphological (Emery 1925, Wheeler & Wheeler 1953, Trager 1984) and molecular studies (Brady *et al.* 2006), However, in the study of Moreau *et al.* (2006), the genus *Plagiolepis, Pseudolasius* and *Prenolepis* emerges first, followed by *Lasius* along with other two genera. Besides the study by LaPolla *et al.* (2010) in which *Prenolepis* was treated as being paraphyletic to the group. In addition, monophyly of the genus *Lasius* was strongly supported (0.99 PP, 90% NJ bootstrap, 99% MP bootstrap).

The results for clade II are consistent with those of previous studies (Bolton 1994, 2003) (Figs. 5E, 5F and 5B). Genera of the tribe Formicini share the following morphological features (Bolton 1994): 12-segmented antennae, antennal sockets situated close to the posterior clypeal margin. Orifices of propodeal spiracle oval, elliptical, or as elongated slits and near-vertical or inclined from the vertical. All of these analyses provided strong support for the two sister-group relationships of (*Formica* + *Polyergus*) and (*Proformica* + *Cataglyphis*), which is consistent with the molecular studies of Moreau *et al.* (2006).

In clade III, the genus *Oecophylla* was separated as a distinct lineage. This result is well supported by previous morphological studies (Wheeler 1922, Wheeler & Wheeler 1985, Bolton 1994, 2003) (Fig. 5), which showed *Oecophylla* as the tribe Oecophllini. In our molecular phylogeny, *Oecophylla* appears to be a sister of Formicini but with low bootstrap support (0.58 PP, 0.54% NJ bootstrap, 16% MP bootstrap). However, this topology is in agreement with that of Moreau *et al.* (2006). Wilson and Taylor (1964) also suggested that *Oecophylla* and clade II cannot be given much credence considering the separate placement in morphologically and parsimony-based phylogenies, as well as its current geographical separation. However, fossil evidence indicate that *Oecophylla* previously occurred in Europe, suggesting that these genera may have shared a common ancestor.

Clade IV is a well supported clade consisting of members from the tribe Plagiolepidini (*Anoplolepis* + (*Plagiolepis* + *Lepisiota*)) (0.95 PP, 82% NJ bootstrap, 53% MP bootstrap). Bolton (1994) had previously placed the three genera into the tribe Plagiolepidini based on a morphological study (Fig. 5F) and the current study is the first to arrive at the same placement based on molecular phylogenetics. This tribe is distinguished by the following features: worker with 11-segmented antennae, antennal sockets fused with the posterior clypeal margin, and palp formula of 6,4. Surprisingly, Bolton (2003) proposed the genus *Plagiolepis* and *Lepisiota* to be included in the tribe Plagiolepidini (Fig. 5B). Although Bolton (2003) represents a more comprehensive summary of ant morphological characters assembled to date than his previous treatment (Bolton 1994), it is likely that this reflects a genuine conflict between morphology and molecular data.

Clade V is strongly supported in all trees (1.0 PP, 100% NJ bootstrap, 87% MP bootstrap) and consists of *Camponotus* and *Polyrhachis*. This result is in agreement with previous morphological (Wheeler 1922, Emery 1925b, Wheeler & Wheeler 1985, Bolton 1994, 2003) (Figs. 5) and molecular studies (Astruc *et al.* 2004, Brady *et al.* 2006, Moreau *et al.* 2006). The tribe Camponotini can be characterized by its 12-segmented antennae, with antennal sockets situated far behind the posterior clypeal margin, and a palp formula of 6,4. *Camponotus* is however a paraphyletic group, as is noted in other studies (Brady *et al.* 1999, Astruc *et al.* 2004, Brady *et al.* 2006). *Camponotus yiningensis* has been placed outside of the genus *Camponotus*, which has been confirmed not to be monophyletic (Brady *et al.* 1999, 2000; Astruc *et al.* 2004, Brady *et al.* 2006). Morphological characters also reflected close, and sometimes overlapping, relationships between

Camponotus and *Polyrhachis*. For instance, many species of *Camponotus* acquired distinctive spines, and many species of *Polyrhachis* have camber-shaped alitrunks. The species *Polyrhachis paracamponota* was first described by Wang and Wu in 1991 based on a single holotype worker which possesses pronotal spines, and was placed in the genus *Polyrhachis*. But having pronotal spines is very common in *Camponotus* and *Polyrhachis,* this morphological character could not be used for distinguishing between the two genera. The original descriptions exact match with the morphological character of the genus *Camponotus*. In our opinion, the authorships also had the same idea, so this species be named "*paracamponota*". Besides, this species has polymorphic workers, and they have been observed to tunnel into the soil for subterranean nesting. In contrast, the workers of *Polyrhachis* are exclusively monomorphic, and can only use existing cavities in the soil or under stones for nesting, but never excavate tunnels themselves. Our phylogenetic reconstruction indicated that this species is associated with *Camponotus*, and is clearly separated from *Polyrhachis*. As such, there is strong evidence from morphological, behavioristic and molecular data that *Polyrhachis paracamponota* should be placed as a member of *Camponotus*.

Conclusion

In conclusion, our study of the phylogenetic relationship of Formicinae from China based on sequences from three protein-coding mitochondrial genes (*Cyt b, COI, COII*) confirms and reinforces the findings of previous morphological studies (Bolton 1994). The tribes Lasiini (*Pseudolasius, Prenolepis, Paratrechina, Lasius*), Formicini (*Formica, Cataglyphis, Proformica, Polyergus*), Plagiolepidini (*Lepisiota, Plagiolepis, Anoplolepis*), and Camponotini (*Camponotus, Polyrhachis*) are strongly supported, while Oecophyllini has moderate support despite being consistent across all analyses. We have also established that the genus *Camponotus* and *Polyrhachis* are indeed not monophyletic. Additionally, evidence from molecular, morphological and behavioral data indicates that *Polyhachis paracamponota* should be corrected as *Camponotus*.

Acknowledgments

We sincerely thank Professor Yu-Feng Xu (National Taiwan Normal University, Taiwan), Dr. Jun-Hao Tang (National University of Singapore) and Dr. John R. Fellowes (Kadoorie Farm and Botanic Garden, Hong Kong) for reviewing the English text. We thank two anonymous reviewers for helpful comments on the manuscript. Thanks also to Chao-Tai Wei (Guangxi Normal University) for providing us with some ant materials, De-Long Zeng (Guangxi Normal University) for helpful assistance and comments on phylogeny analysis. This study was supported by the National Natural Science

Foundation of China (Project Nos. 30770258 and 31071971), Foundation of the Key Laboratory of Ecology of Rare and Endangered Species and Environmental Protection, Ministry of Education, Guangxi Normal University.

References

Agosti, D. (1991). Revision of the oriental ant genus *Cladomyrma*, with an outline of the higher classification of the Formicinae (Hymenoptera: Formicidae). Syst. Entomol., 16: 293-310.

Astruc, C., Julien, J. F., Errard, C. & Lenoir, A. (2004). Phylogeny of ants (Formicidae) based on morphology and DNA sequence data. Mol. Phylog. Evol., 31: 880-893. doi: 10.1016/j.ympev.2003.10.024

Baroni Urbani, C., Bolton, B. & Ward, P. S. (1992). The internal phylogeny of ants (Hymenoptera: Formicidae). Syst. Entomol., 17: 301-329.

Bolton, B. (1994). Identification guide to the ant genera of the world. *Harvard University Press, Cambridge, Massachusetts*, 222 pp.

Bolton, B. (2011). Catalogue of species-group taxa. http://gap.entclub.org/contact.html. *(accessed date: 1 March, 2011)*.

Bolton, B. (2003). Synopsis and classification of Formicidae. Mem. Am. Entomol. Inst., 71: 1-370.

Brady, S.G., Gadau, J. & Ward, P.S. (1999). Is the ant genus *Camponotus* paraphyletic? *4th International Hymenopterists Conference*, Glen Osmond, South Australia, Canberra, Australia.

Brady, S.G., Gadau, J. & Ward, P.S. 2000. Systematics of the ant genus *Camponotus* (Hymenoptera: Formicidae): a preliminary analysis using data from the mitochondrial gene cytochrome oxidase I. *In: Austin, A.D., Dowton, M. (Eds.), Hymenoptera. Evolution, Biodiversity and Biological Control.* CSIRO Publishing, Collingwood, Victoria, pp. 131-139, xi+ 468 pp.

Brady, S.G., Fisher, B.L., Schultz, T.R. & Ward, P.S. (2006). Evaluating alternative hypotheses for the early evolution and diversification of ants. Proc. Nat. Acad. Sci. USA., 103: 18172-18177. doi: 10.1073/pnas.0605858103

Chiotis, M., Jermiin, L.S. & Crozier, R.H. (2000). A Molecular Framework for the Phylogeny of the Ant Subfamily Dolichoderinae. Mol. Phylog. Evol., 17(1): 108-116. doi: 10.1006/mpev.2000.0821

Dlussky, G.M. (1999). The first find of the Formicoidea (Hymenoptera) in the lower Cretaceous of the northern hemisphere. [In Russian.] Paleontol. Zhurnal, 3: 62-66.

Emery, C. (1925). Hymenoptera. Fam. Formicidae. Subfam. Formicinae. Gen. Insectorum, 183: 1-302.

Folmer, O., Black, M., Hoeh, W., Lutz, R. & Vrijenhoek, R. (1994). DNA primers for amplification of mitochondrial cytochrome C oxidase subunit I from diverse metazoan invertebrates. Mol. Mar. Biol. and Biotech., 3(5): 294-299.

Grimaldi, D., Agosti, D. & Carpenter, J.M. (1997). New and rediscovered primitive ants (Hymenoptera: Formicidae) in Cretaceous amber from New Jersey, and their phylogenetic relationships. Am. Mus. Nov., 3208: 1-43.

Huelsenbeck, J.P. & Ronquist, F. (2001). MRBAYES: Bayesian inference of phylogeny. Bioinformatics, 17: 754-755.

Johnson, R.N., Agapow, P.M. & Crozier, R.H. (2003). A tree island approach to inferring phylogeny in the ant subfamily Formicinae, with especial reference to the evolution of weaving. Mol. Phylog. Evol., 29: 317-330. doi: 10.1016/S1055-7903(803)00114-3

LaPolla, J.S, Brady, S.G. & Shattuck, S.O. (2010). Phylogeny and taxonomy of the *Prenolepis* genus-group of ants (Hymenoptera: Formicidae). Syst. Entomol., 35: 118-131. doi: 10.1111/j.1365-3113.2009.00492.x

Moreau, C.S.., Bell, C.D., Vila, R., Archibald, S.B. & Pierce, N.E. (2006). Phylogeny of the ants: diversification in the age of angiosperms. Science, 312: 101-104. doi: 10.1126/science.1124891

Navarro E., Jaffre T., Gauthier D., Gourbiere F., Rinaudo G., Simonet P. & Normand P. (1999). Distribution of Gymnostoma spp. microsymbiotic Frankia strains in New Caledonia is related to soil type and to host-plant species. Mol. Ecol., 8: 1781-1788.

Nylander, J.A.A., Ronquist, F., Huelsenbeck, J.P. & Nieves-Aldrey, J.L. (2004). Bayesian phylogenetic analysis of combined data. Syst. Biol., 53: 47-67 doi: 10.1080/10635150490264699

Nylander, J.A.A. (2004). MrModeltest v2, Program distributed by author. *Evolutionary Biology Centre, Uppsala University, Uppsala.*

Ouellette, G.D., Fisher, B.L. & Girman, D.J. (2006). Molecular systematics of basal subfamilies of ants using 28S rRNA (Hymenoptera: Formicidae). Mol. Phylog. Evol., 40: 359-369. doi: 10.1016/j.ympev.2006.03.017

Posada, D. & Buckley, T.R. (2004). Model selection and model averaging in phylogenetics: advantages of Akaike information criterion and Bayesian approaches over likelihood ratio tests. Syst. Biol., 53: 793-808. doi: 10.1080/10635150490522304

Ronquist, F. & Huelsenbeck, J.P. (2003). MRBAYES 3: Bayesian phylogenetic inference under mixed models. Bioinformatics, 19: 1572-1574.

Saux, C., Fisher, B.L. & Spicer, G.S. (2004). Dracula ant phylogeny as inferred by nuclear 28S rDNA sequence and implications for ant systematics (Hymenoptera: Formicidae). Mol. Phylog. Evol., 33: 457-468. doi: 10.1016/j. ympev.2004.06.017

Swofford, D.L. (2002). PAUP*. Phylogenetic Analysis Using Parsimony (*and Other Methods), Vol. 4. *Sinauer Associates, Sunderland, MA*.

Thompson, J.D., Gibson, T.J., Plewniak, F., Jeanmougin, F. & Higgins, D.G. (1997). The ClustalX windows interface: flexible strategies for multiple sequence alignment aided by quality analysis tools. Nuc. Acids Res., 25: 4876-4882.

Trager, J.C. (1984). A revision of the genus *Nylanderia* (Hymenoptera: Formicidae) of the continental United States. Sociobiology, 9: 49-162.

Villesen, P., Mueller, U.G., Schultz, T.R., Adams, R.M.M. & Bouck, A.C. (2004). Evolution of ant-cultivar specialization and cultivar switching in *Apterostigma* fungus-growing ants. Evolution, 58: 2252-2265. doi: 10.1111/j.0014-3820.2004. tb01601.x

Vogler, A.P. & Pearson, D.L. (1996). A molecular phylogeny of the tiger beetles (Cicindelidae): congruence of mitochondrial and nuclear rDNA data sets. Mol. Phylog. Evol., 6: 321-338.

Wang, C.L. & Wu, J. (1991). Taxonomic Studies on the Genus *Polyrhachis* Mayr of China (Hymenoptera: Formicinae). For. Res., 4(6): 596-601.

Ward, P.S. & Brady, S.G. (2003). Phylogeny and biogeography of the ant subfamily Myrmeciinae (Hymenoptera: Formicidae). Invert. Syst., 17, 361-386. doi: 10.1071/IS02046

Ward, P.S. & Downie, D.A. (2005). The ant subfamily Pseudomyrmecinae (Hymenoptera: Formicidae): phylogeny and evolution of big-eyed arboreal ants. Syst. Entom., 30: 310-335. doi: 10.1111/j.1365-3113.2004.00281.x

Ward, P.S., Brady, S.G., Fisher, B.L. & Schultz, T.R. (2005). Assembling the ant "Tree of Life" (Hymenoptera: Formicidae). Myrmecol. Nachrichten, 7: 87-90.

Ward, P.S. (2007). Phylogeny, classification, and species-level taxonomy of ants. (Hymenoptera: Formicidae). Zootaxa, 1668: 549-563.

Wheeler, W.M. (1922). Ants of the American Museum Congo expedition. A contribution to the myrmecology of Africa. VII. Keys to the genera and subgenera of ants. Bul. Am.. Mus. Nat. Hist., 45: 631-710.

Wheeler, G.C. & Wheeler, J. (1953). The ant larvae of the subfamily Formicinae. Ann. Entom. Soc. Am., 46: 126-171.

Wheeler, G.C. & Wheeler, J. 1985. A simplified conspectus of the Formicidae. *Trans. Am. Entom. Soc.,* 111: 255-264.

Wilson, E.O. (1955). A monographic revision of the ant genus *Lasius*. Bul. Mus. Compar. Zool., 113: 1-201.

Wilson, E.O, & Taylor, R.W. (1964). A fossil ant colony: new evidence of social antiquity. Psyche, 71: 93-103.

Wilson, E.O. & Holldobler, B. (2005). The rise of the ants: a phylogenetic and ecological explanation. Proc. Nat. Acad. Sci. USA, 102: 7411-7414.

Appendix 1

Species	Collection locality	Voucher specimen	GenBank accession numbers		
			Cyt b	COI	COI & COII
Lepisiota xichangensis	Jingxi, Guangxi	GXJX0006	JQ681097	JQ681046	JQ680992
Plagiolepis manczshurica	Helan Mt, Inner Mongolia	NMHL0422	JQ681098	JQ681047	JQ680993
Plagiolepis rothneyi	Xiangtou Mt, Guangdong	GDXT0122	JQ681099	JQ681048	JQ680994
Anoplolepis gracilipes	Beiliu, Guangxi	GXBL0001	JQ681100	JQ681049	JQ680995
Anoplolepis sp.	Bohai, Yunnan	YNBH0003	JQ681101	JQ681050	JQ680996
Pseudolasius cibdelus	Jingxi, Guangxi	GXJX0031	JQ681102	JQ681051	JQ680997
Pseudolasius similus	Jingxi, Guangxi	NMHL0269	JQ681103	JQ681052	JQ680998
Prenolepis sphingthorax	Jingxi, Guangxi	GXJX0144	JQ681104	JQ681053	JQ680999
Cataglyphis aenescens	Heze, Shandong	Shandong_70	JQ681105	HQ619705	JQ681000
Cataglyphis sp.	Yangling, Shanxi	SXYL0007	JQ681106	JQ681054	JQ681001
Formica candida	Xiaowutai Mt, Hebei	Hebei_50	JQ681107	HQ619704	JQ681002
Formica longicepes	Helan Mt, Inner Mongolia	NMHL0227	JQ681108	JQ681055	JQ681003
Formica cunicularia	Xiaowutai Mt, Hebei	Hebei_307	JQ681109	HQ619714	JQ681004
Formica lemani	Xiaowutai Mt, Hebei	Hebei_251	JQ681110	HQ619712	JQ681005
Proformica mongolica	Helan Mt, Inner Mongolia	NMHL0045	JQ681111	JQ681056	JQ681006
Proformica jacoti	Xiaowutai Mt, Hebei	HBXW0039	JQ681112	JQ681057	JQ681007
Nylanderia flavipes	Heze, Shandong	SDHZ0104	JQ681113	JQ681058	JQ681008
Nylanderia vividula	Guilin, Guangxi	GXGL0111	JQ681149	JQ681093	JQ681044
Nylanderia bourbonica	Jingxi, Guangxi	GXJX0022	JQ681114	JQ681059	JQ681009
Lasius niger	Xiaowutai Mt, Hebei	HBXW0263	JQ681115	JQ681060	JQ681010
Lasius flavus	Helan Mt, Inner Mongolia	NMHL0320	JQ681116	JQ681061	JQ681011
Lasius fuliginosus	Xiaowutai Mt, Hebei	HBXW0266	JQ681117	JQ681062	JQ681012
Lasius alienus	Helan Mt, Inner Mongolia	NMHL0316	JQ681118	JQ681063	JQ681013
Oecophylla smaragdina	Xiangtou Mt, Guangdong	GDXT0104	JQ681119	JQ681064	JQ681014
Polyrhachis illaudata	Jingxi, Guangxi	GXJX0141	JQ681120	JQ681065	JQ681015
Polyrhachis halidayi	Jingxi, Guangxi	GDJX0024	JQ681121	JQ681066	JQ681016
Polyrhachis rastellata	Rong'an, Guangxi	GXRA0045	JQ681122	JQ681067	JQ681017
Polyrhachis dives	Beiliu, Guangxi	GXGL0099	JQ681123	JQ681068	JQ681018
Polyrhachis jianghuaensis	Beiliu, Guangxi	GXBL0006	JQ681124	JQ681069	JQ681019
Polyrhachis paracampponota	Jingxi, Guangxi	GXJX0009	JQ681125	JQ681070	JQ681020
Camponotus variegatus	Jingxi, Guangxi	GXJX0155	JQ681126	JQ681071	JQ681021
Camponotus herculeanus	Helan Mt, Inner Mongolia	NMHL0273	JQ681127	JQ681072	JQ681022
Camponotus albosparsus	Jingxi, Guangxi	GXJX0130	JQ681128	JQ681073	JQ681023
Camponotus vanispinus	Jingxi, Guangxi	GXJX0007	JQ681129	JQ681074	JQ681024
Camponotus wasmanni	Xiangtou Mt, Guangdong	GDXT0102	JQ681130	JQ681075	JQ681025
Camponotus dolendus	Jingxi, Guangxi	GXJX0036	JQ681131	JQ681076	JQ681026
Camponotus jianghuaensis	Rong'an, Guangxi	GXRA0010	JQ681132	JQ681077	JQ681027
Camponotus mitis	Bohai, Yunnan	YNBH0111	JQ681133	JQ681078	JQ681028
Camponotus helvus	Jingxi, Guangxi	GXJX0015	JQ681134	JQ681079	JQ681029
Camponotus yiningensis	Jingxi, Guangxi	GXJX0013	JQ681135	JQ681080	JQ681030
Camponotus albivillosus	Helan Mt, Inner Mongolia	NMHL2122	JQ681136	JQ681081	JQ681031
Camponotus lasiselene	Jingxi, Guangxi	GXJX0012	JQ681137	JQ681082	JQ681032
Camponotus parius	Beiliu, Guangxi	GXBL0009	JQ681138	JQ681083	JQ681033
Camponotus singularis	Beiliu, Guangxi	GXBL0008	JQ681139	JQ681084	JQ681034
Camponotus sp. 1	Jingxi, Guangxi	GXJX0017	JQ681140	JQ681085	JQ681035
Camponotus sp. 2	Jingxi, Guangxi	GXJX0123	JQ681141	JQ681086	JQ681036
Polyergus samurai	Beiliu, Guangxi	GXBL0212	JQ681142	JQ681087	JQ681037
Out-group					
Cerapachys sulcinodis	Beiliu, Guangxi	GXBL0095	JQ681145	JQ681090	JQ681040
Radoszkowskius oculata	From GenBank		NC_014485	NC_014485	NC_014485

12

Studies on an Enigmatic *Blepharidatta* Wheeler Population (Hymenoptera: Formicidae) from the Brazilian Caatinga

JC Pereira[1], JHC Delabie[3,4], LRS Zanette[1], Y Quinet[1,2]

1 - Universidade Federal do Ceará, Fortaleza, Ceará, Brazil
2 - Universidade Estadual do Ceará, Fortaleza, Ceará, Brazil
3 - Universidade Estadual de Santa Cruz, Ilhéus-Itabuna, Bahia, Brazil
4 - CEPLAC/CEPEC Centro de Pesquisas do Cacau, Itabuna, Bahia, Brazil

Keywords: northeastern Brazil, semi-arid environment, ants, Blepharidattini, ergatoid queen

Corresponding author:
Yves Quinet
Laboratório de Entomologia
Instituto Superior de Ciências Biomédicas – ISCB
Universidade Estadual do Ceará
Fortaleza, CE, Brazil, 60714-903
Email: yvesq@terra.com.br

Abstract

Blepharidatta is a rare Neotropical ant genus formed by predatory species whose small colonies nest in soil or leaf-litter. A population of *Blepharidatta* that presents affinities with *Blepharidatta conops* Kempf was found in the Caatinga biome, at the "Reserva Particular do Patrimônio Natural Serra das Almas" (RPPNSA), in Crateús (State of Ceará, Brazil). The aim of our study was to obtain data on the nest architecture, size and composition of colonies, foraging behavior, and female castes morphology for this newly found population, and to compare it with other *Blepharidatta* species, particularly with *B. conops*. The results show that *Blepharidatta* sp. and *B. conops* share key features of their biology such as their basic nest architecture, diet and foraging behavior, and the presence of a single ergatoid queen with a phragmotic head. However, marked differences were also found in head and mesosoma morphology of the queen, nest architecture, colony size, and queen location in the nest. Two alternative hypotheses are presented. The newly found *Blepharidatta* population represents a new species, possibly endemic to the Caatinga biome or it represents an extreme of the phenotypic variations observed among the populations forming *B. conops*.

Introduction

The myrmicine ant genus *Blepharidatta* is a strictly Neotropical group that was described by Wheeler (1915) from workers of *Blepharidatta brasiliensis* Wheeler collected near Belém (State of Pará, Brazil), in the Amazon Forest. Together with the genus *Wasmannia*, it forms the monophyletic tribe Blepharidattini (Wheeler & Wheeler, 1991; Bolton, 1995), which is considered to be a close relative of the fungus-growing ant tribe Attini (Schultz & Meier, 1995).

Up to seven species are currently recognized (Silva, 2007), but most of them are waiting for a formal taxonomic treatment or confirmation. Based on morphological as well as behavioral data, only three species are formally recognized: *B. brasiliensis* found in the Amazonian forest, *Blepharidatta conops* Kempf, an inhabitant of savanna-like formation from central Brazil (Cerrado), and an undescribed species (*Blepharidatta* sp-ba, hereafter) known from the Atlantic rainforest

of the State of Bahia, eastern Brazil (Rabeling et al., 2006; Brandão et al., 2008; Cassano et al., 2009) (Fig 1). All are ground-dwelling predatory species that nest in the ground or in the leaf-litter, with small monogynous or polygynous colonies (Brandão et al., 2001; Rabeling et al., 2006; Silva, 2007). Queens are ergatoid (*sensu* Peeters, 1991, i.e. permanently wingless and worker-like) and, at least in *B. conops*, it is believed that the foundation of new colonies is by fission of established colonies (Brandão et al., 2001).

The best studied species is *B. conops* (Diniz & Brandão, 1997; Diniz et al., 1998; Brandão et al., 2001, 2008). *B. conops* colonies (up to 250 workers) live in simple nests with a single opening and a vertical tunnel with 2 cm of diameter and 20 cm deep that ends in a cone-shaped widening at the bottom. Furthermore, all mature nests have a horizontal subsidiary chamber connected to the vertical tunnel through a narrow tunnel that opens at the mid-length of the vertical tunnel. It serves as a refuge for queen and brood when

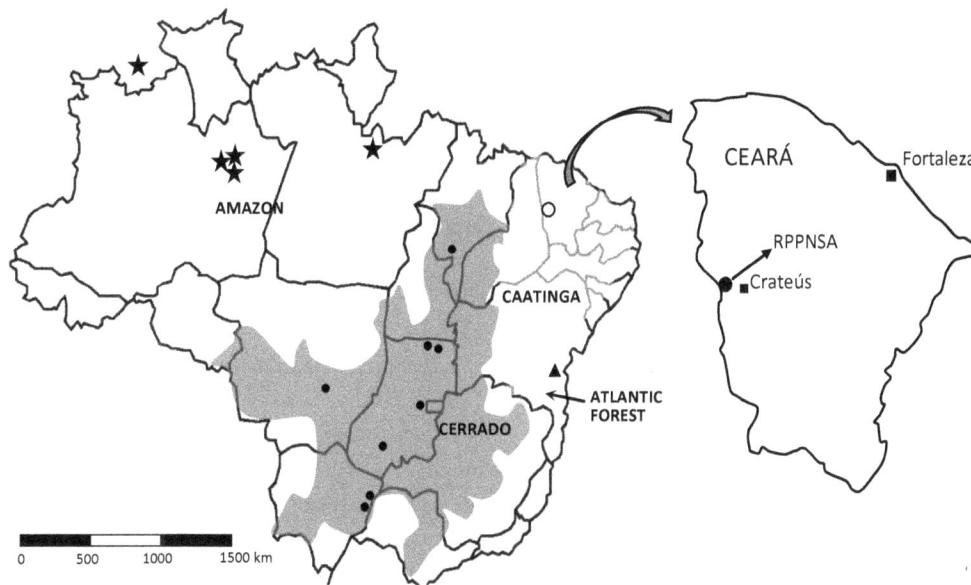

Fig. 1. Distribution of known *Blepharidatta* populations and study site location in the state of Ceará (Brazil) (RPPNSA: Reserva Particular do Patrimônio Natural Serra das Almas).
(★) *Blepharidatta brasiliensis;* (●) *Blepharidatta conops*; (▲) *Blepharidatta* sp-ba; ○ studied *Blepharidatta* new population

nests are visited or inhabited by myrmecophiles or predators. Queens of *B. conops* have a characteristic phragmotic head that together with the anterior slope of pronotum forms a frontal disk whose shape and dimensions fit the subsidiary chamber entrance and is used by the queen to block that entrance (Brandão et al., 2001). The foragers patrol a roughly circular area around the nest opening, where they collect live and dead arthropods (mainly other ants) to feed their larvae (Diniz & Brandão, 1997). In the bottom area of the nest, prey is dismembered, chewed and fed to larvae via trophallaxis. The discarded remains are arranged in a ring around the nest opening and workers concealed in the nest entrance ambush arthropods, including other ants, attracted by the carcasses ring (Diniz et al., 1998).

The other known species of *Blepharidatta* have different nesting habits: their smaller colonies (mean number of workers varying from 112 to 132) are found within the leaf-litter, in natural cavities between leaves or in rotting branches (Rabeling et al., 2006; Silva, 2007). Contrary to *B. conops*, the queens of *B. brasiliensis* and *Blepharidatta* sp-ba do not have a phragmotic head (Silva, 2007).

Colonies of *B. conops* occur in locally dense but widely scattered populations, with large areas of Cerrado devoided of *B. conops* (Diniz & Brandão, 1997; Brandão et al., 2001). According to Brandão et al. (2001), this distribution pattern may be explained by the limited dispersal mode of ergatoid queens and by the type of nest foundation (fission of established colonies). Such nest distribution may be characteristic of all *Blepharidatta* species and probably explains why ant species of this genus are considered rare (Brandão et al., 2001).

Here we present data on nest architecture, size and com-position of colonies, foraging behavior, morphology (female castes) of a Caatinga population of an unidentified species of *Blepharidatta* (Quinet & Tavares, 2005). We compare our findings to the data available for other species of *Blephari-datta*, particularly *B. conops* and discuss the identity of this potential new taxon

Material and methods

Study site

The study was conducted from November 2011 to June 2012 in a 4.5ha area of the "Reserva Particular do Patrimônio Natural Serra das Almas - RPPNSA" (5º08'S, 40º51'W), a 6146ha protected area of deciduous thorny woodland vegetation (Caatinga) in Crateús (State of Ceará, northeastern Brazil, 5°10'S 40°40'W) (Fig 1).

Nest architecture, size and composition of colonies

Twenty nine nests of *Blepharidatta* sp. were located in the study area, using freshly killed termites as baits. Any *Blepharidatta* sp. worker that picked a termite was then followed back to its nest.

Nineteen nests were excavated to study nest architecture as well as the size and composition of colonies. Before initiating a nest excavation, the maximum diameter of the nest opening and of the carcasses ring around it was recorded (see Diniz & Brandão, 1997). The carcasses forming the ring were collected, and a 30-cm deep trench was dug in order to obtain a 20-cm side cube with the nest opening in the middle of the upper face. Starting from one lateral side, the cube was sliced with a spatula until a nest chamber or a tunnel was found. The depth, maximum height and maximum diameter of each chamber were recorded, as well as the direction, diameter and length of each tunnel leading to a chamber.

All biological material (workers, queen(s), male(s) and brood; invertebrate and vegetal fragments; myrmecophiles) found in chambers or tunnels was collected, and, whenever possible, its exact location in the nest was recorded. Workers, queen(s), male(s), brood and myrmecophile organisms found in each nest were counted and fixed in ethanol 90%.

Diet and foraging behavior

In order to obtain information on *Blepharidatta* sp. diet, all invertebrate and vegetal fragments found in the carcasses ring and chambers of 10 excavated nests were analyzed. Fragments were first separated in three categories: ants, other invertebrates and seeds. Ant fragments were identified at least to genus level. Fragments of other invertebrates were identified to order level.

The diel foraging activity pattern was investigated by monitoring three colonies for a 24h period: the first from 12/16/2011 (10 a.m.) to 12/17/2011 (9 a.m.), the two others from 06/16/2012 (9 a.m.) to 06/17/2012 (8 a.m.). The nest opening of each colony was observed for 10min every hour and all ants that left or entered the nest were counted. In total, four hours of foraging activity were monitored per nest. Soil surface temperature was recorded each time ant activity was measured.

The density and foraging area size of *Blepharidatta* sp. nests were assessed using a 144 m^2 area (12 x 12 m) with a grid of 1m^2 quadrats (Fig 6). Each quadrat was baited with sardine and checked 40 min later. Baiting was repeated three times in three successive days. Each quadrat whose bait was visited by *Blepharidatta* sp. workers at least one time was then baited four times, on three consecutive days, with dead workers (or soldiers) of *Nasutitermes* sp. All workers carrying termites back to the nest were followed and their path marked with pieces of plastic straw.

Body measurements

The maximum transverse diameter of the frontal disk (phragmotic head plus anterior slope of the pronotum) of 10 *Blepharidatta* sp. queens (one per nest) and the maximum head width of 54 workers (from 18 nests) were measured under a stereomicroscope with an ocular micrometer. A Petri dish filled with fine white sand was used to correctly position the ants under a microscope.

Total body length of 10 queens (from the foremost part of the frontal disk to the tip of gaster, in individuals with outstretched body) and 20 workers (from the middle part of the clypeus to the tip of the gaster, in individuals with outstretched body) were also measured.

Voucher specimens of *Blepharidatta* sp. are deposited at the Myrmecological Collection of the Laboratório de Entomologia, Universidade Estadual do Ceará, in Fortaleza, CE, Brazil, at the Myrmecological Collection of the Museu de Zoologia of the Universidade de São Paulo [MZSP] in São Paulo, SP, Brazil, and at the Myrmecological Collection of the Centro de Pesquisas do Cacau [CPDC], CEPLAC, in Itabuna, BA, Brazil.

Results

Nest architecture, size and composition of colonies

All excavated nests (n=19) had a carcasses ring that surrounded the nest opening, with invertebrate fragments sometimes partially obstructing it. In three nests, however, the carcasses ring was inconspicuous, with very few invertebrate or vegetal fragments. Carcasses rings were circular or elliptic, with maximum diameter varying from 5 to 20cm. Mean fragment density in the carcasses ring was 8 (± 6.3) fragments/cm^2 (n=10; range = 1.7-19.1 fragments/cm^2).

All nests had only one circular (sometimes elliptic) nest opening whose mean diameter was 0.73 (± 0.17) cm (n=14; range: 0.5-1 cm). Each nest entrance was connected to a tunnel with approximately the same mean diameter (0.74 ± 0.23 cm; n=15; range = 0.5–1.3 cm), that led to a first chamber (carcasses chamber), and ended in a bottom chamber (Fig 2). In six nests, the tunnel was a straight and vertical structure connecting the nest opening to the carcasses and the bottom chambers (Fig 2). In other seven nests, the tunnel had sloped parts, with, sometimes, a branching pattern (Fig 2).

The bottom chamber was located at a mean depth of 26.5 (± 6.3) cm (n=19; range = 15-40 cm). It housed the

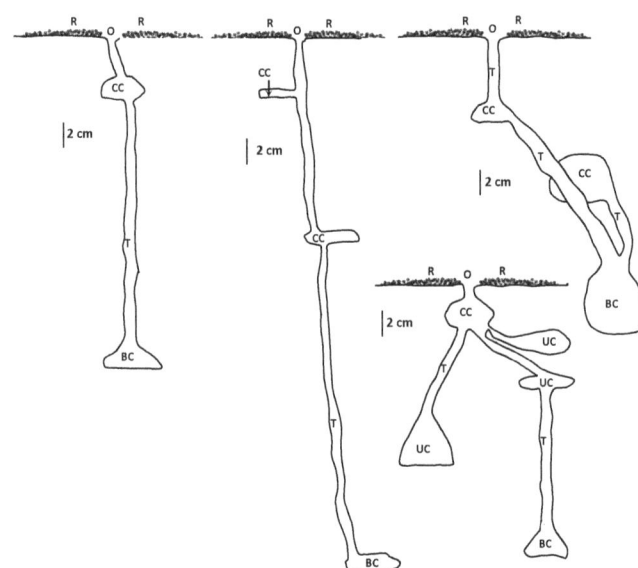

Fig 2. Schematic drawing of four *Blepharidatta* sp. nests excavated from 11/2011 to 06//2012, at the "Reserva Particular do Patrimônio Natural Serra das Almas", CE, Brazil. R: carcasses ring; O: nest opening; T: tunnel; CC: carcasses chamber; BC: bottom chamber; UC: undefined chamber.

queen and brood in most nests (n=8) where a queen was found (n=12). Its mean maximum height and diameter were 2.3 (± 1.54) cm (n=12; range: 0.7-2.5 cm) and 3.3 (± 1.07) cm (n=12; range: 2-4 cm) respectively.

A carcasses chamber, so called because it was full of carcasses, was found in most (14 out 19) nests. It was located very close to the nest opening (Fig 2), at a mean depth of 4.9 (± 2.4) cm (n=14, range: 1-11 cm), and its mean maximum height and diameter were 1.9 (± 0.95) cm (n=12; range: 0.5-2.5 cm) and 2.8 (± 0.8) cm (n=12; range: 2-4 cm), respectively. In six nests, one to two additional carcasses chambers were found at a depth varying from 10 to 25 cm (Fig 2). In one nest, up to five chambers were found (a carcasses chamber, a bottom chamber and three extra chambers), in addition to branching tunnels (Fig 2).

A possible subsidiary chamber, as defined by Brandão et al. (2001) for *B. conops*, was found in one nest. This chamber was located at the mid-length of the vertical tunnel leading to the bottom chamber, at a depth of 13 cm, and contained a queen, brood and 15 males. Queens were missing in seven nests, possibly because they escaped during the excavation process. All other nests (n=12) had a single queen, except one that had two queens. Most queens (8 out 13) were found in the bottom chamber together with brood. In three remaining nests, the queen was found in the carcasses chamber (n=2) or in the subsidiary chamber (*sensu* Brandão et al., 2001) (n=1). Two queens could not be assigned to a specific nest location. A total of twenty five males were found, in chambers of four nests, together with the queen and brood. Fifteen of them were found in the subsidiary chamber of one nest; in the three other nests, from one to five males were found in the bottom chamber.

Three myrmecophiles were frequently found in *Blepharidatta* sp. nests, almost always in the bottom chamber: two species of the crustacean genus *Trichorhina* Budde-Lund (Oniscidea, Platyarthridae) and one cockroach species (Corydiidae); we also noticed the frequent occurrence of the pseudoscorpion *Petterchernes brasiliensis* Heurtault (Cher-

netidae), up to now recorded only from burrows of small mammals in the state of Pernambuco, Brazil (Heurtault, 1986).

Average colony size (number of workers) was 193 (± 107.4) (n=19; range: 30-437). Most colonies (13 out 19) ranged from 110 and 220 workers; a few (n=4) reached more than 300 workers (one of them had more than 400 workers) (Fig 3).

Queen and worker morphology

Blepharidatta sp. queens are ergatoid and have an enlarged phragmotic head that, together with the anterior slope of the pronotum, form a nearly circular disk (Fig 4) whose mean maximum transverse diameter is 1.68 (± 0.04) mm (n=10; range: 1.60-1.76 mm) (Table 1). Frontal disk cuticle is discretely reticulate-punctate, nearly smooth; head margin is strongly curved upwards, and stiff hairs protrude laterally from the perimeter of the disk (Fig 4). Total mean body length of queens is 4.96 (± 0.12) mm (n=10; range: 4.8-5.20 mm). *Blepharidatta* sp. workers are monomorphic (mean body length: 3.79 ± 0.22 mm; n=20; range: 3.50-4.48 mm), and their mean maximum head width is 0.96 (± 0.05) mm (n=54; range: 0.84-1.04 mm) (Table 1).

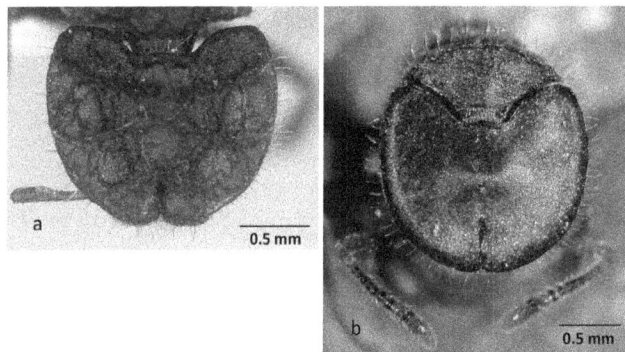

Fig 4. Phragmotic head of *Blepharidatta conops* (a) (photo by April Nobile – www.antweb.org) and *Blepharidatta* sp. (b) queen.

Table 1. Mean maximum transverse diameter (mm) of queen (Q) frontal disk and mean workers (W) maximum head width (mm) in different *Blepharidatta conops* populations (SR, BA, SEL, CH-G) and in the *Blepharidatta* population found at the "Reserva Particular do Patrimônio Natural da Serra das Almas" (RPPNSA), in Crateús (State of Ceará). SR: Serranópolis (State of Goiás); BA: Balsas (State of Maranhão); SEL: Selvíria (State of Mato Grosso do Sul); CH-G: Chapada dos Guimarães (State of Mato Grosso). Data within parenthesis refer to standard errors of the means. Data source for *B. conops*: Brandão et al. (2001).

	Blepharidatta conops				*Blepharidatta* sp.
	SR	BA	SEL	CH-G	RPPNSA
Q	1.55 (± 0.03) *n*=12	1.2 (± 0.06) *n*=5	1.45 *n*=1	1.51 *n*=1	1.68 (± 0.04) *n*=10
W	0.87 (± 0.04) *n*=5	0.75 (± 0.05) *n*=5	0.78 *n*=1	---	0.96 (± 0.05) *n*=54

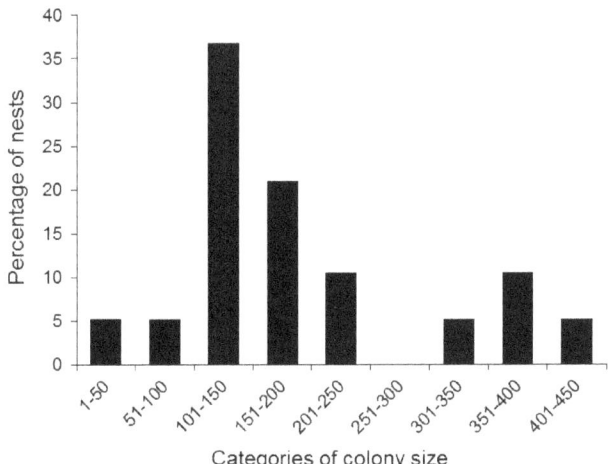

Fig 3. Distribution (%) of colony size categories for 19 nests of *Blepharidatta* sp. excavated from 11/2011 to 06/2012, at the "Reserva Particular do Patrimônio Natural Serra das Almas", CE, Brazil.

Diet and foraging behavior

In total, 13,576 fragments were found in the carcasses ring and chambers of 10 *Blepharidatta* sp. nests. Of these, 8,385 (62%) were from ants, 4,660 (34%) from others invertebrates, and 522 (4%) were whole seeds or pieces of seeds. Most invertebrate and seed items were found in the carcasses ring (77% and 85% respectively); the rest was found in nest chambers, mostly in carcasses chambers. Most ant fragments (93%) could be identified; 61% of the other invertebrate fragments and none of the seeds could be identified.

Ant carcasses included 41 species from 18 (or 19) genera belonging to seven subfamilies (Table 2). Three species accounted for 50% of all ant carcasses. The most common species was *Ectatomma muticum* Mayr (Ectatomminae) (23%), followed by *Camponotus crassus* Mayr (Formicinae) (16%) and *Labidus* sp. nr. *coecus* (Ecitoninae) (11%).

Other eight species (*Acromyrmex rugosus* (Smith), *Gnamptogenys striatula* Mayr, *Camponotus* sp., *Cephalotes pusillus* (Klug), *Eciton* sp., *Crematogaster* sp., *Pheidole* sp., *Odontomachus bauri* Emery) were frequent, accounting each for 3 to 7% of all ant carcasses. It is also worth mentioning the Ecitoninae group since 18% of all ant carcasses were from *Eciton* (3 species), *Labidus* or *Neivamyrmex* (4 species), *Nomamyrmex* (1 species) and an unidentified genus (1 species).

Other invertebrate carcasses were from eight insect groups (Blattodea [including termites], Coleoptera, Dermaptera, Hemiptera, Hymenoptera, Lepidoptera, Neuroptera, Orthoptera), and two arachnid groups (Araneae, Scorpiones). However, the vast majority (92%) of carcasses were from four groups: Hemiptera (37.5%), Coleoptera (22.4%), Araneae (19.4%) and termites (13%) (all termite fragments were heads of *Nasutitermes* sp. soldiers).

Table 2. List of subfamilies and genera identified in ant carcasses found in the carcasses ring and nest chambers of 10 *Blepharidatta* sp. nests, with number of species found in each genus.

Subfamily	Genus	Species number
Dolichoderinae	*Dolichoderus*	1
	Dorymyrmex	1
Ectatomminae	*Ectatomma*	1
	Gnamptogenys	1
Ecitoninae	*Eciton*	3
	Labidus/Neivamyrmex	4
	Nomamyrmex	1
	unidentified genus	1
Formicinae	*Brachymyrmex*	1
	Camponotus	5
Myrmicinae	*Acromyrmex*	1
	Cephalotes	1
	Crematogaster	4
	Cyphomyrmex	1
	Pheidole	10
	Solenopsis	2
	Strumigenys	1
Ponerinae	*Odontomachus*	1
Pseudomyrmecinae	*Pseudomyrmex*	1

Foraging activities of *Blepharidatta* sp. are predominantly crepuscular, with two peaks of activity, the first one corresponding to the night/day transition period (5 a.m.-9 a.m.), the second one to the day/night transition period (4 p.m.-7 p.m.) (Fig 5). At other periods, especially during the warmest period of the day (± 10 a.m.-3 p.m.), foraging activity stops almost entirely, at least at the nest opening (Fig 5). However, at night, from 7 p.m. to 2 a.m., intense cleaning activity was observed in all nests, with workers carrying carcasses or sand particles out of the nest and depositing them on the carcasses ring, while others removed carcasses obstructing the nest opening and organized them on the carcasses ring.

During the period when nests foraging activity was monitored (up to 12 hours of nest opening observation for the three nests), it was observed that the carcasses ring was frequently visited by invertebrates, including by ants that robbed carcasses and sometimes inspected *Blepharidatta* sp. nest opening. However, none of these visitors was ambushed by *Blepharidatta* sp. workers, contrary to what was observed in *B. conops*, whose workers, concealed inside the nest entrance, frequently ambush visitors (Diniz et al., 1998).

Collective transport of large prey items was observed in *Blepharidatta* sp. in one occasion. More than 12 workers

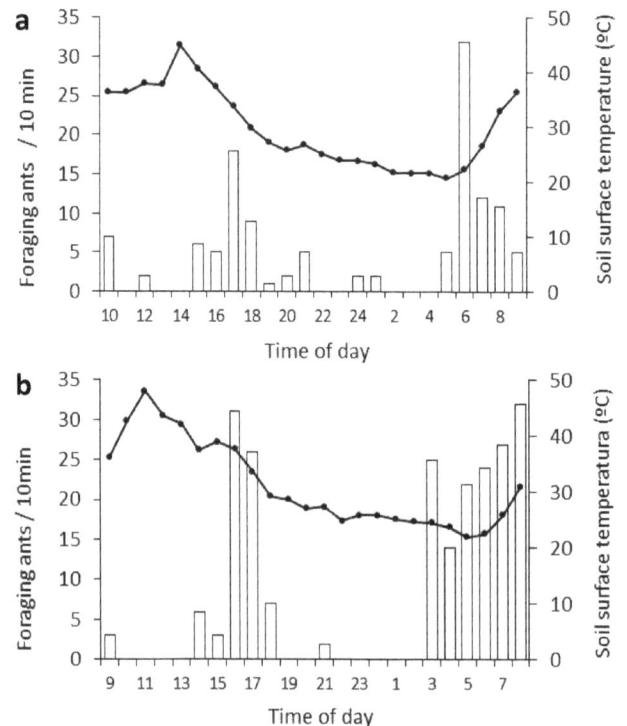

Figure 5. Daily cycle of activity of two *Blepharidatta* sp. colonies (a, b) at the "Reserva Particular do Patrimônio Natural Serra das Almas", CE, Brazil. All ants exiting/entering the nest were counted during 10 minutes, during a 24-hour observation period (a: 16.12-17 .12.2011; b: 16.06-17.06.2012). Bars: foraging ants/10 min; line with dots: soil surface temperature. Sunrise: 5.30 h (onset of daylight: 6.00 h); sunset: 17.30 h (beginning of darkness: 18.00 h). The daily cycle of activity of the third monitored nest (16.06-17.06.2012), not shown here, had similar pattern.

were observed grasping and pulling the appendages of two still alive and fighting *Ectatomma muticum* workers, trying to transport them to the nest.

Only two nests were found in the 144 m² area mapped (three if a nest close to the area border is included) (Fig 6), giving a nest density of 0.020 nest/m². Although some foragers were observed foraging at a distance of seven meters from the nest, the most common foraging distances observed were two to three meters (Fig 6). *Blepharidatta* sp. nest foraging area can therefore be described as a nearly circular area with a radius of ± 2.5m around the nest opening.

Discussion

There is no doubt that *B. conops* and *Blepharidatta* sp. belong to a group of closely related taxa since they share key features of their behavior, ecology and morphology, while at the same time both show fundamental differences with the other *Blepharidatta* species living in tropical humid environments.

B. conops and *Blepharidatta* sp. are soil-dwelling species with nest opening surrounded by a carcasses ring (Diniz et al., 1998), while *B. brasiliensis* and *Blepharidatta* sp-ba nest in leaf-litter, and do not form carcasses ring (Rabeling et al., 2006; Silva, 2007). *B. conops* and *Blepharidatta* sp. are monogynous species whose ergatoid queens have a phragmotic head (Brandão et al., 2001), while *B. brasiliensis* and *Blepharidatta* sp-ba have polygynous colonies, and queens do not have a phragmotic head (Rabeling et al., 2006; Silva, 2007). It is worth mentioning, however, that polygyny or mo-

nogyny of *Blepharidatta* sp-ba remains a controversial question (Silva, 2007).

Frontal disk diameter in queens, head width in workers and body length in queens and workers are almost identical in both species (Table 1) (Diniz et al., 1998; Silva, 2007). On the other hand, body length in queens and workers of *B. conops* and *Blepharidatta* sp. (± 5 mm for queens; ± 3 mm for workers) is almost three times that observed in *B. brasiliensis* and *Blepharidatta* sp-ba (± 2mm and ± 1mm for queens and workers, respectively) (Silva, 2007).

Blepharidatta sp. and *B. conops* have similar diet and foraging activity patterns, with two daily and mostly diurnal peaks of foraging activity separated by periods of inactivity, and a protein diet predominantly made of other ant species (Diniz et al., 1998; Brandão et al., 2008). In *B. brasiliensis*, foraging activity is predominantly nocturnal (Rabeling et al., 2006).

Finally, the two groups of populations are found in completely different environments: rainforests (Amazon and Atlantic Rainforest) for *B. brasiliensis* and *Blepharidatta* sp-ba (Rabeling et al., 2006; Silva, 2007); drier and savanna-like environments (Cerrado and Caatinga) for *B. conops* and *Blepharidatta* sp. (Diniz et al., 1998; Brandão et al., 2001).

According to Silva (2007), *B. conops* possesses more derived character states (e.g. phragmotic head of queens) when compared to other *Blephadatta* species. Considering the number of key behavioral and morphological features shared by *B. conops* and *Blepharidatta* sp., it can be stated that *Blepharidatta* sp. shares with *B. conops* this apomorphic condition.

However, our study also shows the existence of many differences between the populations living in the Cerrado biome (*B. conops*) and the population found in the Caatinga biome (*Blepharidatta* sp.). The most noticeable are those related to queen morphology. In *Blepharidatta* sp., frontal disk cuticle is discretely reticulate-punctate, nearly smooth, while it is rugose with a strong relief composed of closely packed polygonal units delimited by ridges in *B. conops* (Brandão et al., 2001) (Fig 4). Furthermore, other morphological differences, not detailed above, can be seen in the mesosoma sculpture, much stronger in the *B. conops* queen than in *Blepharidatta* sp., in the humeral angle of the pronotum which is sharply angulated in *Blepharidatta* sp. but rounded in *B. conops*. Finally, the pilosity on the first gastral tergite of the queen is erect and short in *Blepharidatta* sp. while it is subdecumbent and a little longer in *B. conops*.

Blepharidatta sp. and *B. conops* also differ in relation to nest architecture, colony size, location of queens and brood in nests and foraging behavior. *Blepharidatta* sp. nests are deeper (± 26.5 cm; up to 40 cm in some nests) than in *B. conops* (up to 20 cm in mature nests) (Diniz et al., 1998). Conversely, the mean diameter of tunnels and nest opening is much greater in *B. conops* (2cm *versus* 0.7cm in *Blepharidatta* sp.) (Brandão et al., 2001). In most excavated nests of *Blepharidatta* sp.,

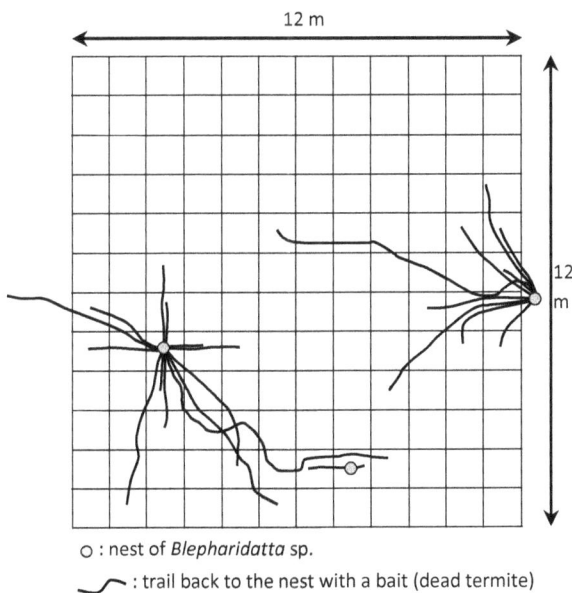

o : nest of *Blepharidatta* sp.

⌒ : trail back to the nest with a bait (dead termite)

Fig 6. Mapping of *Blepharidatta* sp. nests and of the trails used by foragers to go back to the nest with a bait (dead termite) in a 144 m² (12 x 12 m) area with a grid of 1 m² quadrats, at the "Reserva Particular do Patrimônio Natural Serra das Almas", CE, Brazil, in 01/2012.

there was one, or more, chambers located close to the nest opening and full of carcasses. The function of those carcasses chambers in *Blepharidatta* sp. is probably to store prey carcasses before discarding them on the carcasses ring. No such chamber was found in *B. conops* nests. One of the most important differences in nest architecture between *B. conops* and *Blepharidatta* sp. is the so-called subsidiary chamber. In *B. conops*, all mature nests have a subsidiary chambers used as refuge by the queen and the brood (Diniz et al., 1998; Brandão et al., 2001). Only one *Blepharidatta* sp. nest had a chamber full with brood, queen and males and whose features suggest a subsidiary chamber.

More generally, *Blepharidatta* sp. nests are more variable in structure than *B. conops* nests. In *B. conops* nests, there is only one tunnel, always vertical (Diniz et al., 1998). In *Blepharidatta* sp., the main tunnel varies from vertical to inclined. Furthermore, the main tunnel can have branchings, the result being nests three-dimensionally more complex that those of *B. conops*. It cannot be excluded that the characteristics of the soil where the nests of *Blepharidatta* sp. were built have a strong influence on nest architecture. However, such characteristics were not analyzed in the present study, nor in studies with *B. conops* (Diniz et al., 1998).

As in other *Blepharidatta* ants, *Blepharidatta* sp. colonies are small. However, the average number of workers in *Blepharidatta* sp. colonies (193 ± 107, n=19) is higher than in *B. conops* (142 ± 57, n=19) (Diniz et al., 1998), *B. brasiliensis* (132 ± 96, n=13) (Rabeling et al., 2006) and *Blepharidatta* sp-ba (112 ± 13, n=2) (Silva, 2007). One of the *Blepharidatta* sp. nests had up to 437 workers. In *B. conops*, the largest known colony size is 248 workers (Diniz et al., 1998). Interestingly, two queens were found in a single *Blepharidatta* sp. nest. This observation supports the hypothesis that in *B. conops*, the foundation of new colonies is through fission of established colonies (Brandão et al., 2001). If this hypothesis is correct, there must be a time when two queens are present in the nest: the old resident queen and a young virgin queen.

The location of queen and brood in nests is another difference between *B. conops* and *Blepharidatta* sp. In most nests of *Blepharidatta* sp., the queen and brood were found in the bottom chamber. In *B. conops*, they are generally found in the subsidiary chamber, but never in the bottom chamber of the nest, which is used as a place for prey dismembring and to temporarily store carcasses, before they are taken out of the nest (Diniz et al., 1998; Brandão et al., 2001).

Foraging behavior of *Blepharidatta* sp. and *B. conops* is very similar. However, some differences were observed. According to Diniz et al. (1998), the ants and other arthropods visiting the carcasses ring of nests "are frequently ambushed by *B. conops* workers concealed immediately inside the nest entrance". Such behavior was never observed in *Blepharidatta* sp. Another significant difference is the size of foraging area. In *Blepharidatta* sp. the mean radius of foraging area (2.5 m) is nearly twice that observed in *B. conops* (1.5 m), while nest density seems to be much lower in *Blepharidatta* sp. (0.02 nests/m² against 0.2 nests/m² in *B. conops*), at least in the investigated area (Diniz et al., 1998). However, the influence of local factors on nest density, such as prey density or competition with other ant species that use similar resources can not be excluded.

Although it cannot be excluded that *Blepharidatta* sp. is a species different from *B. conops*, it is obvious that both taxa are closely related and constitute a single evolutionary unit, with a branch that spread into the Cerrado biome (savanna) and another into the Caatinga biome (semi-arid). The traits that strongly suggest the occurrence of two distinct species are seen in the queen morphology, mainly on the frontal disk and mesosoma. On the other hand, it has been suggested that *B. conops* distribution in the Cerrado biome is fragmented in a range of local populations, with size and sculpture of queen frontal disk varying from one population to the other (Brandão et al., 2001). Therefore, *Blepharidatta* sp. could be a local variation of *B. conops* adapted to semi-arid conditions and consequently represent one extreme of the phenotypic variations observed in populations of *B. conops*. Ant populations in species that have lost mating flight have low levels of gene flow and can evolve independently, as it could happen in fragmented populations of *B. conops* or its ancestor since the early Quaternary.

Future studies should use cytogenetic and molecular tools to compare these *Blepharidatta* populations (e.g. Mariano et al., 2008; Resende et al., 2010 with *Dinoponera lucida* Emery 1901).

Finally, our study extends the known geographic range of the *Blepharidatta* group and the range of biomes where it is found, as it is the first time that the genus, until now found only in wet forests and savannas (Cerrado), is reported for the Neotropical semi-arid region.

Acknowledgments

The authors are grateful to Rodrigo dos Santos Machado Feitosa (Universidade Federal do Paraná) for the identification of ant carcasses. We are also grateful to Leila Aparecida Souza (Universidade Estadual do Ceará [UECE]), Sonia Maria Lopes (Museu Nacional – Universidade Federal do Rio de Janeiro) and Mark Harvey (Western Australian Museum – Australia) for the identification of the myrmecophile fauna (crustaceans, cockroaches and pseudoscorpions, respectively) found in *Blepharidatta* sp. nests. We also thank UECE for transport facilities as well as the "Associação Caatinga" and the staff of the RPPN "Serra das Almas" for allowing us excellent field work conditions. JCP acknowledges her CAPES fellowship during her MSc, and JHC his CNPq research grant.

References

Bolton, B. (1995). Taxonomic and zoogeographical census of the extant ant taxa (Hymenoptera). Journal of Natural History, 29: 1037-1056. doi: 10.1080/00222939500770411.

Brandão, C.R.F., Diniz, J.L.M., Silva, P.R., Albuquerque, N.L. & Silvestre, R. (2001). The first case of intranidal phragmosis in ants: the ergatoid queen of *Blepharidatta conops* (Formicidae, Myrmicinae) blocks the entrance of the brood chamber. Insectes Sociaux, 48: 251-258. doi: 10.1007/PL00001774.

Brandão, C.R.F., Silva, P.R. & Diniz, J.L.M. (2008). O "mistério" da formiga lenta *Blepharidatta*. In E.F. Vilela, I.A. Santos, J.H. Shoereder, J.E. Serrão, L.A.O. Campos & J.L. Lino-Neto (Eds.), Insetos sociais: da biologia à aplicação (pp. 38-46). Viçosa: Editora da UFV.

Cassano, C.R., Schroth, G., Faria, D., Delabie, J.H.C. & Bede, L. (2009). Landscape and farm scale management to enhance biodiversity conservation in the cocoa producing region of southern Bahia, Brazil. Biodiversity and Conservation, 18: 577-603. doi: 10.1007/s10531-008-9526-x.

Diniz, J.L.M. & Brandão, C.R.F. (1997). Competition for carcasses and the evolution of fungus-ants symbiosis. In J.H.C. Delabie, S. Campiolo, I.C. Nascimento & C.S. Ferreira Mariano (Eds.), Anais do VI International Pest Ant Symposium & XIII Encontro de Mirmecologia (pp. 81-87). Ilhéus: Universidade Estadual de Santa Cruz, Ilhéus.

Diniz, J.L.M., Brandão, C.R.F. & Yamamoto, C.I. (1998). Biology of *Blepharidatta* ants, the sister group of the Attini: a possible origin of fungus-ant symbiosis. Naturwissenschaften, 85: 270-274. doi: 10.1007/s001140050497.

Heurtault, J. (1986). *Petterchernes brasilienses*, genre et espèce nouveaux de Pseudoscorpions du Brésil (Arachnides, Pseudoscorpionida, Chernetidae), Bulletin du Museum National d'Histoire Naturelle, Paris, 2: 351-355.

Mariano, C.S.F., Pompolo, C.G., Barros, L.A.C., Mariano-Neto, E., Campiolo, S. & Delabie, J.H.C., (2008). A biogeographical study of the threatened ant *Dinoponera lucida* Emery (Hymenoptera: Formicidae: Ponerinae) using a cytogenetic approach. Insect Conservation and Diversity, 1: 161-168. doi: 10.1111/j.1752-4598.2008.00022.x.

Peeters, C. (1991). Ergatoid queens and interacaste in ants: two distinct adult forms, which look morphologically intermediate between workers and winged queens. Insectes Sociaux., 38: 1-15.

Quinet, Y.P. & Tavares, A.A. (2005). Formigas (Hymenoptera: Formicidae) da área Reserva Serra das Almas, Ceará. In F.S. Araújo, M.J.N. Rodal & M.R.V. Barbosa (Eds.), Análise das variações da Biodiversidade do Bioma Caatinga (pp. 329-349). Brasília: Ministério do Meio Ambiente.

Rabeling, C., Verhaagh, M. & Mueller, U.G. (2006). Behavioral ecology and natural history of *Blepharidatta brasiliensis* (Formicidae, Blepharidattini). Insectes Sociaux, 53: 300-306. doi: 10.1007/s00040-006-0872-y.

Resende, H.C., Yotoko, K.S.C., Delabie, J.H.C., Costa, M.A., Campiolo, S., Tavares, M.G., Campos, L.A.O. & Fernandes-Salomão, T.M. (2010). Pliocene and Pleistocene events shaping the genetic diversity within the Central Corridor of Brazilian Atlantic Forest. Biological Journal of the Linnean Society, 101(4): 949-960. doi: 10.1111/j.1095-8312.2010.01534.x.

Schultz, T.R. & Meier, R. (1995). A phylogenetic analysis of the fungus-growing ants based on morphological characters of the larvae. Systematic Entomology, 20: 337-370. doi: 10.1111/j.1365-3113.1995.tb00100.x.

Silva, P.R. (2007). Biologia de algumas espécies de *Blepharidatta*. Biológico, 69 (suplemento): 161-164.

Wheeler, W.M. (1915). Two new genera of myrmicine ants from Brazil. Bulletin of the Museum of Comparative Zoology, 59: 483-491.

Wheeler, C.G. & Wheeler, J. (1991). The larva of *Blepharidatta* (Hymenoptera: Formicidae). Journal of the New York Entomological Society, 99: 132-137.

13

The genetic Characterization of *Myrmelachista* Roger Assemblages (Hymenoptera: Formicidae: Formicinae) in the Atlantic Forest of Southeastern Brazil

MA Nakano[1], VFO Miranda[2], RM Feitosa[3], MSC Morini[1]

1 - Universidade de Mogi das Cruzes, Mogi das Cruzes, SP, Brazil.

2 - Universidade Estadual Paulista, Jaboticabal, SP, Brazil.

3 - Universidade Federal do Paraná, Curitiba, PR, Brazil.

Keywords

twigs, ISSR molecular markers, Formicinae, UPGMA, similarity

Corresponding author

Maria Santina de Castro Morini
UMC – Núcleo de Ciências Ambientais
Laboratório de Mirmecologia
Av. Dr. Cândido Xavier de Almeida
Souza, 200
Mogi das Cruzes - SP - Brazil
08780-911
E-mail: mscmorini@gmail.com

Abstract

Arboreal ants of the genus *Myrmelachista*, which have ecologically important relationships with different vegetable species, are found exclusively in the Neotropical region. These ant species are difficult to identify, and their taxonomy remains controversial; moreover, little is known regarding their biology. The objective of the present work is to assess the genetic similarities and dissimilarities between and within *Myrmelachista* species, with the goal of expanding knowledge of the relationships among the taxa of this genus. Sample collection in selected regions of the dense ombrophile forest of southeastern Brazil yielded 256 nests, which were found in vegetation or among scattered twigs in the leaf litter; eight species were recorded. A total of 180 specimens were analyzed, producing 123 polymorphic fragments. Data analyses revealed similarity relationships that allowed the examined species to be classified into the following groups: (1) *Myrmelachista* sp. 4, *M. nodigera*, *M. ruszkii* and *M. gallicola*; (2) *M. catharinae* and *M. arthuri*; (3) *M. reticulata*; and (4) *Myrmelachista* sp. 7. The study results also revealed the existence of two morphological variants of *M. catharinae*; *M. arthuri* was more closely related to one of these *M. catharinae* variants than to the other variant. The present work provides important information regarding genetic variation among *Myrmelachista* species that may contribute to interpreting the complex morphology of this genus.

Introduction

Myrmelachista Roger is a genus of the Formicinae subfamily. The geographical distribution of this genus is restricted to the Neotropical region, and 41% of the species in this genus can be found in Brazil (Kempf, 1972; Fernández & Sendoya, 2004). The species in this genus are arboreal (Longino, 2006) and engage in the specialized practice of nesting in trunk cavities and among twigs (Stout, 1979; Brown, 2000; Longino, 2006; Edwards et al., 2009; Nakano et al., 2012, 2013). These ant species may also form complex mutual associations with certain myrmecophytes (Renner & Ricklefs, 1998; Frederickson, 2005; Edwards et al., 2009) or with Coccidae and Pseudococcidae species (Kusnezov, 1951; Stout, 1979; Ketterl et al., 2003; Longino, 2006). Little information is available regarding the biology of *Myrmelachista* species

(Brown, 2000); however, it is known that these species generally feed on extrafloral nectaries (Haber et al., 1981) and on animal-derived proteins (Torres, 1984; Amalin et al., 2001; McNett et al., 2010).

At present, 69 *Myrmelachista* species have been described, with a few recognized subspecies (Bolton et al., 2006; Bolton, 2013); the diversity of this genus has most likely been underestimated (Longino, 2006) due to the limited taxonomic knowledge available regarding *Myrmelachista* (Snelling & Hunt, 1975). Previously published reports, such as studies by Wheeler (1934) and Longino (2006), are important sources of taxonomic knowledge regarding *Myrmelachista*; in particular, the report by Longino (2006) is the most recent taxonomic review of the genus, although this review was restricted to *Myrmelachista* species found in Costa Rica. In the most recent molecular data-based phylogenetic proposals for ants (Brady et al., 2006; Moreau et al., 2006), *Myrmelachista* is a sister group of *Brachymyrmex*, and these groups constitute the most basal and closely related formicine groups.

Longino (2006) emphasized the difficulty of separating *Myrmelachista* species based on morphological characters but suggested that the combined use of worker, queen and male characters could improve the accuracy of species distinctions. Because the collection of alate individuals is more difficult than the collection of workers (Nakano et al., 2013), molecular techniques may be utilized to help establish the systematics of the *Myrmelachista* genus. These techniques have been applied for similar analyses of other insect groups (Reineke et al., 1998). In particular, ISSRs (inter-simple sequence repeats) may serve as a very important tool for these analyses. ISSR-based techniques enable the development of co-dominant microsatellite markers that may be used in population studies (Gupta et al., 1994; Bornet & Branchard, 2001; Wolfe, 2005). These techniques have allowed genetic distinctions to be drawn among Hymenoptera groups (Borba et al., 2005; León & Jones, 2005; Al-Otaibi, 2008) and among other Insecta groups (Souza et al., 2008; Luque et al., 2002; Dusinsky et al., 2006). However, at present, no known published studies in the literature have applied these techniques exclusively for the examination of ants. Thus, the present work sought to identify relationships and levels of genetic similarity (and dissimilarity) between and within *Myrmelachista* species found in the Atlantic Forest of southeastern Brazil, with the goal of providing additional knowledge regarding the relationships among different taxa of this genus.

Material and methods

Samples for molecular analysis

The collection of *Myrmelachista* nests occurred in three different regions of dense ombrophile forest in southeastern Brazil (Fig 1). A total of 256 nests were collected; these nests were located either in the forest vegetation or among scattered twigs in the leaf litter (for details about biological material collecting see Nakano et al., 2012, 2013). The nests housed eight *Myrmelachista* species: *M. arthuri* Forel, 1903; *M. catharinae* Mayr, 1887; *M. gallicola* Mayr, 1887; *M. nodigera* Mayr, 1887; *M. reticulata* Borgmeier, 1928; *M. ruszkii* Forel, 1903; *Myrmelachista* sp.4; and *Myrmelachista* sp.7.

Whole nests containing living workers were stored at -80°C. To minimize the degradation of genomic DNA, the screening and identification of individuals were performed after specimens were frozen in ice-immersed Petri dishes. Workers from each nest were also collected in the field and stored in 70% ethanol for subsequent morphological identification.

A total of 180 specimens were assessed. For most of the examined species, five individuals per nest were selected from five nests (randomly) for molecular identification (total

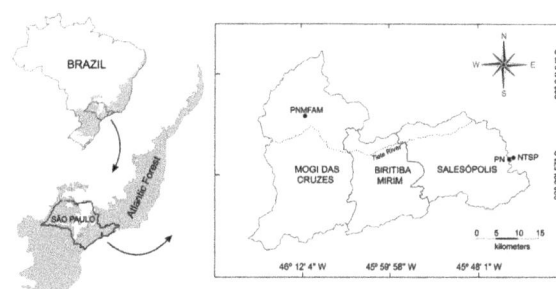

Figure 1. The locations of the Atlantic Forest areas in which *Myrmelachista* species were collected. (PNMFAM: Francisco Affonso de Mello Municipal Natural Park; PN: Ponte Nova Dam; PENT: Tietê Springs State Park).

of 25 individuals and 5 nests per specie), except for *M. reticulata* (5 individuals and 1 nest) and *Myrmelachista* sp.7 (20 individuals and 4 nests). Alate queens (n = 5) were found only in *Myrmelachista* sp.7 nests; these queens were also examined by molecular analyses.

Species (or morphospecies) were identified by comparing the examined specimens with specimens deposited in the reference collection of the Museum of Zoology of the University of São Paulo. Vouchers were deposited in the myrmecofauna collection of the Alto Tietê Myrmecology Laboratory of the University of Mogi das Cruzes and in the Museum of Zoology of the University of São Paulo.

Genomic DNA extraction

Genomic DNA from ants preserved at -80°C was extracted using the protocol described by Sambrook et al. (1989). DNA extraction was performed independently for each examined specimen. Sample integrity and purity were analyzed by 0.8% agarose gel electrophoresis, and spectrophotometric observations were used to assess DNA quantity and purity.

Amplification of ISSRs

Twenty primers (UBC kit, University of British Columbia) were tested; among these primers, the 10 primers that generated the most distinguishable bands were selected, and optimal annealing temperatures were determined for each of these 10 primers (Table 1). Each amplification reaction was performed in a tube containing a total volume of 20 μL. The reaction mixture included the following reagents: buffer (1.6X), deoxynucleotide triphosphates (dNTPs) (0.2 mM), primer (0.8 μM), MgCl$_2$ (1.5 mM), DNA (8 ng), Taq DNA polymerase (1 U) (GoTaq® Flexi DNA Polymerase, Promega) and autoclaved Milli-Q H$_2$O. The amplifications were performed in a thermal cycler (Peltier Thermal Cycler PTC-200, MJ Research). The following temperature conditions were utilized for the amplification reactions: an initial denaturation at 94°C for 3 min; 40 cycles of denaturation at 92°C for 1 min, annealing (at an optimized temperature for each primer) for 2 min and extension at 72°C for 2 min; and a final extension at 72°C for 7 min.

Table 1. The identifying codes, sequences and optimal annealing temperatures of the ISSR primers utilized in this study.

Code	5'-3' Sequence	Annealing temperature (°C)
UBC 842	GAG AGA GAG AGA GAG AYG	54
UBC 888	BDB CAC ACA CAC ACA CA	57
UBC 889	DBD ACA CAC ACA CAC AC	58
UBC 890	VHV GTG TGT GTG TGT GT	58
UBC 808	AGA GAG AGA GAG AGA GC	54
UBC 816	CAC ACA CAC ACA CAC AT	54
UBC 836	AGA GAG AGA GAG AGA GYA	54
UBC 841	GAG AGA GAG AGA GAG AYC	54
UBC 848	CAC ACA CAC ACA CAC ARG	54
UBC 811	GAG AGA GAG AGA GAG AC	54

Y = C or T; R = A or G; B = G, T or C; D = G, A or T; H = A, T or C; V = A, C or G.

Bands were visualized on a 1.8% agarose gel subjected to 2.5 hours of electrophoresis at 120 amperes in a horizontal electrophoresis unit (Gibco). Amplified DNA was visualized by ethidium bromide staining, dissolved in 1X TBE (Tris-borate-EDTA (ethylenediaminetetraacetic acid)) buffer, pH 8.3, and photographed under UV light using a transilluminator equipped with a DC40 digital camera (Kodak).

Data analysis

A binary matrix of molecular data was generated by encoding the bands present in the agarose gel (presence = 1; absence = 0). The CP ATLAS 2.0 software program (Lazar Software, 2009) was used to standardize the criteria for determining band detection and band intensity.

To assess intra and interspecific genetic similarity, cluster analysis was performed using the unweighted pair-group method with arithmetic average (UPGMA) method and Nei and Li's coefficient (Nei & Li, 1979). Analyses using this coefficient only examine bands that are present, comparing the number of bands shared among individual specimens. Principal component analysis (PCA) was used to generate

graphical representations of interspecies genetic variation and to determine the characters that provide the greatest contribution to genetic differentiation. All statistical analyses were performed using the MVSP 3.1 statistical software package (Kovach, 2007).

Results

DNA concentrations in samples from the examined species ranged from 3.3 to 55.4 ng/µL, and the OD_{260}/OD_{280} absorbance ratios of these samples ranged from 1.17 to 2.6. A total of 123 different bands were obtained using the 10 selected primers, and a total of 22,140 bands were analyzed. The sizes of the examined DNA fragments ranged from 100 bp to 1000 bp. The number of bands amplified by each primer ranged from seven (for the UBC 848 primer) to 15 (for the UBC 889 primer), with an average of 12 bands generated per primer. The resulting amplification pattern is depicted in Fig 2.

The largest values of intraspecific genetic similarity obtained in the ISSR analyses were found among workers in *M. arthuri* nests (which exhibited similarity values of up to 1.00). The smallest values of intraspecific genetic similarity were found among *M. ruszkii* (with a minimum similarity value of 0.51) and *Myrmelachista* sp.4 (with a minimum similarity value of 0.56) specimens; these two species exhibited the highest rates of intraspecific genetic diversity.

With respect to interspecific genetic similarity, high similarity between *M. arthuri* and *M. catharinae* (with similarity values of up to 0.84) and between *Myrmelachista* sp.4 and *M. nodigera* (with similarity values of up to 0.77) was observed. The pairs of species with the lowest similarity were *Myrmelachista* sp.4 and *M. reticulata* (with a similarity value of only 0.33) and *Myrmelachista* sp.4 and *M. catharinae* (with a similarity value of only 0.37) (Table 2).

Using the UPGMA-based dendogram, seven main groups were defined. Notably, *Myrmelachista* sp.7 (group one) and *M. reticulata* (group two) were the most genetically dissimilar of the examined species; *Myrmelachista* sp.4 (group three) was more closely related to *M. gallicola* (group

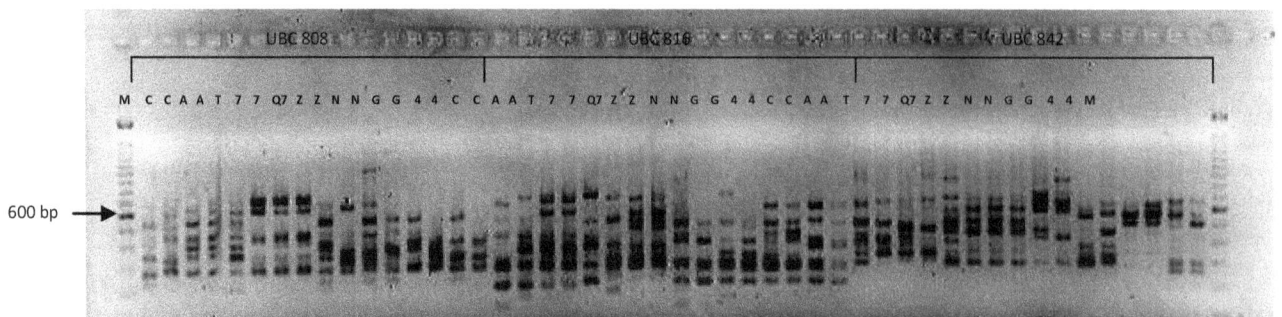

Figure 2. An agarose gel (N%) depicting the ISSR bands obtained from *Myrmelachista* genomic DNA amplified with the UBC 808, UBC 816 and UBC 842 primers. M: 100 bp marker; C: *M. catharinae*; A: *M. arthuri*; T: *M. reticulata*; 7: *Myrmelachista* sp.7; Q7: *Myrmelachista* sp.7 queen; Z: *M. ruszkii*; N: *M. nodigera*; G: *M. gallicola* and 4: *Myrmelachista* sp.4.

Table 2. Variation of intraspecific (underlined) and interspecific molecular similarity for *Myrmelachista* species using ISSR molecular markers and based on the Nei and Li coefficient (Nei & Li, 1979).

Species	*M. catharinae*	*M. arthuri*	*M. reticulata*	*M. sp.7*	*M. ruszkii*	*M. nodigera*	*M. gallicola*	*M. sp.4*
M. catharinae	<u>0.61 – 099</u>							
M. arthuri	0.52 – 0.84	<u>0.81 – 1.00</u>						
M. reticulata	0.47 – 0.66	0.52 – 0.63	<u>0.90 – 0.97</u>					
M. sp.7	0.45 – 0.64	0.44 – 0.63	0.54 – 0.69	<u>0.78 – 0.99</u>				
M. ruszkii	0.46 – 0.72	0.51 – 0.71	0.44 – 0.61	0.47 – 0.63	<u>0.51 – 0.99</u>			
M. nodigera	0.40 – 0.70	0.51 – 0.75	0.42 – 0.62	0.38 – 0.59	0.48 – 0.72	<u>0.64 – 0.99</u>		
M. gallicola	0.38 – 0.68	0.46 – 0.69	0.38 – 0.58	0.38 – 0.59	0.45 – 0.72	0.46 – 0.72	<u>0.64 – 0.97</u>	
M. sp.4	0.37 – 0.64	0.45 – 0.71	0.33 – 0.57	0.37 – 0.62	0.42 – 0.66	0.44 – 0.77	0.43 – 0.74	<u>0.56 – 0.97</u>

four), *M. nodigera* (group five) and *M. ruszkii* (group six) than to the other examined species; and *M. catharinae* (with two morphological variants) and *M. arthuri*, which are both in group seven, were the most similar of the examined species (Fig 3).

A similar result was derived from PCA, which allowed the examined species to be classified into four groups (Fig 4). The PCA findings confirmed that *Myrmelachista* sp.7 (group one) and *M. reticulata* (group two) were the least similar of the examined species. *M. catharinae* and *M. arthuri* were categorized into group three, whereas *M. nodigera*, *M. ruszkii*, *M. gallicola* and *Myrmelachista* sp.4 were categorized into group four.

Discussion

Myrmelachista species possess between nine and 10 antennal segments. Most nine-segmented *Myrmelachista* species are found in Central America and the Caribbean (with only two known nine-segmented *Myrmelachista* species in South America), whereas 10-segmented *Myrmelachista* species are mostly found in South America (with only three known 10-segmented *Myrmelachista* species found in Mexico and Central America) (Longino, 2006). In the present study, only 10-segmented species were examined; as noted by Snelling and Hunt (1975), these species form an extremely heterogeneous group.

The circumscription of *Myrmelachista* species is a complex task because the morphological differences between individuals of a single species that originate from different colonies can be sufficient to cause these individuals to be erroneously regarded as members of different species (Wheeler, 1934; Snelling & Hunt, 1975). Our results show that the ISSR can be powerful markers identification of *Myrmelachista* and facilitates the identification and morphological interpretation. The

ISSR is more advantageous in showing intra and interspecies differences than morphology. According to Nakano (2010), worker ants of the *M. catharinae*, *M. arthuri*, *M. reticulata* and *Myrmelachista* sp.7 species exhibit extremely similar morphologies. However, the molecular analysis results of this study reveal interspecies differences among these morphologically similar workers. *M. catharinae* and *Myrmelachista* sp.7 queens exhibit morphological differences (Nakano, 2010), and our molecular analysis results confirm the interspecies differences between these queens. Longino (2006) reported that analyses of reproductive ants can facilitate the identification of *Myrmelachista* species; however, these types of ants have rarely been examined, and literature data regarding the reproductive biology of *Myrmelachista* species remain scarce (Nakano et al., 2012, 2013).

Two different *M. catharinae* groups are readily identifiable in the dendogram (Fig 3) of this study, and *M. arthuri* is more closely related to one of these *M. catharinae* groups than to the other *M. catharinae* group. Field observations indicated that *M. catharinae* workers that were more closely related to *M. arthuri* exhibited lighter-colored mesosoma than *M. catharinae* workers that were more distantly related to *M. arthuri*. When disturbed, both *M. arthuri* workers and *M. catharinae* workers with lighter-colored mesosoma exhibit the same aggressive behavior of immediately elevating their gasters. However, in contrast to *M. catharinae* with lighter-colored mesosoma, *M. arthuri* produces large nests and forms trails of intensely foraging workers (MA Nakano, personal observation).

Our results demonstrated that *M. nodigera*, *M. gallicola* and *Myrmelachista* sp.4 are genetically similar species. Previous reports have noted morphologic and morphometric similarities between *M. gallicola* and *M. nodigera* with respect to both workers (Quirán & Martínez, 2006) and queens. Field observations have also revealed that when disturbed, ants of these

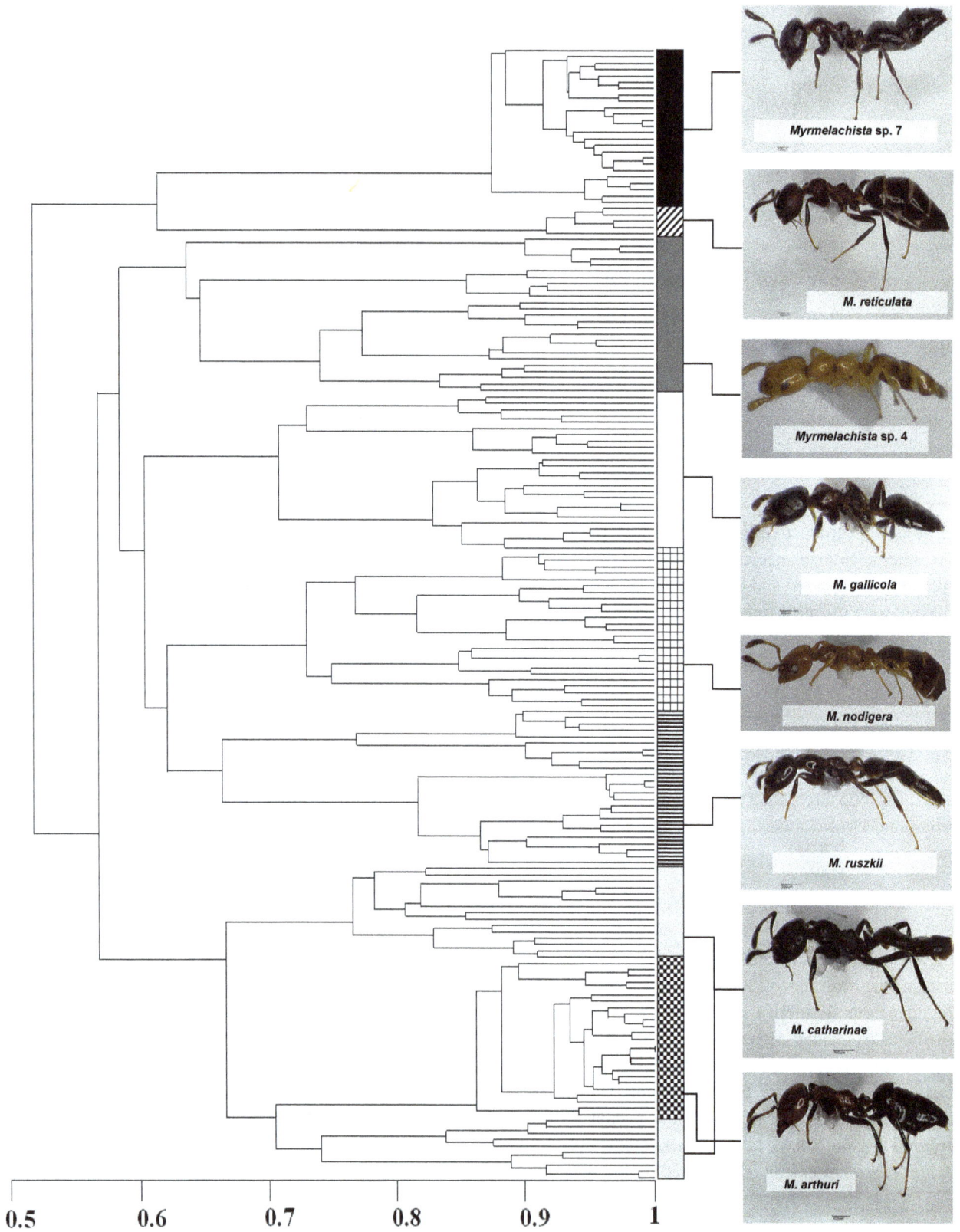

Figure 3. The UPGMA dendrogram generated using Nei and Li's coefficient (Nei & Li, 1979) for eight *Myrmelachista* species.
■ *Myrmelachista* sp.7; ▨ *M. reticulata*; ▧ *Myrmelachista* sp.4; ☐ *M. gallicola*; ⊞ *M. nodigera*; ☰ *M. ruszkii*; ▨ *M. catharinae* and ▨ *M. arthuri*.

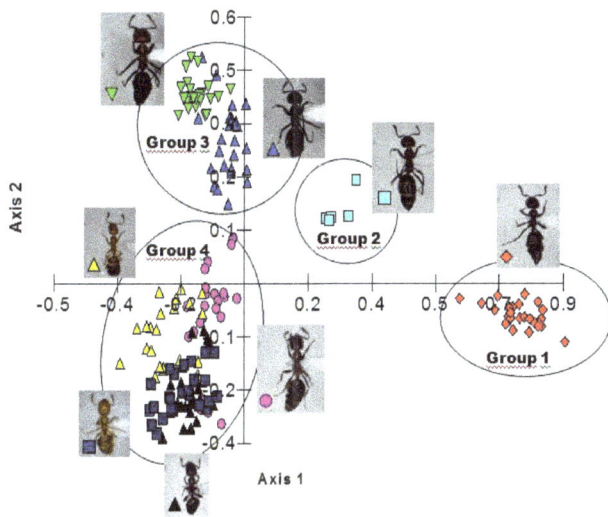

Figure 4. Principal component analysis results for eight *Myrmelachista* species (n = 175) based on 123 bands obtained from ISSR primers. Circles indicate the four different groups into which these species have been classified. PC1 and PC2 account for 16% and 10%, respectively, of the observed variance. ▲ *M. catharinae*; ▽ *M. arthuri*; □ *M. reticulata*; ◆ *Myrmelachista* sp.7; ● *M. ruszkii*; △ *M. nodigera*; ▲ *M. gallicola*; ■ *Myrmelachista* sp.4.

species hastily retreat to nest orifices (Nakano et al., 2013).

The molecular analysis results of this study reveal an interesting scenario that could result in the taxonomic reclassification of the examined groups; that should be studied further with sequencing. For now, the ISSR allowed us to recognize taxonomic unities for *Myrmelachista*. In particular, if less conservative criteria are utilized to interpret intraspecific morphological variations in the examined species (and consequently to delineate the boundaries between species), *M. catharinae* in which workers have lighter-colored mesosoma could be regarded as a new species that is more closely related to *M. arthuri* than to other *M. catharinae*. Similarly, in a more widespread scenario, the superposition of molecular characters among *M. nodigera, M. gallicola, M. ruszkii* and *Myrmelachista* sp.4 could lead to the unification of these species under a single specific name that would represent all of these as a single specie with high intraspecific biological and morphological variation.

Prior to the implementation of the suggested nomenclature changes, individuals of the examined species should be sampled in a more widespread manner across a broad swath of the geographic areas in which these species are distributed. Nonetheless, the present work contributes important information regarding the intra and interspecific genetic variations among the examined species. These contributions facilitate the interpretation of morphological variations in *Myrmelachista*, a conspicuous genus that is ecologically important due to the relationship between *Myrmelachista* species and different vegetable species; despite this importance, this biologically diverse and morphologically complex genus remains insufficiently studied.

Acknowledgments

The authors would like to thank Dr. Regina Lúcia Batista Costa de Oliveira for providing access to the Laboratory of Structural Genomics (University of Mogi das Cruzes); the CNPq for providing a research grant to MA Nakano and MSC Morini (grant numbers 132485/2008-7 and 301151/2009-1, respectively); and the São Paulo Research Foundation for providing support to RM Feitosa (grant number 11/24160–1).

References

Al-otaibi, S. A. (2008). Genetic variability in mite-resistant honey bee using ISSR molecular markers. Arab Journal of Biotecnology, 11: 241-252.

Amalin, D. M., Pena, J. E., Mcsorley, R., Browning, H. W. & Crane, J. H. (2001). Comparison of different sampling methods and effect of pesticide application on spider populations in lime orchards in south florida. Environmental Entomology, 30: 1021-1027. doi: 10.1603/0046-225X-30.6.1021.

Bolton, B., Alpert, G., Ward, P. S. & Naskrecki, P. (2006). Bolton's catalogue of ants of the world. Cambridge: Harvard University Press. CD-ROM.

Bolton, B. (2013). An online catalog of the ants of the world. Available from http://antcat.org. (accessed [November, 30, 2013])

Borba, R. S., Garcia, M. S., Kovaleski, A., Oliveira, A. C., Zimmer, P. D., Castelo-Branco, J. S. & Malone, G. (2005). Dissimilaridade genética de linhagens de *Trichogramma* Westwood (Hymenoptera: Trichogrammatidae) através de marcadores moleculares ISSR. Neotropical Entomology, 34: 565-569.

Bornet, B. & Branchard, M. (2001). Nonanchored Inter Simple Sequence Repeat (ISSR) markers: reproducible and specific tools for genome fingerprinting. Plant Molecular Biology Reporter, 19: 209-215.

Brady, S. G., Schultz, T. R., Fisher, B. L. & Ward, P. S. (2006). From de cover: evaluating alternative hypotheses for the early evolution and diversification of ants. Proceedings of the National Academy of Sciences USA, 103: 18172-18177. doi: 10.1073/pnas.0605858103.

Brown, W. L. (2000). Diversity of ants. In: Agosti, D., Majer, J. D., Alonso, L. T. & Schultz, T. R. (eds.). Ants: standard methods for measuring and monitoring biodiveristy. Washington: Smithsonian Institution Press 45-46.

Dusinsky, R., Kudela, M., Stloukalova, V. & Jedlicka, L. (2006). Use of inter-simple sequence repeat (ISSR) markers for discrimination between and within species of blackflies (Diptera: Simuliidae). Biologia, Bratislava, 61: 299-304.

Edwards, D. P., Frederickson, M. E., Shepard, G. H. & Yu, D. W. (2009). A plant needs ants like a dog needs fleas: *Myrmelachista schumanni* ants gall many trees species to create housing. American Naturalist, 174: 734-740. doi: 10.1086/606022.

Fernández, F. & Sendoya, S. (2004). Lista de las hormigas Neotropicales (Hymenoptera: Formicidae). Biota Colombiana, 5: 5-27.

Frederickson, M. E. (2005). Ant species confer different partner benefits on two neotropical myrmecophytes. Oecologia, 143: 387-395. doi:10.1098/rspb.2006.0415.

Gupta, M., Chyiy, S., Romero-Severson, J. & Owen, J. I. (1994). Amplification of DNA markers from evolutionarily diverse genomes using single primers of simple-sequence repeats. Theoretical and Applied Genetics, 89: 998-1006.

Haber, W. A., Frankie, G. W., Baker, H. G., Baker, I. & Koptur, S. (1981). Ants like flower nectar. Biotropica, 13: 211-214.

Kempf, W. (1972). Catálogo abreviado das formigas da Região Neotropical (Hymenoptera: Formicidae). Studia Entomologica, 15: 1-4.

Ketterl, J., Verhaagh, M., Bihn, J. H., Brandão, C. R. F. & Engels, W. (2003). Spectrum of ants associated with *Araucaria angustifolia* trees and their relations to hemipteran trophobionts. Studies on Neotropical Fauna and Environment, 38: 199-206.

Kovach, W. L. (2007). MVSP 3.1: multivariate statistical packkage. http://kovcomp.co.uk/mvsp.html. (accessed date: 13 March, 2010).

Kusnezov, N. (1951). *Myrmelachista* en la Patagônia (Hymenoptera, Formicidae). Acta Zoologica Lilloana, 11: 353-365.

Lazar Software. CP ATLAS 2.0. (2009). http://lazarsoftware.com/index.html. (accessed date: 1 March, 2010).

León, J. H. & Jones, W. A. (2005). Genetic differentiation among geographic populations of *Gonatocerus ashmeadi*, the predominant egg parasitoid of the glassy-winged sharpshooter, *Homalodisca coagulata*. Journal of Insect Science, 5: 1-9.

Longino, J. T. (2006). A taxonomic review of the genus *Myrmelachista* (Hymenoptera: Formicidae) in Costa Rica. Zootaxa, 1141: 1-54.

Luque, C., Legal, L., Saudter, H., Gers, C. & Wink, M. (2002). Brief report ISSR (Inter Simple Sequence Repeats) as genetic markers in Noctuids (Lepidoptera). Hereditas, 136: 251-253.

Mcnett, K., Longino, J., Barriga, P., Vargas, O., Phillips, K. & Sagers, C. L. (2010). Stable isotope investigation of a cryptic ant-plant association: *Myrmelachista flavocotea* (Hymenoptera: Formicidae) and *Ocotea* spp. (Lauraceae). Insectes Sociaux, 57: 67-72. doi: 10.1007/s00040-009-0051-z.

Moreau, C. S., Bell, C. D., Vila, R., Archibald, B. & Piercel, N. (2006). Phylogeny of the ants: diversification in the age of Angiosperms. Science, 312: 101-104.

Nakano, M. A. (2010). Caracterização morfológica, morfométrica e molecular de espécies de *Myrmelachista* Roger, 1865 (Formicidae: Formicinae) do Alto Tietê. Dissertação de Mestrado em Biotecnologia, Universidade de Mogi das Cruzes, SP, 137p.

Nakano, M. A., Feitosa , R. M., Moraes, C.O., Adriano, L. D. C., Hengles, E. P., Longui, E. L. & Morini, M.S.C. (2012). Assembly of *Myrmelachista* Roger (Formicidae: Formicinae) in twigs fallen on the leaf litter of Brazilian Atlantic Forest. Journal of Natural History, 46: 2103-2115. doi:10.1080/0022 2933.2012.707247.

Nakano, M. A., Miranda, V. F. O., Souza, D. R., Feitosa, R. M. & Morini, M. S. C. (2013). Ocurrence and natural history of *Myrmelachista* Roger (Formicidae: Formicinae) in the Atlantic Forest of southeastern Brazil. Revista Chilena de Historia Natural, 86: 169-179.

Nei, M. & Li, W. H. (1979). Mathematical model for studying genetic variation in terms of restriction endonucleases. Proceedings of the National Academy of Sciences USA, 76: 5269-5273.

Quirán, E. M. & Martínez, J. J. (2006). Redescripción de la obrera de *Myrmelachista gallicola* (Hymenoptera: Formicidae) y primera cita para la provincial de La Pampa (Argentina). Revista de la Sociedad Entomológica Argentina, 65: 89-92.

Reineke, A., Karlovsky, P. & Zebitz, P. W. (1998). Preparation and purification of DNA from insects for AFLP analysis. Insect Molecular Biology, 7: 95-99.

Renner, S. S. & Ricklefs, R. E. (1998). Herbicidal activity of domatia-inhabiting ants in patches of *Tococa guianensis* and *Clidemia heterophylla*. Biotropica, 30: 324-327.

Sambrook, J., Fritsch, E. F. & Maniatis, T. (1989). Molecular Cloning: a laboratory manual. 2. ed. Cold Spring: Harbor Laboratory Press.

Snelling, R. R. & Hunt, J. H. (1975). The ants of Chile (Hymenoptera: Formicidae). Revista Chilena de Entomologia, 9: 110-114.

Souza, G. A., Carvalho, M. R. O., Martins, E. R., Guedes, R. N. C. & Oliveira, L. O. (2008). Diversidade genética estimada com marcadores ISSR em populações brasileiras de *Zabrotes subfasciatus*. Pesquisa Agropecuária Brasileira, 43: 843-849.

Stout, J. (1979). An association of an ant, a mealy bug, and an understory tree from a Costa Rican Rain Forest. Biotropica, 11: 309-311.

Torres, J. A. (1984). Niches and coexistence of ant communities in Puerto Rico: Repeated Patterns. Biotropica, 16: 284-295.

Wheeler, W. M. (1934). Neotropical ants collected by Dr. Elisabeth Skwarra and others. Bulletin of the Museum of Comparative Zoology, 77: 157-240.

Wolfe, A. D. (2005). ISSR techniques for evolutionary biology. Methods in Enzymology, 395: 134-144.

14

Inquilinitermes johnchapmani, a New Termite Species (Isoptera: Termitidae: Termitinae) from the Llanos of North Central Bolivia

RH Scheffrahn

University of Florida, IFAS, Davie, FL, Unites States of America

Keywords

Inquilinitermes johnchapmani, new species, Llanos de Mojos, Bolivia, *Cornitermes snyderi, Constrictotermes, Grigiotermes metoecus*

Corresponding author

Rudolf H. Scheffrahn
University of Florida
Institute of Food and Agricultural Sciences
Fort Lauderdale Res. and Education Center
3205 College Avenue
Davie, Florida, 33314, U.S.A.
rhsc@ufl.edu

Abstract

Inquilinitermes johnchapmani is described from soldiers and workers collected in the Llanos de Mojos of Bolivia. This is the fourth and the smallest species of the genus. Unlike its congeners, *I. johnchapmani* is not an inquiline of *Constrictotermes* spp. nests although an association with *Cornitermes snyderi* nests is likely.

Introduction

Until now, the genus *Inquilinitermes* Mathews (1977) consisted of three neotropical species: *Inquilinitermes fur* (Silvestri, 1901), *I. microcerus* (Silvestri, 1901), and *I. inquilinus* (Emerson, 1925). Previously included in the genus *Termes* (Snyder, 1949), Mathews (1977) consented to erect the genus *Inquilinitermes* for these species on the basis of the absence of the second marginal tooth on the left worker/imago mandible, the poorly developed frontal process of the soldier, and the species' apparent obligate colonization of *Constrictotermes* spp. (Termitidae: Nasutitermitinae) nests (Cunha et al., 2003; Constantino & Acioli, 2006). Apolinário & Martius (2004) report that they found a colony of *Inquilinitermes* cf. *microcerus* in a hollow tree in central Amazonia unassociated with *Constrictotermes*. In all cases however, *Inquilinitermes* soldiers also differ from those of *Termes* by the former's long slender mandibles which are much longer than the head capsule. Herein, I described the fourth and smallest species of *Inquilinitermes, I. johnchapmani*.

Materials and Methods

The distribution map (Fig 1) was created using ArcGIS desktop ver. 10.1 (ESRI, Redlands, CA). Photos in Figs 3, 4, 6, and 8 were taken as multi-layer montages using a Leica M205C stereomicroscope controlled by Leica Application Suite version 3 software. Preserved specimens were taken from 85% ethanol and suspended in pool of Purell® hand sanitizer to position the specimens over a transparent plastic Petri dish background.

The worker enteric valve (Fig 5) and mandible photographs (Fig 7) were taken from slide mounts using PVA medium (BioQuip Products Inc.) and a Leica CTR 5500 compound microscope with phase-contrast optics using the same montage software. Measurements were obtained using an Olympus SZH stereomicroscope fitted with an ocular micrometer. Worker mandible dentition terminology follows that of Sands (1972), and gut sections are named per criteria of Noirot (2001).

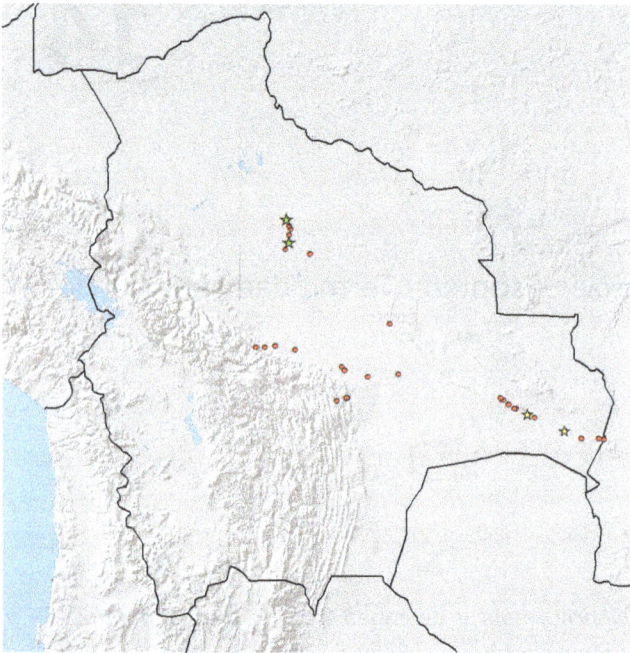

Fig 1. Map of *Inquilinitermes johnchapmani* collection site (green stars), *Constrictotermes cyphergaster* sites (yellow stars), and other 2013 Bolivia termite survey localities (red dots).

Inquilinitermes johnchapmani, new species

Holotype: soldier in UF sample no. BO423 from BOLIVIA: Hwy. 9 N. Trinidad (14.70207, 64.89097) 155 m elev., 29MAY2013, University of Florida Termite collection, Fort Lauderdale Research and Education Center, Davie, Florida.

Paratypes: Twelve soldiers and many workers in holotype colony; BO422 (1 soldier, workers) same data as holotype colony. BOLIVIA: N. San Pedro on Hwy. 9 (14.21260, 64.94026) 147 m elev., 29MAY2013, BO570, (1 soldier, workers) BO573 (2 soldiers, workers). BO422 and BO573 collected from galleries associated with *Cornitermes snyderi* Emerson, and BO423 collected with *Grigiotermes metoecus* Mathews, a common nest inquiline of *Cornitermes* spp. All samples taken collectively by those mentioned in the acknowledgments.

Etymology. This species is named after John Chapman

(1949-2010). John spent his 38-year career with Terminix International and his last 25 years as the company's manager of technical services for wood-destroying organisms. The author interacted with John on many occasions regarding termite identification and control technologies and his absence remains a loss to the pest control industry.

Imago. Unknown.

Soldier. (Figs 3 and 4; Table 1). Head, in dorsal view, rectangular with rounded corners; about 1.5 times as long as broad with lateral margins parallel and straight or very slightly concave. In dorsal view, frontal tubercle triangulate ending in a nipple-like point; tubercle not projecting beyond genal condyles. Antennae with 14 articles, relative length formula 2=3=4<5. Antennae about same length as mandibles. Pronotum angle about 110°. Each tibia with two proximal spurs although a weaker distal spur may occur on fore tibia of some specimens.

Mandibles about 1.7 times as long as head measured at the genal condyles; becoming very thin at distal two-thirds; slightly bending inwards at midpoint, recurving and straightening in distal one-third; tips hooked about 45° inward. Mandibles nearly symmetrical, left mandible slightly longer with more pronounced basal hump than right; tiny obtuse tooth present on inner basal margin of left mandible, occasionally also present on right mandible.

In lateral view, head capsule ovoid. Frontal tubercle rising about 30° from plane of vertex to midpoint of tubercle, then slanting downward about 10° to form ca. 140° angle; nipple-like tip angled upward; angle formed by frontal face of tubercle and postclypeus near 90°. Labrum with sharp triangular lateral points, frontal margin rather deeply incised.

Diagnosis. *Inquilinitermes johnchapmani* is the smallest of the four described congeners and has the most robust frontal process. The nipple-like point of the frontal process is distinctive. Compared with *Termes* soldiers, this and other *Inquilinitermes* spp. have proportionally much longer mandibles relative to head capsule length.

Worker (Figs 5-8; Table 1). Head, thorax, and legs nearly white. Postclypeus moderately inflated. Gut contents dark brown; lighter in P3. P1 long; wrapping 360° around rest of digestive coil and covering median of P4 when viewed on

Table 1. Measurements of *Inquilinitermes johnchapmani* soldiers and workers

Measurement in mm	Range	Mean ± S.D.	Holotype
SOLDIERS (n=11 from 4 colonies)			
Head length to apex of tubercle	1.19-1.48	1.37 ± 0.08	1.38
Head length to genal condyle	1.04-1.31	1.23 ± 0.07	1.23
Left mandible length to genal condyle	1.93-2.22	2.11 ± 0.09	2.22
Head length with mandibles*	2.96-3.48	3.34 ± 0.15	3.46
Head height (max. w/o gula)	0.64-0.72	0.68 ± 0.03	0.67
Head width (max.)	0.79-0.86	0.85 ± 0.02	0.84
Pronotum width (max.)	0.57-0.64	0.61 ± 0.03	0.64
WORKERS (n=8 from 4 colonies)			
Head width (max.)	0.67-0.77	0.73 ± 0.03	
Pronotum width (max.)	0.40-0.54	0.45 ± 0.05	

* = HL to genal condyle + LML to genal condyle.

right side; posterior 1/5 of P1 greatly inflated in volume until reaching constriction at P2. In ventral view, P3 occupying 2/3 of abdomen. Enteric valve armature composed of six spiny cushions, three larger and three smaller, in symmetrical arrangement. Spines consist of hollow cones with pointed tips projecting into lumen of P2; when mounted, spines pushed flat by cover slip.

Mandibles dominated by prominent apical teeth; left mandible with 2nd marginal tooth absent along cutting edge formed by first and third; right mandible with second marginal tooth reduced and forming a weak concavity along posterior margin of first marginal.

Fig 2. Ground cover where *I johnchapmani* was collected. Mounds are those of *Cornitermes snyderi*.

Fig 3. Dorsal, lateral and ventral view of *Inquilinitermes johnchapmani* soldier head.

Diagnosis. Presumed to be the smallest workers among three other congeners based on size of soldier caste in each. Unlike *Termes* spp., this and other *Inquilinitermes* spp. lack a second marginal tooth on the left mandible, however enteric valve armature is very similar to that of *Termes* (Miller (1991), Sands (1998)) with the exception of *T. hispaniolae* (Banks) in which pads are adorned with thickly sclerotized spines (Scheffrahn unpubl. obs.). Like *I. johnchapmani*, Constantino (2002) reports a 2:2:2 tibial spur formula which he uses to distinguish the genus from *Termes* (3:3:2 or 3:2:2).

Discussion

The currently known habitat for *I. johnchapmani* is the Mamoré subregion of the Llanos de Mojos in Depto. Beni, Bolivia (Fig 1). This biome is characterized by a pronounced wet/dry season (1300-2000 mm precipitation/yr) and encompasses a patchwork of floodplains, savannas with palm stands, and gallery forests (Hamilton et al. 2004). Our two roadside collection sites of *I. johnchapmani* were dominated by epigeal mounds built by *Cornitermes snyderi* (Fig 2). Some trees had

Fig 4. Soldier of *Inquilinitermes johnchapmani* showing the frontal process (antennae removed).

Fig 5. Enteric valve armature of *Inquilinitermes johnchapmani*. Inset showing arrangement of all six pads.

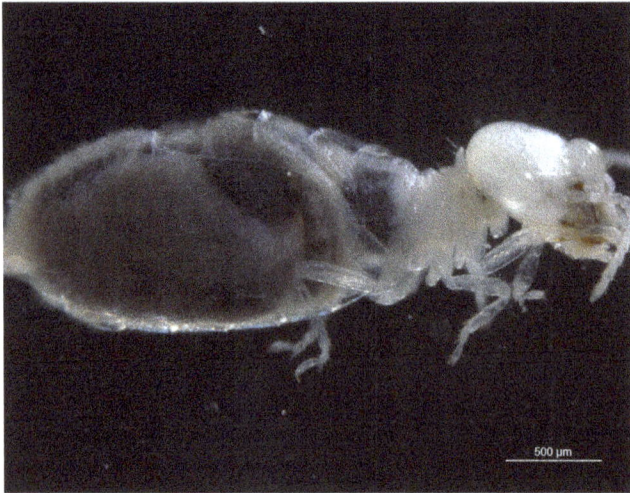

Fig 6. Lateral view of worker of *Inquilinitermes johnchapmani*.

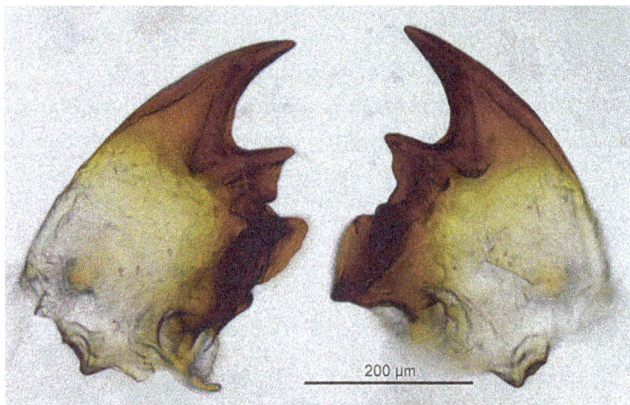

Fig 7. Worker mandibles of *Inquilinitermes johnchapmani*.

been cleared from the area in recent time. Much of the habitat, including our collection sites, were grazed by cattle. We collected 128 colony samples of termites (some with multiple species) from the two *I. johnchapmani* sites. During our survey expedition (Fig 1), the nearest collection of *Constrictotermes, C. cyphergaster* (Silvestri), was taken 712 km to the southeast (-18.46413, -59.47732, Aguas Calientes) in the Llanos de Chiquitos, a humid subregion of the Gran Chaco (Fig 1). The workers of *I. johnchapmani* resemble soil-feeding termites as they lack fat bodies found in sympatric cellulose feeders like *Termes*, presumably because of its feeding habit inside *C. snyderi* nests.

Acknowledgments

The 2013 Bolivian termite survey team members included the author, J. Krěcěk, and A. Mullins (Univ. of Florida), J.A. Chase and J. R. Mangold (Terminix International), Tom Nishimura (BASF Corp.), and Reginaldo Constantino (Universidade de Brasília). We are indebted to Terminix International for travel support. We also thank BASF Corp. for partial funding. We thank Estado Plurinacional de Bolivia, Ministerio de Medio Ambiente y Agua for issuance of collecting permit MMAyA-VMABCCGDG-DGBAP No. 1052/2013.

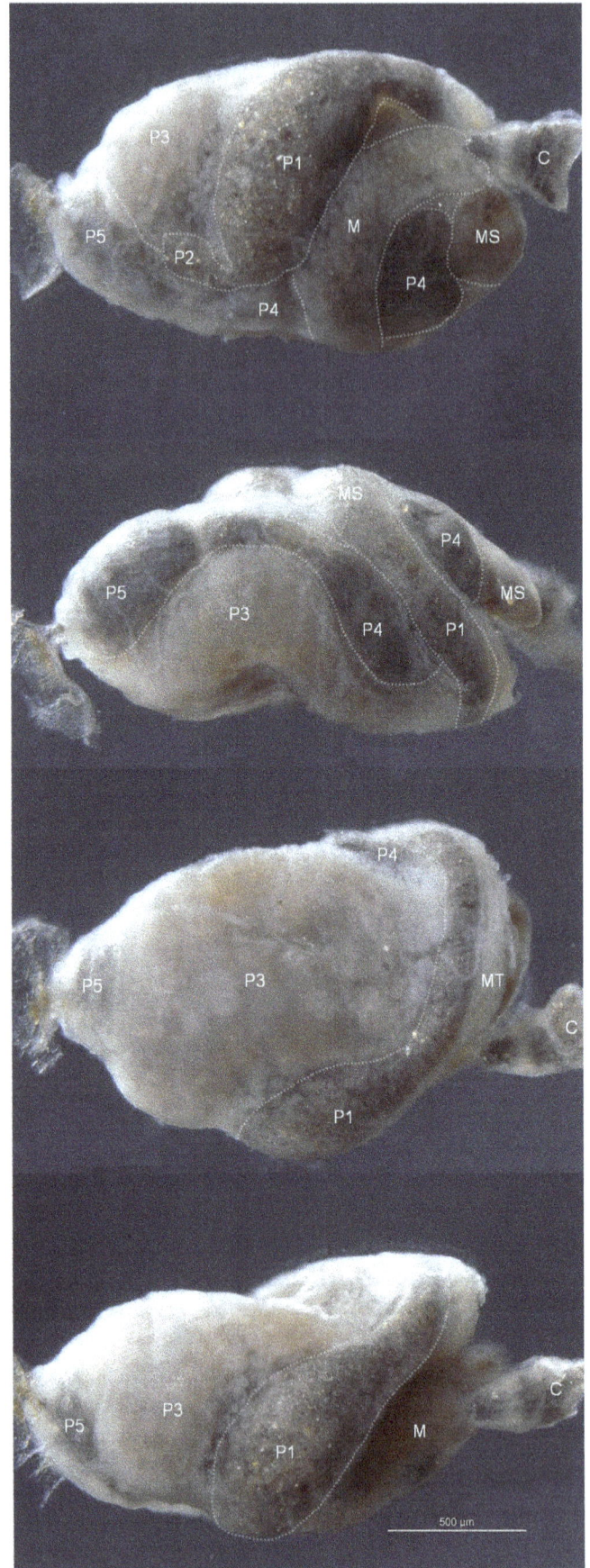

Fig 8. Worker abdomen of *Inquilinitermes johnchapmani* with integument removed. C = crop, M = mesenteron, MS = mixed segment, MT = malphigian tubules, P1-P5 = protodeal segments 1-5.

References

Apolinário, F.E., & Martius, C. (2004). Ecological role of termites (Insecta, Isoptera) in tree trunks in central Amazonian rain forests. Forest Ecology and Management, 194: 23-28.

Constantino, R. (2002). An illustrated key to Neotropical termite genera (Insecta: Isoptera) based primarily on soldiers. Zootaxa, 67: 1-40.

Constantino, R., & Acioli, A.N.S. (2006). Termite diversity in Brazil (Insecta: Isoptera). Pp 117-128 In: Moreira, F. M., Siqueira, J. O., & Brussaard, L. (Eds.). Soil Biodiversity in Amazonian and Other Brazilian Ecosystems. Wallingforg, CAB International.

Cunha, H.F., Costa, D.A., Filho, K.E., Silva, L.O., & Brandão, D. (2003). Relationship between *Constrictotermes cyphergaster* and inquiline termites in the Cerrado (Isoptera: Termitidae). Sociobiology, 42: 761-770.

Emerson, A.E. (1925). The termites from Kartabo, Bartica District, Guyana. Zoologica, 6: 291-459.

Hamilton, S.K., Sippel, S.J., & Melack, J.M. (2004). Seasonal inundation patterns in two large savanna floodplains of South America: the Llanos de Moxos (Bolivia) and the Llanos del Orinoco (Venezuela and Colombia). Hydrological Processes, 18: 2103-2116.

Mathews, A.G.A. (1977). Studies on Termites from the Mato Grosso State, Brazil. Rio de Janeiro: Academia Brasileira de Ciências, 267 p.

Miller, L.R. (1991). A revision of the *Termes-Capritermes* branch of the Termitinae in Australia (Isoptera: Termitidae). Invertebrate Systematics, 4: 1147-1282.

Noirot, C. (2001). The gut of termites (Isoptera): comparative anatomy, systematics, phylogeny. II. Higher Termites (Termitidae). Annales de la Société Entomologique de France, 37: 431-471.

Sands, W. A. (1972). The soldierless termites of Africa (Isoptera: Termitidae) British Museum (Natural History) Entomology supplement 18. 244pp.

Sands, W.A. (1998). The identification of worker castes of termite genera from soils of Africa and the Middle East. Cab International.

Silvestri, F. (1901). Nota preliminare sui termitidi sud-americani. Bollettino dei Musei di Zoologia e Anatomia Comparata della Università di Torino XVI(389):1-8.

Snyder, T. (1949). Catalog of the termites (Isoptera) of the world. Smithsonian Misc. Coll., ll2: 490pp.

15

Post-embryonic Development of Intramandibular Glands in *Pachycondyla verenae* (Forel) (Hymenoptera: Formicidae) workers

LCB Martins[1,] JHC Delabie[2,3,] JC Zanuncio[1,] JE Serrão[1]

1 - Universidade Federal de Viçosa, Viçosa, MG , Brazil

2 - CEPLAC/CEPEC, Itabuna, BA, Brazil

3 - Universidade Estadual de Santa Cruz. Ilhéus, BA, Brazil

Keywords

histology; histochemistry; Ponerinae; secretory cell; exocrine gland.

Corresponding author

José Eduardo Serrão
Departamento de Biologia Geral
Universidade Federal de Viçosa
36570-000 - Viçosa – MG, Brazil
E-Mail: jeserrao@ufv.br

Abstract

The current knowledge of intramandibular glands in Hymenoptera is focused on occurrence and morphology in adult insects. This is the first report regarding the post-embryonic development of intramandibular glands in a "primitive" ant, *Pachycondyla verenae*. In this study, we analyzed mandibles of prepupae, white-eyed, pink-eyed and black-eyed pupae, pupa of pigmented body pupae, and adults. Adult workers of *P. verenae* have intramandibular glands with epidermal secretory cells of class I and isolated glands of class III, and both glands have onset differentiation in pink-eyed pupae. Some histological sections were submitted to histochemical test for total proteins and neutral polysaccharides. Histochemical tests showed occurrence of polysaccharides and proteins in epidermal secretory cells of class I from the white-eyed pupae, polysaccharides and proteins in pink-eyed pupae to black-eyed pupae in both glands classes I and III and presence of polysaccharides in adult ants also in both gland classes I and III. Intramandibular glands of classes I and III in *P. verenae* workers differentiate during pupation, with onset occurring in pink-eyed pupae, and completion occurring in black-eyed pupae.

Introduction

Hymenoptera have two mandibular gland types, the ectomandibular or mandibular glands, and the mesomandibular or intramandibular glands (Cruz-Landim & Abdalla, 2002). Unlike the mandibular glands that are the most studied, the current knowledge regarding intramandibular glands is restricted to occurrence and descriptive morphology in adult bees (Nedel, 1960; Toledo 1967; Costa-Leonardo, 1978; Cruz-Landim & Abdalla, 2002; Santos et al. 2009a, Cruz-Landim et al., 2011) and ants (Schoeters & Billen, 1994, Billen & Espadaler, 2002, Grasso et al., 2004, Roux et al., 2010, Martins & Serrão, 2011).

The intramandibular glands in ants can be divided in three classes (Martins & Serrão, 2011). First, gland characterized by cubical or columnar epidermal cells, which are described as class I gland cells by Noitot & Quennedey (1974). Second, isolated glands with spherical cells in the inner cavity of the mandible characterized by the presence of canaliculi that open in pores on the mandible surface, this

gland class has a secretory and a conducting cell (Billen 2009) and is classified as class III gland cells by Noitot & Quennedey (1974). The third type occurs in Attini, with hypertrophy of epidermal cells containing a reservoir (Amaral & Caetano, 2006; Billen, 2009; Martins & Serrão, 2011).

In ants, studies of salivary system are restricted to morphology of post-pharyngeal and hypopharyngeal (Gama, 1978; Amaral & Caetano, 2005), mandibular (Gama, 1978; Pavon & Camargo-Mathias, 2005;), labial or salivary glands (Gama, 1978; Meyer et al., 1993; Lommelen, et al., 2002; 2003; Amaral & Machado-Santelli, 2008) and intramandibular glands (Martins & Serrão, 2011) but the intramandibular glands development remains unclear.

Ponerini ants are an interesting model to study gland development since their representatives have features similar to ants ancestor (Kusnezov, 1955; Peeters & Crewe, 1984, Hölldobler & Wilson, 1990).

In this work we describe the histology and histochemistry of intramandibular glands from prepupae to adult workers of the Neotropical ant *Pachycondyla verenae*.

Material and Methods

Five colonies of *P. verenae* were collected in the experimental field of the Department of Fruit Science, Federal University of Viçosa, Viçosa, state of Minas Gerais, Brazil. Worker ants at the following developmental stages were evaluated: prepupae, white-eyed, pink-eyed and black-eyed pupae, body pigmented pupae, and adults (Fig. 1A, 1B). Features that allowed for the identification of the successive post-embryonic developmental stages were the focus of these evaluations (Soares et al., 2004).

Five individuals in each developmental stage were -anesthetized by thermal shock at 0 ºC for 3 min and the mandibles were removed and transferred to Zamboni fixative solution (Stefanini et al., 1967). Mandibles were dehydrated in a graded ethanol series, embedded in JB-4 historesin, and 3 µm slices were stained with hematoxylin and eosin.

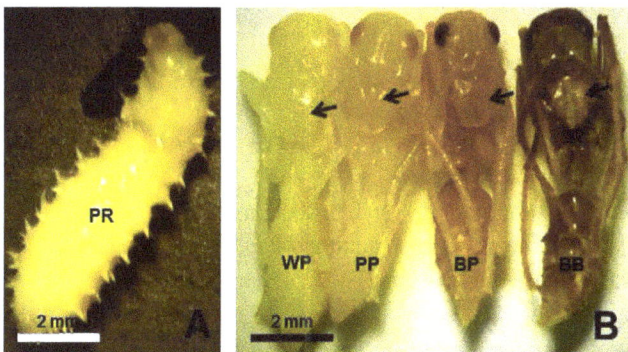

Fig. 1: Different post-embryonic developmental stages in *Pachycondyla verenae* workers. A - prepupae (PR). B - white-eyed (WP), pink eyed (PP) and black-eyed pupae (BP) and body pigmented pupae (BB). Arrows: mandibles.

Some mandible sections were subjected to histochemical analysis of mercury-bromophenol for protein detection and Periodic acid-Schiff (PAS) for neutral polysaccharide detection according to protocols from Pearse (1985) and Bancroft & Gamble (2007). Negative control for PAS test was performed by omitting periodic acid oxidation.

Results

The morphology of the intramandibular glands was characterized in adult workers, followed by monitoring of their development from the prepupae stage.

Adult

Mandibles of *P. verenae* adult workers had two classes of intramandibular glands, class I epidermal secretory cells and class III isolated glands. The class I secretory cells were characterized as columnar epidermal cells with well developed spherical nucleus and a homogeneously acidophilic cytoplasm (Fig. 2A). Class III isolated glands were scat-

tered within the mandible cavity and showed well developed spherical nuclei and strongly basophilic cytoplasm with many vacuoles (Fig. 2B).

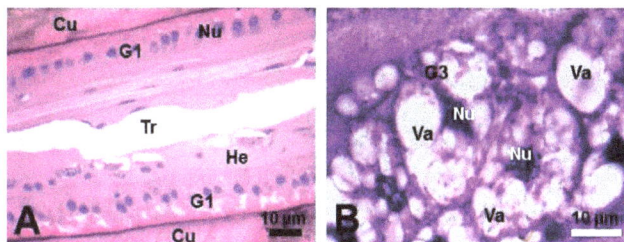

Fig. 2. Histological sections of the mandibles in adult workers of *Pachycondyla verenae*. A - Class I epidermal secretory cells (G1) characterized by cubic cells. Trachea (Tr). B - Class III unicellular glands of (G3) with vacuoles (Va) in the cytoplasm containing and spherical nucleus (Nu) with predominance of uncondensed chromatin. Cu: cuticle. He: hemocoel. Hematoxyline and eosin stained.

Histochemical analyses of both intramandibular gland types were positive for the presence of neutral polysaccharides and proteins.

Prepupae

In the mandibles of *P. verenae* prepupae, free cells were not found within the intramandibular cavity, which was filled with flocculent material. The epidermis was formed by flattened cells (Fig. 3), without features of glandular secretory cells.

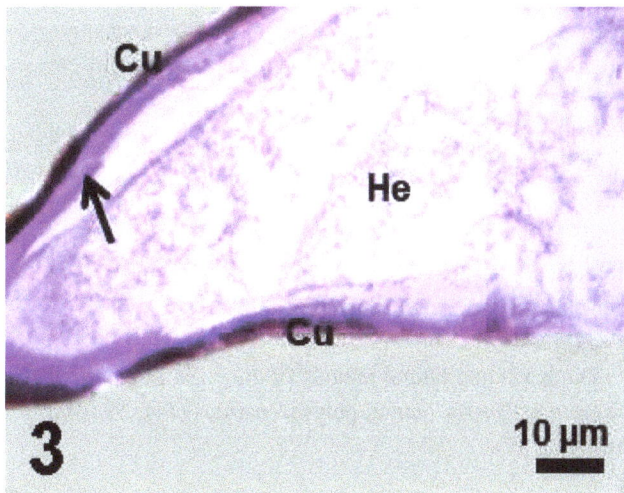

Fig. 3. Histological section of mandible of *Pachycondyla verenae* prepupae. A - flattened epidermis (arrow) and mandibular cavity, hemocoel (He) filled with flocculent material. Cu: cuticle. Hematoxyline and eosin stained.

White-eyed pupae

In white-eyed pupae of *P. verenae*, the mandible cavity was characterized by intense cell reorganization, with cells displaying irregular distribution and filamentous projections.

Table I - Histochemical tests in the intramandibular glands types I (G1) and III (G3) of *Pachycondyla verenae* (Formicidae: Ponerini) during different developmental stage of prepupae (PP), white-eyed pupae (WEP), pink-eyed pupae (PEP), black-eyed pupae (BEP), body pigmented pupae (BPP) and adult.

Histochemical test	Developmental stage											
	PP		WEP		PEP		BEP		BPP		Adult	
	G1	G3	G1	G3	G1	G3	G1	G3	G1	G3	G1	G3
Protein	ga	ga	ga	ga	ga	ga	+	+	+	+	+	+
Neutral polysaccharide	ga	ga	ga	ga	+	+	+	+	+	+	+	+

ga - gland absent, + positive reaction

Cell identification was based on the presence of small nucleus (Fig. 4A) as well as aggregation of protein rich rectangular cells (Fig. 4B). The single layered epidermal cells were cubic (Fig. 4A, 4C), although in some mandible regions they remained flattened. In addition to cells, the mandible cavity displayed an accumulation of flocculent material positive for proteins (Fig. 4C) and neutral polysaccharides (Fig. 4D).

Pink-eyed pupae

In pink-eyed pupae of *P. verenae*, the mandible cuticle was thicker than in previous pupae and mandibular teeth were present (Fig. 5A inset). The mandibular epidermal cells were similar to class I secretory cells found in adult workers, but with variations in size and shape. Specifically, some cells were spherical while others were columnar (Fig. 5A). The epidermis within the mandible was composed of columnar epithelium on one side and a pseudostratified-like epithelium on the opposed side (Fig. 5A). In addition, we found sensitive cell precursors (neuroblasts) of mandibular sensilla closely associated with the epidermis (Fig. 5B).

In the mandible cavity, precursors of class III gland cells were dispersed and had clear cytoplasm, making it difficult to identify the cell boundaries (Fig. 5C, 5D). The nuclei of these cells had uncondensed chromatin (Fig. 5B, 5C). In addition to cells dispersed in the mandible cavity, there were some chromatic bodies, which were small and strongly basophilic (Fig. 5A, 5B).

The basal and lateral regions of the class I epidermal cells were positive for neutral polysaccharides (Fig. 5E) and proteins (Fig. 5F).

Black-eyed pupae

In black-eyed pupae of *P. verenae*, the mandible was almost entirely lined by cubic class I epidermal cells (Fig. 6A). The class III gland cells had well developed nuclei and cytoplasm that was homogeneous, without vacuoles (Fig. 6B). Chromatic bodies were rare within the mandible cavity.

Histochemical analyses were positive for neutral polysaccharides (Fig. 6C) and proteins (Fig. 6D) in both class I and III gland cells.

Body pigmented pupae

Pupae of *P. verenae* with a pigmented body had a mandible epidermis with class I cubic gland cells and class III gland cells with vacuolated cytoplasm (Fig. 7).

The histochemical tests results in different intramandibular glands of different developmental stages of *P. verenae* are summarized in Table I.

Discussion

The morphology and histochemistry of intramandibular glands in *P. verenae* adult workers described here are similar to those of other Ponerini ants reported by Martins & Serrão (2011).

The occurrence of intramandibular glands in *P. verenae* with class I epidermal secretory cells and class III isolated glands was similar to that of adult ants (Schoeters & Billen, 1994, Billen & Espadaler, 2002; Grasso et al., 2004; Roux et al., 2010; Martins & Serrão, 2011).

Both classes I and III of intramandibular glands in *P. verenae* workers began to differentiate during the pink-eyed pupae stage. In bees, intramandibular glands also differentiate in the pupal stage (Cruz-Landim & Abdalla 2002). In addition, mandibular glands appear during the pupal development of the ant *P. obscuricornis* (Lommelen et al., 2003). Intramandibular gland differentiation in *P. verenae* was not observed in prepupa. During that larval stage, ants, likely other holometabolous insects, accumulates food reserves and increases in size (Wheeler & Martinez, 1995), followed by metamorphosis, which consists of the complete structural reorganization of larval organs to adult organs (Wheeler & Buck, 1992; Rosell & Wheeler, 1995; Roma et al., 2006; 2009).

The white-eyed pupae of *P. veranae* also had no intramandibular glands, however, cell reorganization at this stage did occur and was likely due to proliferation and differentiation of epidermal cells. The white-eyed pupae of Hymenoptera is the earliest period of internal organs modifications, because it has yet larval organs and absence of those present in adults (Soares et al., 2004, Azevedo et al., 2008, Santos et al., 2009b; Cruz-Landim et al., 2011).

Intramandibular glands characterized as class I epidermal secretory cells or class III isolated glands can be definitively identified during the pink-eyed pupae stage of *P. verenae* in a manner similar to likely the labial, post-pharyngeal, hypopharyngeal, and mandibular glands that also differentiate in *Camponotus rufipes* pupae (Gama 1978).

In body pigmented pupae of *P. verenae*, both classes I and III of intramandibular glands are almost completely differentiated, similar to that reported from the salivary system of *C. rufipes,* which reaches maximum differentiation during the body pigmented pupae stage, despite the lack of secretory activity within the glands (Gama, 1978). The same may be occurring in the intramandibular glands of *P. verenae*, as the amount of gland cells in body pigmented pupae seems low compared to the adult ants.

The cellular processes that result in tissue and organ changes during metamorphosis include addition of new cells by cell division, cell loss due to programmed cell death, and cell migration (Cruz-Landim, 2009). In the present study,

we are unable to detect some of these processes, but after the black-eyed pupae stage of *P. verenae*, intramandibular glands were similar to those found in adult workers, as is the case for other internal organs of Hymenoptera (Neves et al., 2002, 2003, Soares, 2004, Azevedo et al., 2008; Santos et al., 2009b).

The presence of proteins and neutral polysaccharides in intramandibular glands from white-eyed pupae of *P. verenae* suggests that during this developmental stage, there is a high level of cellular activity derived from the metabolic cell machinery. In the leaf-cutter ant, *Atta sexdens rubropilosa,* class III secretory cells contain carbohydrate and protein releasing glucoconjugate compounds (Amaral & Caetano, 2006).

In conclusion, class I epidermal secretory cells and class III isolated cells of the intramandibular glands of *P. verenae* workers differentiate during pupation, with onset occurring in pink-eyed pupae, and completion occurring in black-eyed pupae.

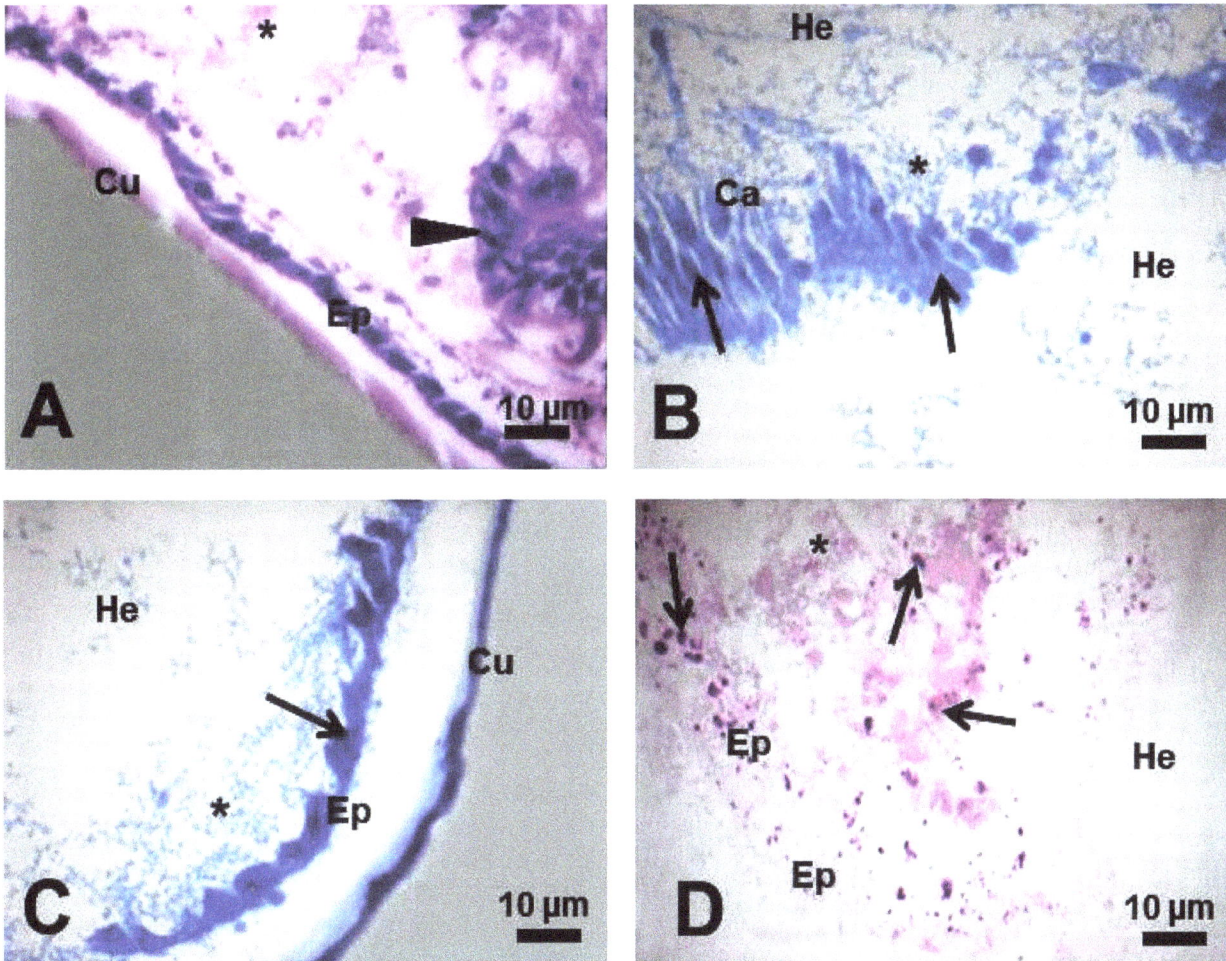

Fig. 4. Histological sections of mandible of *Pachycondyla verenae* white-eyed pupae. A- Epidermis (EP) with flattened cells and irregular cells in the mandible cavity with small nucleus (arrowheads) and flocculent material (*). Hematoxyline and eosin stained. B - Cell aggregate (Ca) and flocculent material (*) into mandible cavity positive (arrows) for protein. Mercury-bromophenol blue stained. C - epidermis (Ep) with cubic cells and flocculent material (*), both positive (arrows) for proteins. Mercury-bromophenol blue stained. D - flocculent material (*) and epidermis (Ep) positive for polysaccharides (arrows). Periodic acid- Schiff stained. He: hemocoel. Cu: cuticle.

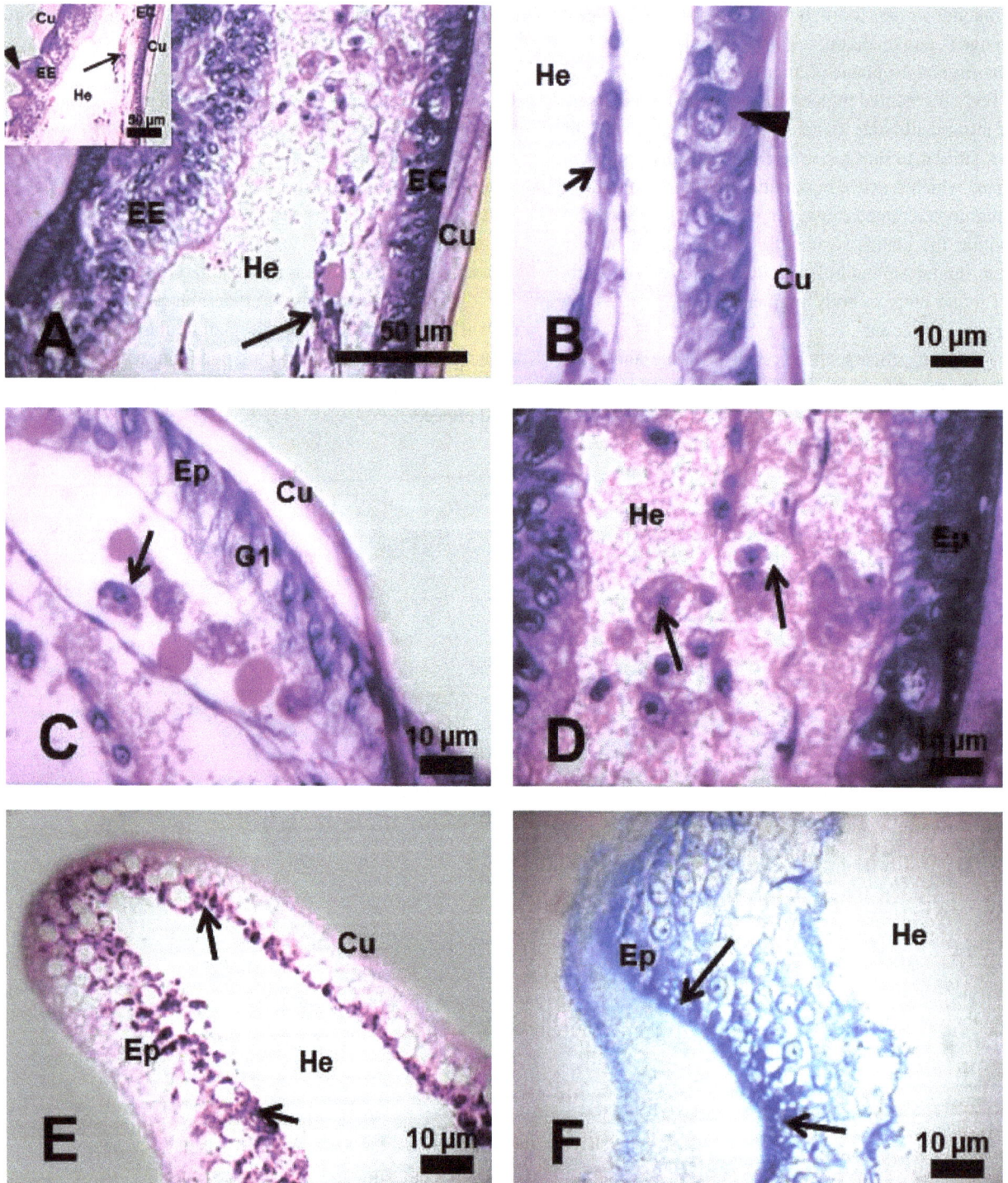

Fig. 5. Histological sections of the mandible of *Pachycondyla verenae* pink- eyed pupae. A - columnar (EC) and pseudostratified-like (EE) epidermis Note chromatic bodies (arrow) into mandible cavity Insert - general view of the mandible showing teeth (arrowhead). Hematoxyline and eosin stained. B - Neuroblast (arrowhead) in the epidermis and chromatic bodies (arrow) into mandible cavity (He). Hematoxyline and eosin stained. C - Class I epidermal gland showing columnar cells (GI). Hematoxyline and eosin stained. D - Class III precursor cells (arrows) with light cytoplasm into mandible cavity. Hematoxyline and eosin stained. E - positive reaction for polysaccharides (arrows) in the class I secretory cells. Periodic acid- Schiff stained. F - positive reaction for proteins (arrows) in class I secretory cells. Mercury-bromophenol blue stained. Cu: cuticle. Ep: epidermis. He: hemocoel.

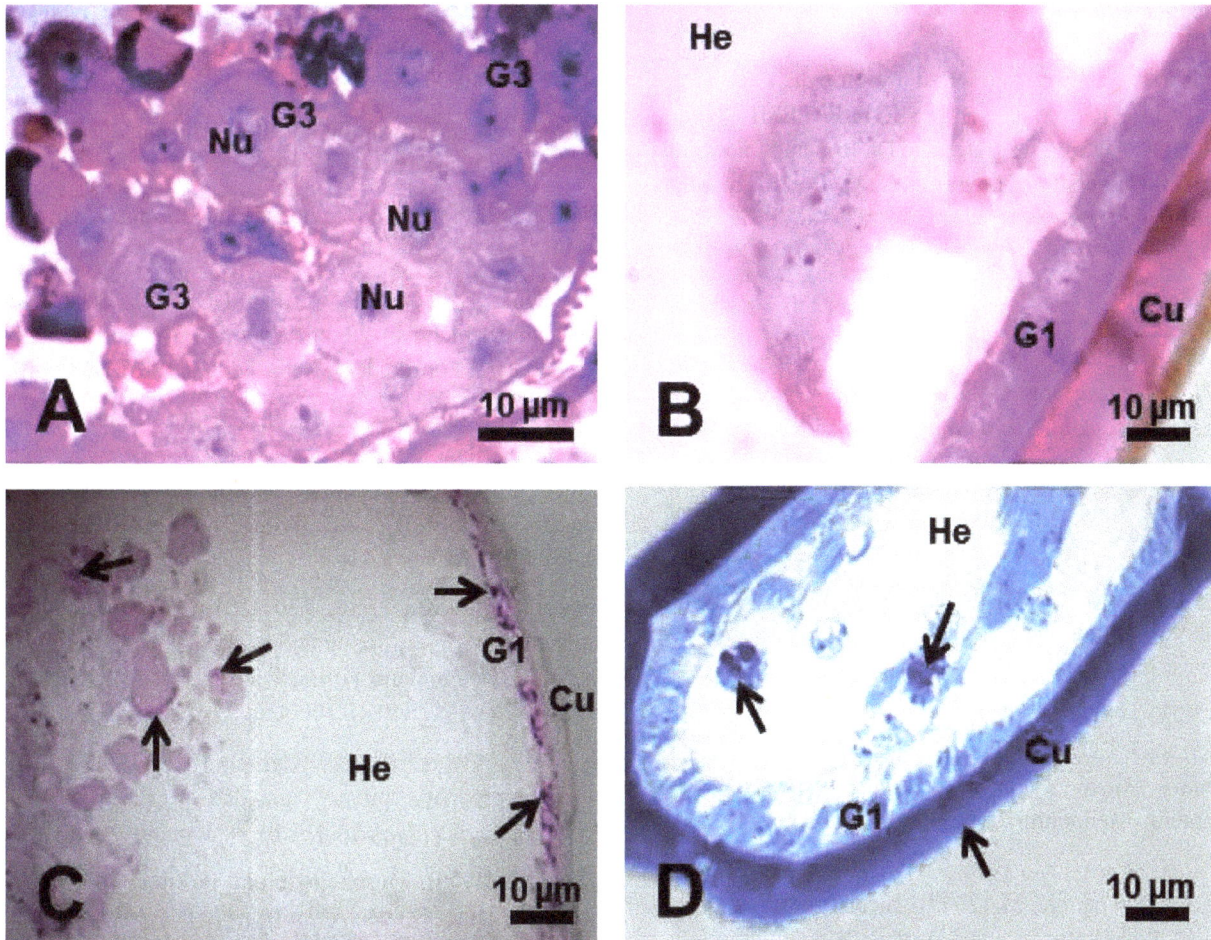

Fig. 6. Histological sections of the mandibles of *Pachycondyla verenae* black-eyed pupae. A - Class III unicellular glands of (GIII) showing well developed nucleus (Nu) and homogeneous acidophilic cytoplasm. Hematoxyline and eosin stained. B - Class I epidermal secretory cells (GI) showing nuclei with predominance of uncondensed chromatin. Hematoxyline and eosin stained. C - Positive reaction for polysaccharides (arrows) in the class I epidermal secretory cells (G1) and class III glands (G3). Periodic acid- Schiff stained.   D - Positive reaction for proteins in the class I epidermal secretory cells (G1) and class III glands (G3). Mercury-bromophenol blue stained. Ep: epithelium. He: hemocoel. Cu: cuticle.

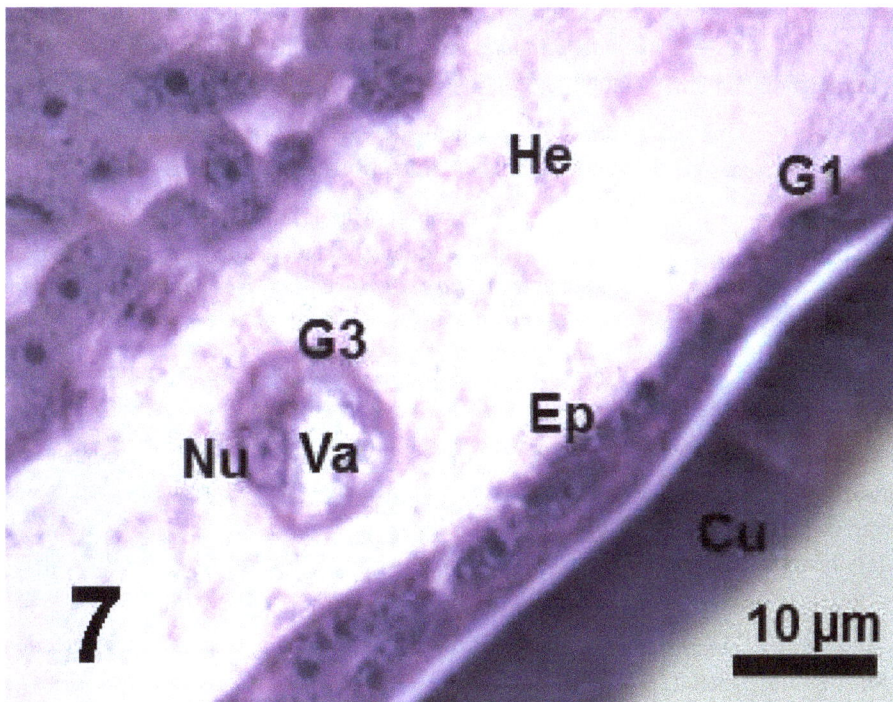

Fig. 7. Histological section of the mandible of body pigmented pupa of *Pachycondyla verenae*. A - well developed class I epidermal cells (G1) and Class III unicellular glands of (G3) with cytoplasm vacuoles (Va). Ep: epithelium. Hematoxyline and eosin stained. He: hemocoel. Cu: cuticle. Nu: nuclei.

Acknowledgements

This research was supported by Brazilian research agencies FAPEMIG, CNPq and SECTI/FAPESB-CNPq-PNX0011/2009. Authors are grateful to Dr. Ronara S. Ferreira of the Federal University of Lavras (UFLA), Brazil, for help in ant collection.

References

Amaral, J.B. & Caetano, F.H. (2005). The hypopharyngeal gland of leaf-cuttting ants (*Atta sexdens rubropilosa*) (Hymenoptera: Formicidae). Sociobiology, 46, 515-524.

Amaral, J.B. & Caetano, F.H. (2006). The intramandibular gland of leaf-cutting ants (*Atta sexdens rubropilosa* Forel 1908). Micron, 37, 154-160.

Amaral, J.B. & Machado-Santelli, G.M. (2008). Salivary system in leaf-cutting ants (*Atta sexdens rubropilosa* Forel, 1908) castes: A confocal study. Micron, 39, 1222-1227.

Azevedo, D.O., Gus-Matiello, C.P., Ronnau, M., Zanuncio, J.C. & Serrao, J. E. (2008). Post-embryonic development of the antennal sensilla in *Melipona quadrifasciata anthidioides* (Hymenoptera: Meliponini). Microsc. Res. Tech., 71, 196-200.

Bancroft, J.D. & Gamble, M. (2007). Theory and Practice of Histological Techniques (6 edition). Philadelphia: Churchill Livingstone Elsevier.

Billen J. (2009). Occurrence and structural organization of the exocrine glands in the legs of ants. Arthropod Struct. Dev., 38, 2-15. doi: 10.1016/j.asd.2008.08.002

Billen, J. & Espadaler, X. (2002). A novel epithelial intramandibular gland in the ant *Pyramica membranifera* (Hymenoptera, Formicidae). Belgian J. Zool., 132, 175-176.

Costa-Leonardo, A.M. (1978). Glândulas intramandibular em abelhas sociais. Cienc. Cult., 30, 835-838.

Cruz-Landim, C. (2009). Abelhas: morfologia e funções de sistemas. São Paulo: UNESP.

Cruz-Landim, C. & Abdalla, F.C. (2002). Glândulas exócrinas das abelhas. Ribeirão Preto: FUNPEC-RP.

Cruz-Landim, C., Gracioli-Vitti, L.F. & Abdalla, F.C. (2011). Ultrastructure of the intramandibular gland of workers and queens of the stingless bee, *Melipona quadrifasciata* (Meliponini). J. Insect Sci., 11, 1-9.

Gama, V. (1978). Desenvolvimento pós-embrionário das glândulas componentes do sistema salivar de *Camponotus* (*Myrmothrix*) *rufipes* (Fabricius, 1775) (Hymenoptera: Formicidae). Arq. Zool., 29, 133-183.

Grasso, D.A., Romani, R., Castracani, C., Visicchio, R., Mori, A., Isidoro, N. & Le Moli, F. (2004). Mandible associated glands in queens of the slave-making ant *Polyergus rufescens* (Hymenoptera, Formicidae). Insectes Soc., 51, 74-80. doi: 10.1007/s00040-003-0700-6

Hölldobler, B. & Wilson, E.O. (1990). The ants. Cambridge: Harvard University Press.

Kusnezov, N. (1955). Evolución de las hormigas. Dusenia, 6, 1-34.

Lommelen E., Schoeters, E. & Billen, J. (2002). Ultrastructure of the labieal gland in the ant *Pachycondyla obscuricornis* (Hymenoptera, Formicidae). Neth. J. Zool., 52, 61-68.

Lommelen E., Schoeters, E. & Billen, J. (2003). Development of the labial gland of the ponerine ant *Pachycondyla obscuricornis* (Hymenoptera, Formicidae) during the pupal stage. Arthropod Struct. Dev., 32, 209-217. doi: 10.1016/S1467-8039(803)00052-5

Martins, L.C.B. & Serrão, J.E. (2011). Morphology and histochemistry of the intramandibular glands in Attini and Ponerini (Hymenoptera, Formicidae) species. Microsc. Res. Tech., 74, 763-771. doi: 10.100/jemt.20956

Meyer, W., Beyer, C. & Wissdorf, H. (1993). Lectin histochemistry of salivary glands in the ant eater (*Myrmecophaga tridactyla*). Histol. Histopathol., 8, 305-316.

Nedel, O.J. (1960). Morphologie und Physiologie der Mandibeldruse einiger Bienen Arten (Apidae). Z. Morphol. Ökol Tiere, 49, 139-83.

Neves, C.A., Bhering, L.L., Serrão, J.E. & Gitirana, L.B. (2002). FMRFamide-like midgut endocrine cells during the metamorphosis in *Melipona quadrifasciata anthidioides* (Hymenoptera, Apidae). Micron, 33, 453-460.

Neves, C.A., Gitirana, L.B. & Serrão, J.E. (2003). Ultrastructural study of the metamorphosis in the midgut of *Melipona quadrifasciata anthidioides* (Apidae, Meliponini) worker. Sociobiology, 41, 443-459.

Noirot, C. & Quennedey, A. (1974). Fine structure of insect epidermal glands. Annu. Rev. Entomol., 19, 61-80.

Pavon, L.F. & Camargo-Mathias, M.I. (2005). Ultrastructural studies of the mandibular glands o fhte minima, media and soldier ants of *Atta sexdens rubropilosa* (Forel 1908). Micron, 36, 449-460.

Pearse, A.G.E. (1985). Histochemistry: Theoretical and Applied. Churchill Livingstone, London.

Peeters, C. & Crewe, R.M. (1984). Insemination controls the reproductive division of labour in a Ponerine ant. Naturwissenschaften, 71, 50-51.

Roma, G.C., Camargo-Mathias, M.I. & Bueno, O.C. (2006). Fat body in some genera of leaf-cutting ants (Hymenoptera: Formicidae). Proteins, lipids and polysaccharides detection. Micron, 37, 234-242.

Roma, G.C., Bueno, O.C. & Camargo-Mathias, M.I. (2009). Ultrastructural analysis of the fat body in workers of Attini ants (Hymenoptera: Formicidae). Anim. Biol., 59, 262. doi: 10.1163/157075609X437745

Rosell, R.C. & Wheeler, D.E. (1995). Storage function and ultrastructure of the adult fat body in workers of the ant *Camponotus pestinatus* (Buckley) (Hymenoptera, Formicidae). Int. J. Insect Morphol. Embryol., 24, 413-426.

Roux, O., Billen, J., Orivel, J. & Dejean, A. (2010). An overlooked mandibular- rubbing behavior used during recruitment by the African weaver ant, *Oecophylla longinoda*. PLoS ONE, 5, e8957. doi: 10.1371/journal. pone.0008957.

Santos, C.G., Megiolaro, F., Serrão, J.E. & Blochtein, B. (2009a). Morphology of the head salivary and intramandibular glands of the stingless bee *Plebeia emerina* (Friese) (Hymenoptera, Meliponini) workers associated with propolis. Ann. Entomol. Soc. Am., 102, 137-143.

Santos, C.G., Neves, C.A., Zanuncio, J.C. & Serrão, J.E.

(2009b). Postembryonic development of rectal pads in bees (Hymenoptera, Apidae). Anat. Rec., 292, 1602-1611. doi: 10.1002/ar.20949

Soares, P.A. O., Delabie, J.H.C. & Serrão, J.E. (2004). Neuropile organization in the brain of *Acromyrmex* (Hymenoptera, Formicidae) during the post-embryonic development. Braz. Arch. Biol. Techn., 47, 635-641. doi: 10.1590/S1516-89132004000400017

Schoeters, E. & Billen, J. (1994). The intramandibular gland, a novel exocrine structure in ants (Insecta, Hymenoptera). Zoomorphology, 114, 125-131.

Stefanini, M., Demartino, C. & Zamboni, L. (1967). Fixation of ejaculated spermatozoa for electron microscopy. Nature, 216, 173-174.

Toledo, L.F.A. (1967). Histo-anatomia de glândulas de *Atta sexdens rubropilosa* Forel (Hymenoptera). Arq. Inst. Biol., 34, 321-329.

Wheeler, d.a. & Buck, N.A. (1992). Protein, lipid and carbohydrate use during metamorphosis in the fire ant, *Solenopsis xyloni*. Physiol. Entomol., 17, 397-403.

Wheeler, D.E. & Martinez, T. (1995). Storage proteins in ants (Hymenoptera: Formicidae). Comp. Biochem. Phsyiol. B, 112, 15-19.

16

Reproductive Status of the social wasp *Polistes versicolor* (Hymenoptera, Vespidae)

VO Torres[1], D Sguarizi-Antonio[2], SM Lima[1], LHC Andrade[1] & WF Antonialli-Junior[1]

1 - *Universidade Federal da Grande Dourados, Dourados-MS, Brazil.*

2 - *Universidade Estadual de Mato Grosso do Sul, Dourados-MS, Brazil.*

Keywords

Cuticular hydrocarbons, Polistinae Ovarian development

Corresponding author

Viviana de Oliveira Torres
Progr. de Pós-graduação em Entomologia e Conservação da Biodiversidade
Univ. Federal da Grande Dourados
Dourados, Mato Grosso do Sul, Brazil
79804-970
E-mail: vivianabio@yahoo.com.br

Abstract

A fundamental feature in the evolution of social insects is the separation of castes, and the presence of wide differentiation between castes indicates a more advanced degree of sociability. In this study, we evaluated factors indicating the reproductive status of females in colonies of the social wasp *Polistes versicolor*. The reproductive status of each female was examined by measuring nine morphometric characters, by tracing the cuticular chemical profile, by evidence of insemination and by recording the relative age. We conclude that *P. versicolor* colonies present 3 female groups according to cuticular chemical profile difference. The first group is made of females with filamentous ovarioles, typical of workers; the second one is females with intermediate ovarioles; and the third group is the group of the queens, which are older females, already inseminated and with the greatest degree of ovarian development. No significant external morphological differences were found among these female groups. Therefore, despite the lack of significant morphological differences among females, there are other factors such as the chemical composition of the cuticula, which are indicative of the reproductive physiological condition of the female in the colony.

Introduction

An important feature for the ecological success of social insects is the division of labor among individuals in their colonies (Wilson, 1985). For this reason, many investigators have devoted their efforts to elucidate the parameters that determine this division, especially the distinction and determination of the caste (Robinson, 1992; O'Donnell, 1995; O'Donnell, 1998).

The subfamily Polistinae has characteristics that are important to understand how the social behavior has evolved in the wasps (Ross & Matthews, 1991). The degree of morphological differences among castes in this group can range from total absence (Richards, 1971; Strassmann et al., 2002) to sharp differences among castes (Jeanne, 1991). This, indeed, may be a key feature in the evolution of social insects, since the presence of wide differentiation among castes indicates a higher degree of sociality (Bourke, 1999).

In the basal Polistinae such as *Mischocyttarus* and *Polistes*, females are distinguished by their behavior, dominance hierarchy, degree of ovarian development and/or their reproductive physiology (Röseler et al., 1985). The dominance status of the individual apparently initiates a physiological response that directly affects their ovarian development (Wheeler, 1986). The queen shows the highest degree of ovarian development and, by using behavioral strategies to dominate all of the other females, she largely monopolizes reproduction while avoiding energy-consuming tasks such as foraging (Jeanne, 1972; Strassmann & Meyer, 1983).

Evidently, on the lack of visible external traits, some other kind of detection mechanism, used by each member of the colony, is needed for the establishment and recognition of this hierarchy, and chemical communication is the most effective way to accomplish this recognition. Among the compounds involved in this process are the cuticular hydrocarbons (CHCs), which are a constituent of the lipid layer that coats the cuticle of insects, and have the primary functions of preventing desiccation (Lockey, 1988) and creating a barrier against microorganisms (Provost et al., 2008). CHCs also act as contact pheromones, allowing conspecific individuals to identify each other, thus assisting in maintaining the colony structure, separating individuals according to their function in the colony, their physiological status and their hierarchical rank (Provost et al., 2008), functioning, therefore, as a specific chemical signature of the individuals.

Sledge et al. (2001) and Monnin (2006) noted that there is a strong correlation between reproductive status and CHC profile of each individual in a colony of wasps. According to Dapporto et al. (2005), in colonies of *Polistes dominula* (Christ) founded by a single female, the CHC profile of the queen and its brood are different and, when one of those colonies becomes orphaned, a worker assumes the queen position, its ovaries develop, and acquires a CHC profile similar to that of the original queen.

Maintenance of a reproductive monopoly by a queen is one of the goals reached by many social insects. In independent-founding species, it was believed that the queen maintains her reproductive status by using aggression toward other females; however, in recent decades many studies have demonstrated the importance and role of CHCs in the communication among members of the colonies and in maintaining the status of the queen (Bonavita-Cougourdan et al., 1991; Peeters et al., 1999; Liebig et al., 2000; Sledge et al., 2001; Dapporto et al., 2005).

This study is focused on analyzing the reproductive status of females of *Polistes versicolor* (Olivier) through examining morphological and reproductive physiological features and by tracing the chemical profiles of the cuticula.

Material and Methods

We collected 10 colonies of *P. versicolor* in the southern region of the state of Mato Grosso do Sul, in the cities of Dourados (22°13'16" S 54°48'20" W) and Mundo Novo (23°56'23" S 54°17'25"W). All of the females from each colony were evaluated for morphological, physiological and cuticular chemical profile analyses. The classification of the colonial stage are done according to the system proposed by Jeanne (1972).

After collection, the gaster of each female was individually fixed in an Eppendorf containing absolute ethyl alcohol (99.8% PA) for later analysis of ovarian development, insemination and relative age. The remainder of the body was preserved by freezing, for subsequent morphometric measurements and analysis of the cuticular chemical profile.

We performed nine morphometric measurements, modified from Shima et al. (1994) and Noll et al. (1997), in order to detect morphological diferences: Head: width (HW), minimum interorbital distances (IDx); Mesosoma: width, length and height of mesoscutum (MSW, MSL and MSH, respectively); Metasoma: basal and apical heights of tergite 2 (T2BH and T2AH), length of tergite 2 (T2L); Wing: partial length of the forewing (WL).

The gaster was dissected under a Zeiss binocular stereomicroscope for evaluation of the degree of ovarian development, insemination and relative age. The ovaries were classified according to the stage of development of the ovarioles, based on the observations of Baio et al. (2004).

For each female, the spermatheca was removed and put on a slide in a 1:1 solution of acid fucsina (1%) in order to determine the presence of sperm cells under a light microscope.

The relative age of all adult females was determined, according to the pigmentation of the transverse apodeme across the hidden base of the fifth sternum, as follows: LY (light yellow), LB (light brown), DB (dark brown) and BA (black). According to Richards (1971) and West-Eberhard (1973), this color sequence indicates a progression in the age of individuals, from younger (LY) to older females (BA).

For analysis of the cuticular chemical profile, the thorax of each female was submitted to optical spectroscopy by Fourier Transform Infrared Photoacoustic Spectroscopy (FTIR-PAS), after 48 hours in a vacuum oven, in order to minimize the water content. This technique was used by Antonialli-Junior et al. (2007 and 2008) and Neves et al. (2012) and has proved reliable for assessing the CHCs profiles of ants and wasps, even when compared to gas chromatography (Ferreira et al., 2012).

The FTIR-PAS technique measures the radiation absorbed by the sample. It is advantageous for application on very fragile objects, such as biological materials, because the low-intensity radiation does not destroy the sample. FTIR-PAS uses the infrared spectrum from 400 to 4000 cm^{-1} (Silverstein et al., 2000; Skoog et al., 2002), which is sensitive to the vibrations and rotations of molecular chemical groups, so it can identify and distinguish molecular radicals and some kinds of chemical bonds in the samples (Smith, 1999).

The resulting spectrum for each thorax was obtained from the mean of 64 spectra with a resolution of 8 cm^{-1}, which were separated in absorption lines between 400 and 4000 cm^{-1}, mostly those related to vibrations of CHCs.

The degree of ovarian development, morphometric data and the cuticular chemical profile were evaluated by stepwise discriminant analysis, which reveals the group of variables that better explain the groups evaluated in case of a difference. This is indicated by Wilk's Lambda, a measure of the difference, if any, among the groups (Quinn & Keough, 2002). The chi-square test was performed to test the association between the relative age and the three groups of females (workers 1, workers 2 and queens). For all analyses, the variable was considered significant when the level reached was <0.05.

Results and discussion

Four kinds of ovarian development (Fig. 1) were found in females of this species: type A, filamentous ovarioles without visibly developed oocytes; type B, ovarioles containing some oocytes in the initial stage of development; type C, ovarioles with moderately developed oocytes, some in the final phase of vitellogenesis; and type D, well-developed, longer ovario-

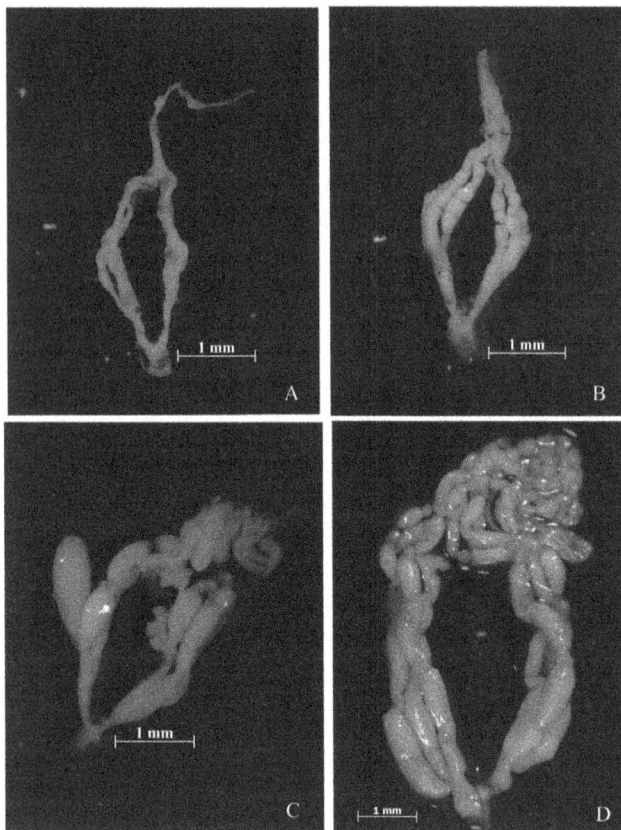

Fig 1. Different degrees of ovarian development found in females of *Polistes versicolor*. A) type A, filamentous ovarioles without visible developed oocytes; B) type B, ovarioles containing some oocytes in the initial stage of development; C) type C, ovarioles with moderately developed oocytes, some in the final phase of vitellogenesis; D) type D, well-developed, longer ovarioles, each containing two to several producing oocytes.

les, each containing two to several producing oocytes. Type D females were always inseminated.

These four physiological conditions were described by Baio et al. (2004) for *Brachygastra augusti* (Saussure). Noll et al. (2004) also described these same conditions in the species *Apoica pallens* (Fabricius), *Charterginus fulvus* (Fox) and *Nectarinella championi* (Dover). In all of these species, the females with the most advanced ovarian deve-

lopment were also the inseminated ones. However, Giannotti and Machado (1999), Gobbi et al. (2006) and Murakami et al. (2009) analyzed several independent founding species, *Polistes lanio* (Fabricius), *P. versicolor* and *Mischocyttarus cassununga* (von Ihering) among them, and found six, five and five ovarian development patterns, respectively.

Other studies evaluating the relationship between the degree of ovarian development and the reproductive position occupied by the females were performed on *Parachartergus smithii* (Saussure) (Mateus et al., 1997), *Pseudopolybia vespiceps* (Ducke) (Shima et al., 1998), *Chartergellus communis* (Richards) (Mateus et al., 1999), *Brachygastra lecheguana* (Latreille) (Shima et al., 2000), *Parachartergus fraternus* (Gribodo) (Mateus et al., 2004) and *Protopolybia chartegoides* (Gribodo) (Felippotti et al., 2007), all of them species of Epiponini.

The colonies of *P. versicolor* contained on average 2.9 ± 1.72 inseminated females, demonstrating the potential of other females to replace the queen; however, only one female had an ovarian condition typical of a laying individual (Fig. 1D). Giannotti & Machado (1999) and Murakami et al. (2009) also reported more than one inseminated female in colonies of *P. lanio* and *M. cassununga*, respectively. In fact, the presence of more than one inseminated female in the colony, capable of reproducing, is a common feature in independent-founding species, as is the case for *Polistes* and *Mischocyttarus*. This fact shows that the distinction between reproductive and non-reproductive females is quite flexible and complex, depending on physiological, behavioral and ecological factors (Murakami & Shima, 2006).

Insemination of two or more females in the same colony can be a strategy to overcome problems encountered during the colony cycle, such as predation or parasitism, as suggested by Murakami et al. (2009) for *M. cassununga*. Gobbi et al. (2006) found that 75% of *P. versicolor* females in the aggregate and 85% in the foundation association were inseminated; therefore, insemination must occur before the formation of aggregates.

The results from the analyzes of the cuticular chemical profile show that the 4 different types of ovarian development

Fig 2. Mean curve for each group of mid-infrared absorption spectra of the thorax of *Polistes versicolor* females, grouped according to cuticular chemical profile and indicating the significant peaks for separation of the groups.

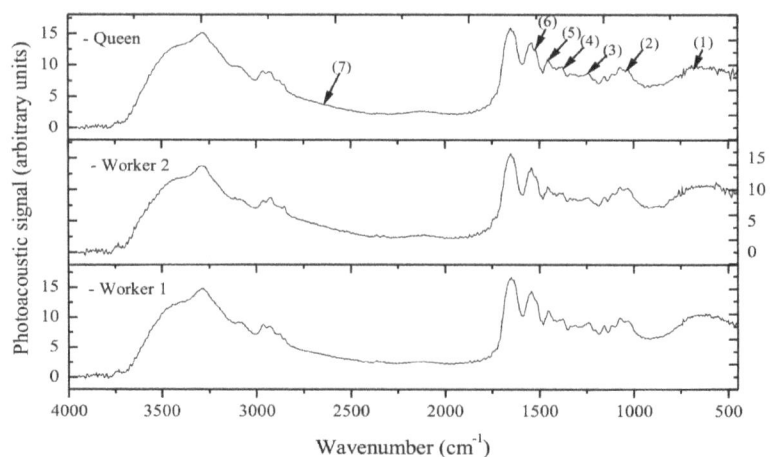

degree are distributed among three distinct categories of females: a) Workers 1, with filamentous ovarioles, type A; b) Workers 2, with partially developed ovarioles, types B and C; and c) Queens, with fully developed ovarioles, type D (Fig. 1 and 2). These differences were significant (Wilks' Lambda = 0.476, F = 6.303, P <0.001) (Fig. 3).

The spectra analyzed by FTIR-PAS showed significant differences among the cuticular chemical profiles, indicating seven significant peaks for the separation of females groups (Fig. 2 and Table 1). These compounds were linked to chitin (1238, 1523 e 2634 cm^{-1}) and CHCs (667, 1030, 1377 and 1450 cm^{-1}) present in the female cuticle (Table I). Antonialli-Junior et al. (2007) and Neves et al. (2012) discuss the importance of these peaks for distinguishing the groups analysed. However, the most significant peaks for those groups were mainly those corresponding to the hydrocarbon band (Fig. 2 and Table 1).

The first canonical root explained 93% of the results, and the second one the reimaning 7%, explaining together 100% of the results. Therefore, it seems that every female within these three groups had a different physiological status within the colony, which leads to a difference in the cuticular chemical profile and probably in the recognition by other females of their position in the colony hierarchy.

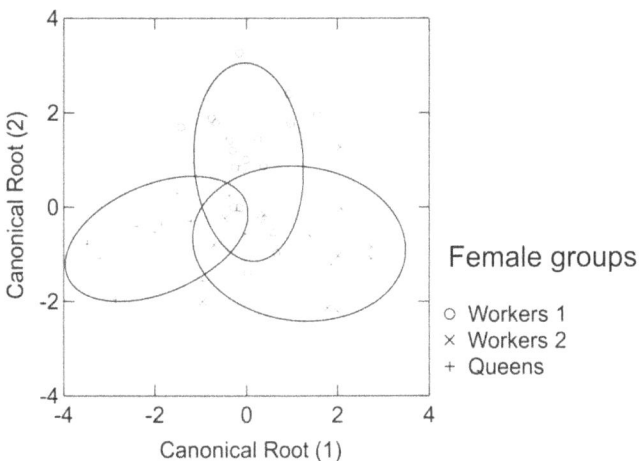

Fig 3. Dispersion diagram of the results of the stepwise discriminant analysis, showing the two canonical roots of differentiation of cuticular chemical profile in 3 different groups of females of *Polistes versicolor.*

According to Monnin (2006), there are correlations between reproductive status and the CHC profile in social insects, and that differentiation is important for the establishment of a hierarchy in independent-founding species, as it is the case for many Neotropical members of Polistinae. In colonies of *P. dominula* founded by a single female, the CHC profiles of the queen and workers were different (Dapporto et al., 2005). Bonckaert et al. (2012) investigated colonies of *Vespula vulgaris* (Linnaeus) and found that laying queens, queens in aggregate, virgin queens, and workers had different degrees of ovarian development, and this was correlated with their respective CHCs profiles.

The discriminant analysis of the 3 groups of females with different cuticular chemical profiles, however, showed no significant morphological differences (Wilks' Lambda = 0.851, F = 3.182, P <0.05), indicating an absence of morphological differences among these groups. The absence of morphological differences among castes was also reported by Giannotti & Machado (1999) for *P. lanio* and by Murakami et al. (2009) for *M. cassununga*, all of them independent-founding species. However, Gobbi et al. (2006) observed that of *P. versicolor* female aggregate are significantly larger than first emerged females, foundress association and workers. Tannure-Nascimento et al. (2005) suggested that the morphological differences between reproductive and non-reproductive females of *Polistes satan* (Bequaert) during the colony cycle is due to seasonal nutritional differences.

In fact, most studies describe the absence of morphological differences among castes in *Polistes* and *Mischocyttarus* (Cumber, 1951; Giannotti & Machado, 1999; Tannure-Nascimento et al., 2005; Murakami et al., 2009), supporting the hypothesis of post-imaginal caste determination. However, studies such as those of Gadagkar et al. (1991), Keeping (2002), Dapporto et al. (2008) and Hunt et al. (2011) performed with *Ropalidia marginata* (Fabricius), *Belonogaster petiolata* (DeGeer), *Polistes metricus* (Say) e *P. dominula*, respectively, show that, at least in part, the distinction of castes may be pre-imaginal.

By analyzing the relative age of the females we found a higher frequency of young females during the post-emergence (pre-male) stage and, as the colony cycle advances, old females become more frequent during the post-emergence

Table 1. Wave numbers, coefficients of the two canonical roots, functional groups, and vibrations of the peaks identified in the infrared absorption spectra of the thorax of the wasps, for analysis of the cuticular chemical profile.

Peak	Wave number (cm^{-1})	Canonical root 1	Canonical root 2	Functional group	Vibration model
(1)	667	1.453	3.036	Out-of-plane C-H (benzene)	Bending
(2)	1030	3.185	-5.432	In plane C-H (benzene)	Bending
(3)	1238	5.766	-5.481	-C-N	Stretching
(4)	1377	-15.713	7.655	C-CH$_3$	Symmetric bending
(5)	1450	2.659	1.669	C-CH$_2$ and C-CH$_3$	Asymmetric bending scissors
(6)	1523	3.216	-1.956	N-H	Bending
(7)	2634	0.173	1.456	C-N and N-H	Overtone bending

(post-male) and decline stages (Fig. 4). The chi-square test ($\chi^2 = 65.594$, df = 6, P <0.05) indicated a relationship between the female age and the degree of ovarian development, in which the queens are always among the older females of the colony. The workers have a range of ages, which is probably related to the stage of the colony cycle.

Corroborating these results, Baio et al. (2003), Murakami et al. (2009) and Felippotti et al. (2010), investigating colonies of *Metapolybia docilis* (Richards), *M. cassununga* and three species of *Clypearia* respectively, reported that queens are among the older females in the colony, and that the presence of young and old females varies according to the stage of the colony cycle. Murakami et al. (2009) observed that females with the most advanced ovarian development are older and are also more aggressive in the hierarchical ranking. All of these results agree with the system of gerontocracy (Strassmann & Meyer, 1983), common in independent-founding species, which means that, as the workers grow older, they are subject to more aggressive acts of from the dominant female.

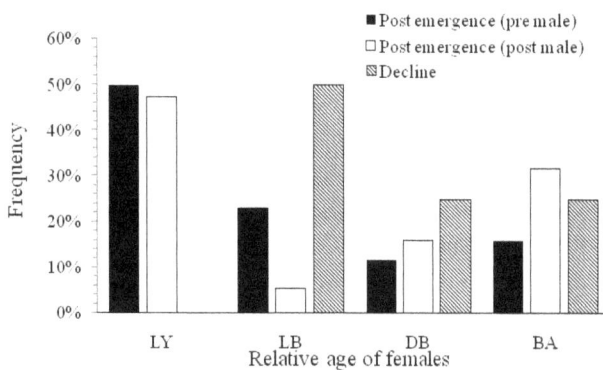

Fig 4. Frequency of the relative ages for females of *Polistes versicolor* from the ten colonies. Separation made according to the color patterns of the transverse apodeme in different colony cycle stages. LY: light yellow; LB: light brown, DB: dark brown; BA: black.

We conclude that there are three groups of females showing different cuticular chemical profiles in *P. versicolor* colonies. The first group is females with filamentous ovarioles, typical of workers; the second one is that of females with intermediate ovarioles; and the third one is the group of the queens, which are older, inseminated females with the greatests degree of ovarian development found among all females. On the other hand, no significant morphological differences were found among these female groups. Therefore, although there are no significant morphological differences among females, there are other factors such as the cuticular chemical composition, which are indicative of the reproductive physiological condition of each female in the colony.

Acknowledgments

The authors thank Janet W. Reid (JWR Associates) for the revision of the English text, and Orlando T. Silveira (Museu Paraense Emílio Goeldi) for the identification of the species. We are grateful to the Conselho Nacional de Desenvolvimento Científico e Tecnológico (CNPq) for an undergraduate fellowship awarded to the first author, and to CAPES for a doctoral fellowship to VO Torres. WF Antonialli-Júnior acknowledges research grants warded by the CNPq.

References

Antonialli-Junior, W.F., Lima, S.M., Andrade, L.H.C. & Súarez, Y.R.. (2007). Comparative study of the cuticular hydrocarbon in queens, workers and males of *Ectatomma vizottoi* (Hyminoptera, Formicidae) by Fourier transform-infrared photoacoustic spectroscopy. Genetics and Molecular Research, 6: 492-499.

Antonialli-Junior, W.F., Andrade, S.M., Súarez, Y.R. & Lima, S.M. (2008). Intra- and interspecific variation of cuticular hydrocarbon composition in two *Ectatomma* species (Hymenoptera: Formicidae) based on Fourier transform infrared photoacoustic spectroscopy. Genetics and Molecular Research, 7: 559–566.

Baio, M.V., Noll, F.B. & Zucchi, R. (2003). Shape differences rather than size differences between castes in the Neotropical swarm-founding wasp *Metapolybia docilis* (Hymenoptera: Vespidae, Epiponini). BMC Evolutionary Biology, 3: 10. doi:10.1186/1471-2148-3-10.

Baio, M.V., Noll. F.B. & Zucchi, R. (2004). Morphological caste differences and non-sterility of workers in *Brachygastra augusti* (Hymenoptera, Vespidae, Epiponini), a Neotropical swarm-founding wasp. Journal of the New York Entomological Society, 111: 243-253. doi: 10.1664/0028-7199-(2003)111[0242:MCDANO]2.0.CO;2.

Bonavita-Cougourdan, A., Theraulaz, G., Bagnères, A.G., Roux, M., Pratte, M., Provost, E. & Clément, J.L. (1991). Cuticular hydrocarbons, social organization and ovarian development in a polistine wasp: *Polistes dominulus* Christ. Comparative Biochemistry and Physiology, 100: 667–680. doi: 10.1016/0305-0491(91)90272-F.

Bonckaert, W., Drijfhout, F.P., D'Ettorre, P., Billen, J. & Wenseleers, T. (2012). Hydrocarbon signatures of egg maternity, caste membership and reproductive status in the common wasp. Journal of Chemical Ecology, 38: 42-51. doi: 10.1007/s10886-011-0055-9.

Bourke, A.F.G. (1999). Colony size, social complexity and reproductive conflict in social insects. Journal of Evolutionary Biology, 12: 245-25. doi: 10.1046/j.1420-9101.1999.00028.x.

Cumber, R.A. (1951). Some observations on the biology of

the Australia wasp *Polistes humilis* Fabr. (Hymenoptera: Vespidae) on North Auckland (New Zeland) with special reference to the nature of work caste. Proceedings of the Royal Entomological Society of London. Series A, General Entomology, 26: 11-16. doi: 10.1111/j.1365-3032.1951.tb00104.x.

Dapporto, L., Sledge, F.M. & Turillazzi, S. (2005). Dynamics of cuticular chemical profiles of *Polistes dominulus* workers in orphaned nests (Hymenoptera, Vespidae). Journal of Insect Physiology, 51: 969-973. doi: 10.1016/j.jinsphys.2005.04.011.

Dapporto, L., Lambardi, D. & Turillazzi, S. (2008). Not only cuticular lipids: First evidence of differences between foundresses and their daughters in polar substances in the paper wasp *Polistes dominulus*. Journal of Insect Physiology, 54: 89-95. doi:10.1016/j.jinsphys.2007.08.005.

Felippotti, G.T., Noll, F.B. & Mateus, S. (2007). Morphological studies on castes of *Protopolybia chartergoides* (Hymenoptera, Vespidae, Epiponini) observed in colonies during male production stage. Revista Brasileira de Entomologia, 51: 494-500. doi: 10.1590/S0085-56262007000400015.

Felippotti, G.T., Mateus, L., Mateus, S., Noll, F.B. & Zucchi, R. (2010). Morphological caste differences in three species of the Neotropical genus *Clypearia* (Hymenoptera: Polistinae: Epiponini). Psyche: A Journal of Entomology, 2010: 1-9. doi: 10.1155/2010/410280.

Ferreira, A.C., Cardoso C.A.L., Neves, E.F., Súarez, Y.R., Antonialli-Junior, W.F. (2012) Distinct linear hydrocarbon profiles and chemical strategy of facultative pasasitism among *Mischocyttarus* wasps. Genetics and Molecular Research, 11: 4351-4359. doi: 10.4238/2012.

Gadagkar, R, Bhagavan, S., Chandrashekara, & Vinutha, C. (1991). The role of larval nutrition in pre-imaginal biasing of caste in the primitively eusocial wasp *Ropalidia maginata* (Hymenoptera: Vespidae). Ecological Entomology, 16: 435-440. doi: 10.1111/j.1365-2311.1991.tb00236.x.

Giannotti E., & Machado, V.L.L. (1999). Behavioral castes in the primitively eussocial wasp *Polistes lanio* Fabricius (Hymenoptera: Vespidae). Revista Brasileira de Entomologia, 43: 185-190.

Gobbi, N., Noll, F.B. & Penna, M.A.H. (2006). "Winter" aggregations, colony cycle, and seasonal phenotypic change in the paper wasp *Polistes versicolor* in subtropical Brazil. Naturwissenschaften, 93: 487-494. doi: 10.1007/s00114-006-0140-z.

Hunt, J.H., Mutti, N.S., Havukainen, H. Henshaw, M.T. & Amdam, G.V. (2011). Development of an RNA interference tool, characterization of its target, and an ecological test of caste differentiation in the eusocial wasp *Polistes*. PLoS ONE 6(11): e26641. doi:10.1371/journal.pone.0026641.

Jeanne, R.L. (1972). Social biology of the Neotropical wasp *Mischocyttarus drewseni*. Bulletin of the Museum of Comparative Zoology, 144: 63-150.

Jeanne, R.L. (1991). The Swarm-founding Polistinae. *In* The Social Biology of wasps (pp.191-231). Cornell University Press, Ithaca, New York, xvii+678 p.

Keeping, M.G. (2002). Reproductive and worker castes in the primitively eusocial wasp *Belonogaster petiolata* (DeGeer) (Hymenoptera: Vespidae): evidence for pre-imaginal differentiation. Journal of Insect Physiology, 48: 867–879. doi.org/10.1016/S0022-1910(02)00156-7.

Liebig, J., Peeters, C., Oldham, N.J., Markstädter, C. & Hölldobler, B. (2000). Are variations in cuticular hydrocarbons of queens and workers a reliable signal of fertility in the ant *Harpegnathos saltator* ?. Proceedings of the National Academy of Sciences, USA, 97: 4124-4131. doi: 10.1073/pnas.97.8.4124.

Lockey, K.H. (1988). Lipids of the insect cuticle: origin, composition and function. Comparative, Biochemistry and Physiology, 89: 595–645.

Mateus, S., Noll, F.B. & Zucchi, R. (1997). Morphological caste differences in the Neotropical swarm-founding Polistinae wasp, *Parachartegus smithii* (Hymenoptera, Vespidae). Journal of the New York Entomological Society, 105(3-4): 129-139.

Mateus, S., Noll, F.B. & Zucchi, R. (1999). Caste differences and related bionomic aspects of *Chartergellus communis*, a neotropical swarm-founding polistinae wasp (Hymenoptera: Vespidae: Epiponini). Journal of the New York Entomological Society, 107: 391–406.

Mateus, S., Noll, F.B. & Zucchi, R. (2004). Caste flexibility and variation according to the colony cycle in the swarm-founding wasps, *Parachartergus fraternus* (Gribodo) (Hymenoptera; Apidae: Epiponi). Journal of the Kansas Entomological Society, 77: 470-483.

Monnin, T. (2006). Chemical recognition of reproductive status in social insects. Annales Zoologici Fennici, 43(5-6): 515-530.

Murakami, A.S.N. & Shima, S.N. (2006). Nutritional and social hierarchy establishment of the primitively eusocial wasp *Mischocyttarus cassununga* (Hymenoptera, Vespidae, Mischocyttarini) and related aspects. Sociobiology, 48: 183-207.

Murakami, A.S.N., Shima, S.N. & Desuó, I.C. (2009). More than one inseminated female in colonies of the independent-founding wasp *Mischocyttarus cassununga* Von Ihering (Hymenoptera, Vespidae). Revista Brasileira de Entomologia, 53: doi: 10.1590/S0085-56262009000400017.

Neves, E.F., Andrade, L.H.C., Súarez, Y.R., Lima, S.M. & Antonialli-junior, W.F. (2012). Age-related changes in the surface pheromones of the wasp *Mischocyttarus consimilis* (Hymenoptera: Vespidae). Genetics and Molecular Research: 11(3), 1891-1898. doi: 10.4238/2012.July.19.8.

Noll, F.B., Simões, D. & Zucchi, R. (1997). Morphological caste differences in the neotropical swarm-founding wasps: *Agelaia m. multipicta* and *Agelaia p. pallipes* (Hymenoptera, Vespidae). Ethology, Ecology and Evolution, 4: 361-372. doi: 10.1080/08927014.1997.9522878.

Noll, F.B., Wenzel, J.W. & Zucchi, R. (2004). Evolution of caste in neotropical swarm-founding wasps (Hymenoptera: Vespidae: Epiponini). American Museum Novitates, 3467: 1–24.

O'Donnell, S. (1995). Necrophagy by Neotropical Swarm-Founding Wasps (Hymenoptera: Vespidae, Epiponini). Biotropica, 27: 133–136. doi: 10.2307/2388911.

O'Donnell, S. (1998). Dominance and polyethism in the eusocial wasp *Mischocyttarus mastigophorus* (Hymenoptera: Vespidae). Behavioral Ecology and Sociobiology, 43: 327-331. doi: 10.1007/s002650050498.

Peeters, C., Monnin, T. & Malosse, C. (1999). Cuticular hydrocarbons correlated with reproductive status in a queen less ant. Proceedings of the Royal Society London B. Biol., 266: 1323–1327. doi:10.1098/rspb.1999.0782.

Provost, E., Blight, O., Tirard, A. & Renucci, M. (2008). Hydrocarbons and insects' social physiology. Pp. 19–72. *In.* Maes R. P. (ed.), Insect Physiology: new research. Nova Science Publishers.

Quinn, G.P. & Keough, M.J. (2002). Experimental design and data analysis for biologists. Cambridge University Press, Cambridge, 520 p.

Richards, O.W. (1971). The biology of social wasps (Hymenoptera, Vespidae). Biological Reviews, 46: 483–528.

Robinson, G.E. (1992). Regulation of division of labor in insect societies. Annual Review of Entomology, 37: 637–665. doi: 10.1146/annurev.en.37.010192.003225.

Röseler, P.F., Röseler, I. & Strambi, A. (1985). Role of ovaries and ecdysteroids in dominance hierarchy establishment among foundresses of the primitively social wasp, *Polistes gallicus*. Behavioral Ecology and Sociobiology, 18: 9-13.

Ross, K.G. & Matthews, R.W. (1991). The social biology of wasps. Cornell University Press, Ithaca, New York, xvii+678 p.

Shima, S.N., Yamane, S. & Zucchi, R. (1994). Morphological caste differences in some Neotropical swarm-founding polistine wasps. I. *Apoica flavissima* Hymenoptera, Vespidae). Japanese Journal of Entomology, 62: 811-822.

Shima, S.N., Noll, F.B., Zucchi, R. & Yamane, S. (1998). Morphological caste differences in the Neotropical swarm-founding Polistine wasps IV. *Pseudopolybia vespiceps*, with preliminary considerations on the role of intermediate females in the social organization of the Epiponini (Hymenoptera, Vespidae). Journal of Hymenoptera Research, 7: 280–295.

Shima, S.N., Noll, F.B. & Zucchi, R. (2000). Morphological caste differences in the neotropical swarm-founding polistine wasp, *Brachygastra lecheguana* (Hymenoptera: Vespidae, Polistinae, Epiponini). Sociobiology, 36: 41–52.

Silverstein, R.M. & Webster, F.X. (2000). Identificação espectrométrica de compostos orgânicos. 6ª edição. Livros Técnicos e Científicos Editora, Rio de Janeiro.

Skoog, D.A., Holler, J.F. & Nieman, T.A. (2002). Princípios de análise instrumental. 5ª edição. Bookman, Porto Alegre.

Sledge M.F., Boscaro, F. & Turillazzi, S. (2001). Cuticular hydrocarbons and reproductive status in the social wasp *Polistes dominulus*. Behavioral Ecology and Sociobiology, 49: 401–409. doi:10.1007/s002650000311.

Smith, B.C. (1999). Infrared. Spect. Interpretation: A Syst. Approach. CRC Press, Boca Raton, 265p.

Strassmann, J.E. & Meyer, D.C. (1983). Gerontocracy in the social wasp, *Polistes exclamans*. Animal Behavior, 31: 431-438.

Strassmann, J.E., Sullender, B.W. & Queller, D.C. (2002). Caste totipotency and conflict in a large-colony social insect. Proceedings of the Royal Society London B. Biol., 269(1488): 263-270. doi: 10.1098/rspb.2001.1880.

Tannure-Nascimento, I.C., Nascimento, F.S. & Zucchi, R. (2005). Size and colony cycle in *Polistes satan*, a Neotropical paper wasp (Hymenoptera, Vespidae). Ethology, Ecology and Evolution, 17: 105-119. doi: 10.1080/08927014.2005.9522601.

West-Eberhard, M.J. (1973). Monogyny in polygynous social wasps. Proceedings of the VII Congress of I.U.S.S.I. London, p. 396-403.

Wheeler, D.E. (1986). Developmental and physiological determinants of caste in social Hymenoptera: evolutionary implications. American Naturalist, 128: 13-34.

Wilson, E.O. (1985). The sociogenesis of insect colonies. Science, 228(4704): 1479-148.

17

Honey Bee Health in Apiaries in the Vale do Paraíba, São Paulo State, Southeastern Brazil

LG Santos[1]; MLTMF Alves[2]; D Message[3]; FA Pinto[1]; MVGB Silva[4]; EW Teixeira[2]

1 - Universidade Federal de Viçosa (UFV), Viçosa, Minas Gerais, Brazil.

2 - Agência Paulista de Tecnologia dos Agronegócios, Pindamonhangaba, São Paulo. Brazil.

3 - Universidade Federal Rural do Semiárido. (PVNS/CAPES), Mossoró/RN, Brazil.

4 - EMBRAPA Gado de Leite, Juiz de Fora, MG, Brazil.

Key words

Apis mellifera, Varroa destructor, Nosema sp., Paenibacillus larvae

Corresponding author

Lubiane Guimarães dos Santos
Graduate Program in Entomology
Federal University of Viçosa(UFV)
Avenida Peter Henry Rolfs s/n.
CEP: 36.571-000. Viçosa, MG, Brazil
E-Mail: lubi.guimaraes@gmail.com

Abstract

Bee health is a growing global concern due to phenomena with as yet undefined causes, such as the sudden population decline of colonies that has been observed in apiaries in many countries, recently including Brazil. The main objective of this study was to assess the presence and/or prevalence of pathogens that afflict Africanized Apis mellifera bees in the Vale do Paraíba region of São Paulo state, Brazil. Three sampling periods were established: period 1 – August and September 2009 (winter/early spring); period 2 – December 2009 and January 2010 (summer); and period 3 – April and May 2010 (autumn). Samples were collected of honeycomb from the brood area, combs containing capped brood, adult bees that cover the brood and foraging bees, to evaluate the presence and prevalence of Paenibacillus larvae, Varroa destructor and Nosema sp. The results indicated that the intensity of infection by Nosema ceranae and infestation rates of V. destructor in the hives were low (mean 637x10^3 spores of Nosema ceranae, 5.41% infestation of Varroa in adult bees and 4.17% infestation of Varroa in brood), with no detection of P. larvae spores in the samples. The prevalence of N. ceranae and V. destructor was high, at respective values of 85.2 and 95.7%. All told, 1,668 samples were collected from 438 hives, in 59 apiaries. These results demonstrate that although mites and microsporidia are widespread in the region's colonies, the Africanized bees are apparently tolerant to pathogens and parasites. However, the mechanisms related to defense against pathogens are not completely clear, and monitoring and prophylactic measures are essential to maintain the health of bee colonies.

Introduction

Although honey bees are the most frequently studied social insects due to their ecological and economic importance (Winston, 1987; Martin, 2001), a set of factors still not well understood has been affecting these arthropods in an increasing number of countries. Various organisms can parasitize honey bees in the immature or adult phase, possibly leading to collapse of the colony depending on the level of virulence (Bailey & Ball 1991; Ellis & Munn, 2005). There have been many reports of colony collapse in the Northern Hemisphere in recent years, prompting strong concern in the scientific community, particularly due to the importance

of bees as pollinizers (Neumann & Carreck, 2010). This phenomenon, called Colony Collapse Disorder (CCD), was first identified in the United States in the winter of 2006, when 23% of apiaries were affected, with average colony losses of 45% (Van Engelsdorp et al., 2007). Besides the risks posed by CCD, another factor that must be considered is the potential problems caused by applying drugs to control diseases, because of the possibility of generating resistant populations and of contamination of bee products, with consequent risks to human health (Lodesani et al., 2008).

Although honey bees are afflicted by numerous parasites and pathogens, including viruses, bacteria, protozoa and mites (Bailey & Ball, 1991), in Brazil there are few reports of diseases causing large-scale mortality. This pattern could be related to a series of biotic and abiotic factors typical of each region, among them the breed of bees used in local apiculture (Moretto, 1997). Africanized honey bees (AHB) have traits that favor resistance to parasites such as the mite *Varroa destructor*, among other pathogens (De Jong, 1996; Rosenkranz, 1999; Rosenkranz et al., 2010).

Nevertheless, in recent years honey production in some Brazilian regions has been declining, accompanied by observations of apparent weakening of colonies, with adult bees and brood in many cases presenting anomalous symptoms (Message et al., 2012). The explanation for this fact can be related both to increased virulence of parasites already present in the country and the introduction of new pathogenic agents, such as viruses (Teixeira et al., 2008).

Preventive monitoring of the levels of infestation and prevalence of harmful agents is important to maintain the health of apiculture in the country. In this study, we investigated the prevalence and incidence of *V. destructor*, *Nosema* sp. and *Paenibacillus larvae* in apiaries of southeastern Brazil. These agents have been indicated as responsible for causing large losses to apiculture worldwide.

Material and Methods

Samples were collected in 438 colonies of 59 apiaries, located in 13 municipalities in the region of Vale do Paraíba, São Paulo state. In this region, the climate is considered subtropical (Koeppen climate classification), with temperatures varying from 18°C to 29°C during the year.

Samples included pieces of honeycomb from the brood area, comb containing capped brood, adult bees covering the breeding area from brood comb, and forager bees collected at the entrance of the hive, to assess the presence and prevalence of *Paenibacillus larvae*, *Varroa destructor* and *Nosema* sp., respectively. We established three collection periods, to obtain samples from different moments of colony development as a function of the natural conditions and availability of food resources in the field (nectar and pollen flow): period 1 – August and September 2009 (winter/early spring); period 2 – December 2009 and January 2010 (summer); and period 3 – April and May 2010 (autumn). All the analyses were carried out in Pindamonhangaba, São Paulo, in the Honey Bee Health Laboratory of the São Paulo State Agribusiness Technology Agency (LASA/APTA). The collection of the samples for analysis of *Varroa destructor*, *Nosema* sp. and *Paenibacillus larvae* was based on Teixeira & Message (2010). Evaluations of *Varroa destructor* infestations were based on De Jong et al. (1982). The protocol for microbiological analyses developed by Schuch et al. (2001) and later considered as the Brazilian method (Brasil, 2003) was used to analyze samples for *P. larvae*, while the method of Cantwell (1970) was used to count the spores for *Nosema*. In order to identify the *Nosema* species, samples from all apiaries were submitted to duplex PCR assay (Martín-Hernández et al., 2007), using the primers 321APIS-FOR/REV, for identification of *N. apis* and 218MITOC-FOR/REV for *N. ceranae*. Positive and negative controls were used.

Analysis of variance was carried out considering a statistical model whose representation is given by: $y_{ijk} = \mu + l_i + c_j + e_{ijk}$, where: y_{ijk} = dependent variables; μ = overall mean; l_i = effect of the ith site, c_j = effect of the jth sample and e_{ijk} is the random effect of the error. The degrees of freedom for the sources of variation studied were decomposed into contrasts and evaluated through the F-test at a significance level of 1%. All analyses were performed using the GLM procedure of the SAS statistical package (2001).

Results

During the three sampling periods, 438 samples of foraging bees were collected at the hive entrances, and for other types of samples it was possible to collect 432 samples of adult bees present in the brood area, 368 comb pieces containing honey from the brood area and 430 samples of comb containing at least 100 older pupae. Table 1 shows the intensity of infection by the microsporidium *Nosema ceranae*, represented by the average number of spores per bee, as well as the average infestation rate (%) by the mite *V. destructor* in adult bees and in brood in apiaries located in 13 locations in the Vale do Paraíba: Cunha, Bananal, Lagoinha, Lorena, Monteiro Lobato, Natividade da Serra, Paraibuna, Pindamonhangaba, Redenção da Serra, São José dos Campos, São Luís do Paraitinga, Santo Antônio do Pinhal, Taubaté e Tremembé. Only the species *N. ceranae* was detected by molecular analysis (Fig. 1), and no *P. larvae* spores were detected in the honey samples.

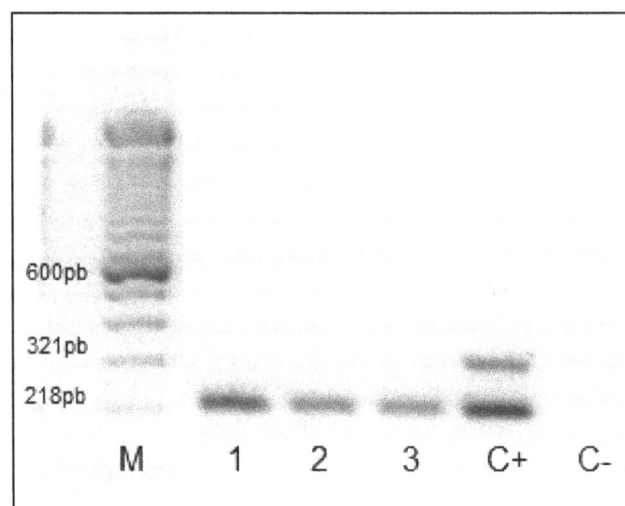

Fig. 1. 2% agarose gel showing duplex *Nosema apis* and *Nosema ceranae* PCR products. M: 100 bp marker. Column 1- 3: samples showing *N. ceranae* amplification. Column C+: Positive control – *Nosema ceranae* (218 bp) and *Nosema apis* (321 bp). Column C-: Negative control.

Table 1. Mean number of spores of *Nosema ceranae* per bee, mean infestation rate (%) by the mite *V. destructor* in brood worker bees and mean infestation rate (%) by *V. destructor* in brood in different municipalities*.

Municipality	No. of *Nosema* sp. spores (x10³)	*V. destructor* in adult bees (%)	*V. destructor* in brood (%)
Bananal	623 ± 188a	5.49 ± 0.97a	6.63 ± 1.34a
Cunha	485 ± 128a	3.53 ± 0.69a	2.93 ± 0.78a
Lagoinha	258 ±106a	5.87 ± 0.56a	5.65 ± 0.66a
Monteiro Lobato	623 ± 145a	5.92 ± 0.76a	5.70 ± 0.88a
Paraibuna	581 ± 99a	4.94 ± 0.51a	3.61 ± 0.61a
Pindamonhangaba (APTA)	1.655 ± 114b	4.74 ± 0.59a	3.81 ± 0.69a
Pindamonhangaba	424 ± 110a	6.75 ± 0.59a	3.30 ± 0.68a
Redenção da Serra	636 ± 153a	4.01 ± 0.79a	2.12 ±0.91a
São Jose dos Campos	459 ± 145a	4.26 ± 0.75a	2.52 ± 0.88a
Santo Antonio do Pinhal	607 ± 275a	4.17 ± 1.42a	1.14 ± 1.66a
São Luis do Paraitinga	369 ± 386a	5.66 ± 1.99a	2.90 ± 2.33a
Taubaté	1.055 ± 153b	4.90 ± 0.79a	3.55 ± 0.93a
Tremembé	658 ± 99a	5.82 ± 0.51a	4.65 ± 0.60a

* In collection period 2, it was not possible to obtain samples in five localities. [†]Means accompanied by different letters differ significantly (P<0.01).

Table 2. Mean number of spores of *Nosema ceranae* per bee, mean infestation rate (%) by the mite *V. destructor* in brood worker bees and mean infestation rate (%) by *V. destructor* in brood in different collection periods (1: winter 2009, 2: summer 2010, 3: autumn 2010).

Collection Periods	No. of *Nosema* sp. spores (x10³)	*V. destructor* in adult bees (%)	*V. destructor* in brood (%)
1	379 ± 62a	6.03 ± 0.32a	4.58 ± 0.37a
2	689 ± 87b	3.57 ± 0.45b	1.95 ± 0.53b
3	879 ± 74b	5.65 ± 0.39a	4.66 ± 0.46a
Mean	637 ± 36	5.41 ± 0.20	4.17 ± 0.23

[†] Means accompanied by different letters differ significantly (P<0.01).

The apiaries in Pindamonhangaba and Taubaté presented the highest infection intensities, differing significantly (P<0.001) from the other municipalities studied.

Table 2 reports the intensities of infection by the microsporidium in the three sampling periods. The first period (winter and early spring 2009) presented the least intense infection by *Nosema ceranae*, differing (P<0.001) from the intensities of periods 2 (summer 2010) and 3 (autumn 2010).

The results for the number of adult *V. destructor* mites as well as descendants in the 13 locations and three collection periods can be observed in Tables 1 and 2, respectively. The overall average infestation of *V. destructor* in adult bees was 5.41 ± 0.20 (Table 1). The infestation rates observed in period 2 were lower in relation to the other two periods (1 and 3), both in adult bees and brood. Therefore, there was a difference (P<0.01) of the infestation rates observed in these periods; but in all cases the infestation levels were low.

Discussion

This is the first comprehensive study (1,668 samples analyzed, collection periods during all seasons) conducted to identify the prevalence of three typical honey bee pathogens (*Nosema* sp., *Varroa destructor* and *Paenibacillus larvae*) in Brazil. Although *Nosema apis* was a highly problematic pathogen in the 1970s in the southern region of Brazil (Teixeira et al., 2013), it was not detected in the present study. In the first collection period (winter and early spring 2009), the number of spores detected per bee was very low, differing significantly (P<0.001) in relation to the other two periods (summer and autumn 2010).

Traver et al. (2011) also observed low levels of this pathogen in the winter, but in the autumn their results were opposite to ours, even though in the region of the United States studied by them the autumn temperatures are near those in the winter in São Paulo. In turn, Higes et al. (2008) observed higher intensities of infection by *N. ceranae* in the coldest months and lower intensities at the beginning of spring, while in Germany, Gisder et al. (2010) found that the intensity of infection by *Nosema* spp. was greater in spring than in autumn.

Although comparisons of infection intensity between different regions are highly contradictory, even considering data from temperate regions where the seasons are well defined, the climatic peculiarities of the regions under analysis must be considered, because they can affect the intensity of infection, even if only indirectly (Le Conte & Navajas, 2008). In Brazil, no pattern in the infection intensity of *Nosema ceranae* has been observed during the year (Teixeira et al., 2013), considering the weather or region.

In the two places (Pindamonhangaba – SP and Taubaté – SP) where infection by *Nosema ceranae* was higher than at the other sites, the experimental apiaries are frequently managed for teaching/research purposes rather than production, so the frequent management practices (change of frames, common use of material between colonies, etc.) might have facilitated dispersion and/or made the bees more susceptible due to stress from frequent human intervention.

The low numbers of spores in our samples and the absence of collapse of any of the colony analyzed, in contrast to the observations of Higes et al. (2008), can possibly be credited to a higher tolerance of Africanized *A. mellifera* honey bees. In fact, many questions can be posed in relation to *Nosema ceranae*, mainly regarding the absence of clinical symptoms in infected colonies and the considerable variation of infection intensity within very short periods (D. Message, unpublished data).

In the state of São Paulo, considering the samples analyzed so far, the presence of the species *Nosema ceranae* has been confirmed in 100% of the cases where *Nosema* has been observed (Fig 1). According to Teixeira et al. (2013), even with the proof of the high prevalence of *N. ceranae* in the country's apiaries, the recommendation by researchers and technicians who work in the area of honey bee health is not to use chemical products, due to the inconsistency of the real effect of the presence of this pathogen in the colonies.

The first report of the species *Nosema ceranae* in Brazil was by Klee et al. (2007); these authors reported this species of microsporidium on four continents. According to various authors, *N. ceranae* jumped from *Apis cerana* to *Apis mellifera* probably in the 1990s, since then dispersing to most regions of the world (Fries et al., 1996; Higes et al., 2006; Klee et al., 2007; Paxton et al., 2007; Fries, 2010). Nevertheless, in Brazil, Teixeira et al. (2013) detected its presence in samples that had been collected in the southern state of Rio Grande do Sul more than three decades ago. In Asia, where it supposedly originated, recent articles have reported the presence of *N. ceranae* in Vietnam (Klee et al., 2007) and Iran (Nabian et al., 2011). In Europe, studies have confirmed its presence since 1998 (Fries et al., 2006; Paxton et al., 2007), in several countries: Spain (Fries et al., 2006; Higes et al., 2006; Klee et al, 2007; Martín- Hernandez et al., 2007), France (Chauzat et al., 2007), Germany (Klee et al., 2007), Sweden (Klee et al., 2007), Finland (Klee et al., 2007; Paxton et al., 2007), Denmark (Klee et al., 2007), Greece (Klee et al., 2007), Hungary (Tapaszti et al., 2009), Holland (Klee et al., 2007), United Kingdom (Klee et al., 2007), Italy (Klee et al., 2007), Serbia (Klee et al, 2007), Poland (Topolska &Kasprzak, 2007), Bosnia (Santrac et al., 2009) and Turkey (Whitaker et al., 2010). In Africa, Higes et al., (2009) reported *N. ceranae* in Algeria in 2008. In North America, this species has been present since 1995, affecting colonies mainly in the United States (Klee et al., 2007; Chen et al., 2008; Williams et al., 2008), as well as Canada (Williams et al., 2008) and Mexico (Guzmán-Novoa et al., 2011). In Central America, *N. ceranae* was found in Costa Rica by Calderón et al. (2010), and in South America it is present in Brazil and Argentina (Klee et al., 2007; Medici et al., 2012), Uruguay (Invernizzi et al, 2009) and Chile (Martínez et al., 2012). In Oceania, recent studies reported this species in New Zealand (Klee et al., 2007) and Australia (Giersch et al., 2009).

With respect to the mite *Varroa destructor*, the infestation rates were low, as also reported by Pinto et al. (2011) in Africanized bees in a region with similar environmental conditions to those in the present study, although with a smaller number of samples. On the other hand, in evaluating the infestation levels in the Forest Zone of the state of Minas Gerais, a region with a tropical climate with temperatures varying from 14 to 26°C during the year (Koeppen climate classification), Bacha Junior et al. (2009) found an average value of 7.8% during the summer. Apparently, there is great variation in the infestation rates of this mite according to the region of the country and the respective climate (Moretto, 1997).

Showing the same tendency as the levels of infestation in adult bees, the rates of infestation by the mite in brood combs were low in comparison with those found in colonies without treatment in Bulgaria, England and the entire United Kingdom (18 to 49%, 15 to 40% and 6 to 42%, respectively) (Martin, 1994; 2001). Usually, when the temperature drops, especially during the autumn and the winter, the brood area shrinks, as a consequence of the reduced availability of food resources in the environment and the number of phoretic mites increases. This can explain the low number of the mites in the capped brood.

The high infertility rate of this mite on Africanized bees in Brazil (Message & Gonçalves, 1995; Rosenkranz, 1999; Calderón et al., 2010; Rosenkranz et al., 2010) could also explain the low infestation rate observed. However, Garrido et al. (2003), analyzing samples from different regions of Brazil, found predominance of the haplotype K, and also an increase in the mite's fertility rate.

The intensity levels of infection by *N.ceranae* and the rates of infestation by *V. destructor* observed in this study are low compared to other regions where chemical treatments are used for control in temperate zones, but the prevalence of these pathogens was high, considering that 85.2% of the hives were infected by the microsporidium and 95.7% presented infestation by mites.

Acknowledgments

To CNPq for the scholarship and research funding (MAPA/CNPq process 578293/2008-0, EWT.), APTA, SP, for the institutional support, and Carmen L. Monteiro for help with the analyses.

References

Bacha Júnior, G.L., Felipe-Silva, A.S., Pereira, P.L.L. (2009). Taxa de infestação por ácaro *Varroa destructor* em apiários sob georreferenciamento. Arquivos Brasileiros de Medicina Veterinária e Zootecnia, 61: 1471–1473.

Bailey, L.,Ball, B.V. (1991). Honey Bee Pathology. London: Academic Press, 208p.

Brasil (2003) Ministério da Agricultura e do Abastecimento. Instrução Normativa n° 62, de 26 de agosto de 2003. Métodos Analíticos Oficiais para Análises Microbiológicas para Controle de Produtos de Origem Animal. Anexo, Capítulo XIX Pesquisa de *Paenibacillus larvae* subsp. *larvae*. Diário Oficial da União, 18/09/2003.

Calderón, R.A., Van Veen J.W., Sommeijer, M.J., Sanchez, L.A. (2010). Reproductive biology of *Varroa destructor* in Africanised honey bees (*Apis mellifera*). Experimental and Applied Acarology, 50: 281–297.

Cantwell, G. R. (1970). Standard methods for counting nosema spores. American Bee Journal, 110: 222-223.

Chauzat, M.P., Higes, M., Martín-Hernandez, R., Meana, A., Cougoule, N., Faucon, J.P., (2007). Presence of Nosema ceranae in French honey bee colonies, Journal of Apicultural Research, 46: 127-128. DOI: 10.3896/IBRA.1.46.2.12.

Chen, Y., Evans, J.D., Smith, I.B., Pettis, J.S., (2008). *Nosema ceranae* is a long-present and wide-spread microspridean infection of the European honey bee *(Apis mellifera)* in the United States. Journal of Invertebrate Pathology, 97: 186–188.

De Jong, D., Roma, D.A., Gonçalves, L.S. (1982). A comparative analysis of shaking solutions for the detection of *Varroa jacobsoni* on adult honey bees. Apidologie, 13: 297–306.

De Jong, D. (1996). Africanized honey bees in Brazil, forty years of adaptation and success. Bee World, 77: 67–70.

Ellis, J.D. & Munn, P. (2005). The worldwide health status of honey bees. Bee World, 86: 88-101.

Fries, I. (2010). *Nosema ceranae* in European honey bees (*Apis mellifera*). Journal of Invertebrate Pathology, 103: 573-579.

Fries, I., Feng, F., Silva, A., Slemenda, S.B., Pieniazek, N.J. (1996). *Nosema ceranae* sp. (Microspora, Nosematidae), morphological and molecular characterization of a microsporidian parasite of the Asian honey bee *Apis cerana* (Hymenoptera, Apidae). European Journal of Protistology, 32: 356-365.

Fries, I., Martín, R., Meana, A., García-Palencia, P., Higes, M. (2006) .Natural infections of *Nosema ceranae* in European honey bees. Journal of Apicultural Research, 45: 230-233.

Garrido C., Rosenkranz, P., Paxton, R.J., Gonçalves, L.S. (2003). Temporal changes in *Varroa destructor* fertility and haplotype in Brazil. Apidologie, 53: 535–541.

Giersch, T., Berg, T., Galea, F., Hornitzky, M. (2009). *Nosema ceranae* infects honey bees (*Apis mellifera*) and contaminates honey in Australia. Apidologie, 40: 117-123.

Gisder, S., Hedtke, K., Mockel, N., Frielitz, M., Linda, A., Genesrch, E. (2010). Five-Year Cohort Study of *Nosema* spp. in Germany: Does Climate Shape Virulence and Assertiveness of *Nosema ceranae*? Applied and Environmental Microbiology, 76:3032–3038. DOI: 10.1128/AEM.03097-09.

Guzmán-Novoa, E., Hamiduzzaman, M, M., Arechavaleta, M. E., Koleoglu, G. (2011). *Nosema ceranae* has parasited Africanized Honey bees in Mexico since at least 2004. Journal of Apicultural Research, 50: 167-169 DOI: 10.3896/IBRA.1.50.2.09.

Higes, M., Martín -Hernández R., Meana, A. (2006). *Nosema ceranae*, a new microsporidian parasite in honeybees in Europe. Journal of Invertebrate Pathology, 92: 93-95. DOI: 10.1016/j.jip.2006.02.005.

Higes, M., García-Palencia, P., Martín-Hernández, R., Botías, C., Garrido-Bailón, E., González-Porto, A.V., Barrios, L., Del Nozal, M.J., Berna, J. L., Jiménez, J. J., Palencia, P. G., Meana, A. (2008). How natural infection by *Nosema ceranae* causes honeybee colony collapse. Environmental Microbiology, 10: 2659-2669. DOI: 10.1111/j.1462-2920.2008.01687.x.

Higes, M., Matin-Hernandez, R., Garrido-Bailón E., Botías C., Meana A. (2009). First detection of *Nosema ceranae* (Microsporidia) in African Honey bees (*Apis melifera intermissa*). Journal of Apicultural Research, 48: 217–219. DOI: 10.3896/IBRA.1.48.3.12.

Invernizzi, C., Abud C., Tomasco, I.H., Harriet, J., Ramallo, G., Campa, J., Katz, H., Gardiol, G., Mendonça, Y. (2009). Presence of *Nosema ceranae* in honeybees (*Apis mellifera*) in Uruguay. Journal of Invertebrate Pathology, 101: 150-153

Klee J., Besana A.M., Genersch E., Gisder S., Nanetti, A., Tam D.Q., Chinh T.X., Puerta F., Ruz, J.M., Kryger P., Message D., Hatjina, F., Korpela, S., Fries, I., Paxton, R.J. (2007). Widespread dispersal of the microsporidian *Nosema ceranae*, an emergent pathogen of the western honey bee, *Apis mellifera*. Journal of Invertebrate Pathology, 96: 1-10.

Le Conte, I.& Navajas, M. (2008). Climate change: impact on honey bee populations and diseases. Revue Scientifique et Technique de L`Office Internationalndes Epizooies, 27 : 499-510.

Lodesani, M., Costa, C., Serra, G., Colombo, R., Sabatini, A. G. (2008). Acaricide residues in beeswax after conversion to organic beekeeping methods. Apidologie, 39: 324-333.

Martin, S.J. (1994) Ontogenesis of the mite *Varroa jacobsoni* Oud. in worker brood of the honeybee *Apis mellifera* L. under natural conditions. Experimental and Applied Acarology, 18: 87–100.

Martin, S.J. (2001). *Varroa destructor* reproduction during the winter in *Apis mellifera* colonies in UK. Experimental and Applied Acarology, 25: 321–325.

Martín-Hernández, R., Meana, A., Prieto, L., Salvador, A.M., Garrido-Bailon, E., Higes, M. (2007). Outcome of colonization of *Apis mellifera* by *Nosema ceranae*. Applied Environmental Microbiology, 73: 6331-6338. DOI: 10.1128/AEM.00270-07.

Martínez, J., Leal, G., Conget, P. (2012). *Nosema ceranae* an emergent pathogen of *Apis mellifera* in Chile. Parasitological Research, 111: 601–607. DOI: 10.1007/s00436-012-2875-0.

Medici, S.K., Sarlo, E.G., Porrini, M.P., Braunstein, M., Eguaras, M.J.(2012). Genetic variation and widespread dispersal of *Nosema ceranae* in *Apis mellifera* apiaries from Argentina. Parasitol. Res., 110: 859-864. DOI: 10.1007/ s00436-011-2566-2.

Message, D., Gonçalves, L.S. (1995). Effect of size of worker brood cells of Africanized honey bees on infestation and reproduction of the ectoparasitic mite *Varroa jacobsoni* Oud. Apidologie, 26: 381- 386.

Message, D., Teixeira, E.W., De Jong, D. (2012). Situação da Sanidade das abelhas no Brasil. In: Imperatriz-Fonseca et al, Polinizadores no Brasil: Contribuição e perspectivas para a biodiversidade, uso sustentável, conservação e serviços ambientais. São Paulo, EDUSP.

Moretto, G. (1997). Defense of Africanized bee workers against the mite *Varroa jacobsoni* in Southern Brazil. American Bee Journal, 137: 746–747.

Nabian, S., Ahmadi, K., Nazem Shirazi, M.H., Gerami Sadeghin, A. (2011). First Detection of Nosema ceranae, a Microsporidian Protozoa of European Honeybees (*Apis mellifera*) In Iran. Iranian Journal of Parasitology, 6: 89-95.

Neumann, P. & Carreck, N.L. (2010). Honey bee colony losses. J Apic. Res., 49: 1-6. DOI 10.3896/IBRA.1.49.1.01

Paxton, R., Klee, J.S., Korpela, S., Fries, I. (2007). *Nosema ceranae* has infected *Apis mellifera* in Europe since at least 1998 and may be more virulent than *Nosema apis*. Apidologie, 38: 558–565. DOI: 10.1051/apido:2007037.

Pinto, F.A., Puker, A., Message, D., Barreto, L.M.R.C. (2011). *Varroa destructor* in Juquitiba, Vale do Ribeira, Southeastern Brazil: Sazonal Effects on the Infestation Rate of Ectoparasitic Mites in Honeybees. Sociobiology, 57: 511–518.

Rosenkranz, P. (1999). Honey bee (*Apis mellifera* L.) tolerance to *Varroa jacobsoni* Oud. in South America. Apidologie, 30: 159–172.

Rosenkranz, P., Aumeier, P., Ziegelmann, B. (2010). Biology and control of *Varroa destructor.* Journal of Invertebrate Pathology, 103: 96-119. DOI: 10.1016/j.jip.2009.07.016.

Santrac, V., Granato, A., Mutinelli, F. (2009). First detection of *Nosema ceranae* in *Apis mellifera* from Bosnia and Herzegovina, Proc. Workshop "*Nosema* disease: lack of knowledge and work standardization" (COST Action FA0803)

Guadalajara.

SAS - Statistical Analysis System (2001). SAS User's Guide: Statistics. SAS Institute Inc., Cary, NC, USA.

Schuch, D.M.T., Madden, R.H.; Sattler, A., 2001. An improved method for the detection and presumptive identification *Paenibacillus larvae* spores in honey. Journal of Apicultural Research, 40: 59-64.

Tapaszti, Z., Forgách, P., Kövágó, C., Békési, L., Bakonyi, T., Rusvai, M. (2009). First detection and dominance of *Nosema ceranae* in Hungarian honeybee colonies. Acta Veterinaria Hungarica, 57: 383-388. DOI: 10.1556/AVet.57.2009.3.4.

Teixeira, E.W., Chen, Y., Message, D., Pettis, J., Evans, J.D. (2008). Virus infections in Brazilian honey bees. Journal of Invertebrate Pathology, 99: 117–119. DOI: 10.1016/j.jip.2008.03.014

Teixeira, E. W., Message, D. (2010). Manual Veterinário de Colheita e Envio de Amostras-Abelhas *Apis mellifera.* São Paulo: Editora Horizonte, OMS/OPAS/MAPA. p. 175-213, 2010.

Teixeira, E. W., Santos, L. G., Sattler, A., Message, D., Alves, M. L. T. M. F., Martins, M. F., Grassi-Sella, M. F., Francoy, T. M. (2013). *Nosema ceranae* has been present in Brazil for more than three decades infecting Africanized honey bees. Journal of Invertebrate Pathology, 114: 250-254. 2013. DOI: 10.1016/j.jip.2013.19.002.

Topolska, G. & Kasprzak, S. (2007). First cases of *Nosema ceranae* infection in bees in Poland. Medycyna Weterynaryjna Suppl., 63: 1504-1506.

Traver, B.E., Williams, M.R., Fell, R.D. (2011). Comparison of within hive sampling and seasonal activity of *Nosema ceranae* in honey bee colonies. Journal of Invertebrate Pathology, 109: 187-93.

VanEngelsdorp, D., Underwood, R. M., Caron, D., Hayes Jr. J. (2007). An estimate of managed colony losses in the winter of 2006-2007: a report commissioned by the Apiary Inspectors of America. American Bee Journal, 147: 599-603.

Whitaker, J., Szalanski, A. L., Kence, M. (2010). Molecular detection of *Nosema ceranae* and *Nosema apis* from Turkish honey bees. Apidologie, 41: 364-374. DOI: 10.1051/apido/2010045

Williams G.R., Shafer, A.B., Rogers, R.E., Shutler, D., Stewart, D.T. (2008). First detection of *Nosema ceranae*, a microsporidian parasite of European honey bees (*Apis mellifera*), in Canada and central USA. Journal of Invertebrate Pathology, 97: 189–192. DOI:10.1016/J.JIP.2008.04.005.

Winston, M. L. (1987). The Biology of the Honey Bee. Harvard University Press. Cambridge. London. UK.

18

Coexistence Patterns Between Ants And Spiders In Grassland Habitats

AM Rákóczi, F Samu

Centre for Agricultural Research, Hungarian Academy of Sciences, Budapest, Hungary.

Keywords
Sas Hill, species co-occurrence, correlation, mimicry, myrmecomorphy, myrmecophagy

Corresponding author
Ferenc Samu
Plant Protection Institute
Centre for Agricultural Research
Hungarian Academy of Sciences
Postal address: PO. Box 102, Budapest
H-1525 Hungary
E-Mail: feri.samu@gmail.com

Abstract
The ecological importance of both ants and spiders is well known, as well as the relationship between certain spiders and ants. The two main strategies - myrmecomorphy (ant-mimicking) and myrmecophagy (ant-eating) - that connect spiders to ants have been mostly studied at the behavioural level. However, less is known about how these relationships manifest at the ecological level by shaping the distribution of populations and assemblages. Our question was how ant-mimicking and ant-eating spiders associate with ant genera as revealed by field co-occurrence patterns. For both spider groups we examined strength and specificity of the association, and how it is affected by ant size and defence strategy. To study spider-ant association patterns we carried out pitfall sampling on the dolomitic Sas Hill located in Budapest, Hungary. Spiders and ants were collected at eight grassland locations by operating five pitfalls/location continuously for two years. To find co-occurrence patterns, two approaches were used: correlation analyses to uncover possible spider-ant pairs, and null-model analyses (C-score) to show negative associations. These alternative statistical methods revealed consistent co-occurrence patterns. Associations were generally broad, not specific to exact ant genera. Ant-eating spiders showed a stronger association with ants. Both ant-mimicking and ant-eating spiders associated more strongly with Formicine ants - species with formic acid or anal gland secretions, and had neutral association with Myrmicine ants - species with stings and cuticle defences.

Introduction

Ants have immense and complex effects on ecosystems because of their sheer abundance, biomass and the complex interactions in which they are involved (Hölldobler & Wilson, 1990). Ants possess various forceful defence mechanisms such as formic-acid, aggressive attack, stings, and social defence (Wilson, 1976; Yanoviak & Kaspari, 2000). Defence makes ants best avoided by most predators, which presents them as ideal models for mimics among arthropods (Schowalter, 2006), or makes them a food best suited for specialist predators. Ant associations that have developed in many arthropod taxa fall into three categories: myrmecomorphy, myrmecophagy and myrmecophily. Myrmecomorphs are ant-mimicking species which have acquired morphological and/or behavioural similarity to ants, myrmecophagous species are ant-eaters that specialise in subduing ant prey. Here - since only those association types occurred in our study area - we only consider the ant-eating and ant-mimicking species and do not deal with the third type of ant associated spiders, the myrmecophils, which are highly integrated into host colonies (Cushing, 2012; Pekar et al., 2012).

Spiders can use one or more of these strategies, making spider-ant relationship a complex system to observe (McIver & Stonedahl, 1993; Cushing, 1997). Ant associates can be found in various spider families (Salticidae, Gnaphosidae, Theridiide, Zodariidae, Liocraniidae, Linyphiidae) (Cushing, 1997; Pekar, 2004b). Ant associated spiders have many morphological and behavioural adaptations. In ant-mimicking species body shape often resembles three body regions, legs are long and slender and there may be cuticle modifications present that resemble mandibles, compound eyes or sting. The movement of ant-mimics frequently becomes ant-like, including holding forelegs like antennae (Reiskind, 1977; Ceccarelli, 2008). Ant association may also manifest in special foraging and predatory strategies, most tangible in specialist ant-eaters, like *Zodarion* spp. (Pekar, 2004).

Spider-ant relationship is also shaped by ants, which are the models of mimicry and/or potential prey. Such a relationship is logically influenced by ant size and also by defence type ants possess (Holway, 1999; Feener, 2000). Ants concerned in the present study fall into two main categories: ants that rely on cuticular structures, sting and ants that mostly rely on the use of formic acid or gland secretions. These coincide with two broader taxonomic groups, being either "myrmicine" (Myrmicinae subfamily) or "formicine" (Formicinae and Dolichoderinae subfamilies) ants (Edwards et al., 1974; Shattuck, 1992; Bolton, 2003). Myrmicine ants have thick cuticle and cuticle structures, such as spines (also present in some Formicinae, but not present in the genera included in the present study); they possess a distinct postpetiole and a functional sting is always present, while in the formicine group, species armour is different, lack both postpetiole and sting; their defence is based on the use of their mandibles and on toxin exuded from the tips of their abdomens (Hermann, 1969; Edwards et al., 1974). We treated these taxonomic groups as representing two different defence types, because such modifications are important selective factors for both predators and mimics.

Although ant associations have been mostly studied through the resulting morphological and behavioural modifications, it also has an ecological context, because ant models should be present in the same microhabitat, and have direct or indirect ecological interactions that are related to co-occurrence (Edmunds, 1978). Direct trophic connection may exist between ants and spiders, but ants may also influence spiders indirectly through their ecological impact, e.g. aphid tending (Renault et al., 2005; Sanders & van Veen, 2012).

In recent years connection between spiders and ants has gained more and more attention in behavioural, morphological and evolutionary studies (Cushing, 1997; Pekar, 2004b; Pekar, 2004a; Nelson & Jackson, 2009; Cushing, 2012; Nelson & Jackson, 2012), but the ecological patterns observable in the field has to be examined for a complex view on ant-spider relationship. Analysing seasonally divided datasets from 40 pitfalls in a grassland ecosystem we tried to answer the following questions: (i) Is there any non-neutral co-occurrence pattern between ant associated spiders and ants? (ii) How specific is the association between ant associated spiders and ants? (iii) Is the strength of the relationship different between spider strategies and is it influenced by spider and ant size and ant defence type?

Material and Methods

Study area

Our field study took place on the top area of Sas Hill Nature Reserve, Budapest (47°28'48.68"N, 19° 1'1.22"E), between 2010 and 2012. This is a grassland covered dolomitic hill, a refuge for many rare spider species (Szinetár et al., 2012), and has been a nature reserve since 1958. Arachnological research at Sas Hill has an especially rich tradition (Balogh, 1935; Samu & Szinetár, 2000; Rákóczi & Samu, 2012; Szinetár et al., 2012). These studies made us notice the especially high number of ant associated spider species, which reaches 14 species with the present study (Szinetár et al., 2012). Contrary to spiders the ant fauna of the hill have not been previously studied and published neither on generic or specific

level. From Hungary 126 species of ants in 34 genera are known (Csősz et al., 2011). Collection of ants and spiders was made in eight dry dolomitic grassland patches scattered on the 35 ha area of the hill. Botanically they belonged to open and closed dolomitic dry grasslands, with *Festuca pallens* Host as a characteristic grass species. Detailed habitat description and co-ordinates are given in Szinetár et al. (2012).

Sampling

We collected spiders and ants by pitfall trapping. Pitfall traps containing 40% ethylene-glycol with a small drop of liquid soap, had 7.5 cm diameter openings and a laminated plate was applied c. 3 cm higher than the surface as a cover (Kádár & Samu, 2006). Pitfall trap sampling lasted from 29 April 2010 to 24 May 2012. Traps were emptied fortnightly, except in winter when, depending on the weather, the traps were emptied c. every four weeks. Each location was sampled with five traps in a linear transect with 2 m between traps.

Collected samples were placed in 70% alcohol; both spiders and ants were sorted and identified under a stereomicroscope. Adult spiders were determined to species, while ants were determined to genera. Voucher specimens were placed in the collection of the Plant Protection Institute, Centre for Agricultural Research, Hungarian Academy of Sciences. We used several determination keys for spiders (Loksa, 1969; Loksa, 1972; Roberts, 1995; Nentwig et al., 2013), and for ants (Somfai, 1959; Czechowski et al., 2012). The nomenclature of spiders followed the World Spider Catalogue (Platnick, 2013).

Data classification and analysis

The co-occurrence of spiders and ants was examined at two different levels, for which two datasets were derived from raw data: 'trap' level dataset contained summarized data of a given pitfall trap over all emptying occasions (n = 40 datasets); 'trap-season' level datasets contained summarized data of a given pitfall trap for a season of a year. In the latter datasets we placed winter catches (that represented fewer animals) into autumn or spring, with the division date 1 January, resulting in 7 seasons: 2010 spring, 2010 summer, 2010 autumn, 2011spring, 2011summer, 2011autumn and 2012 spring (n = 280 datasets). In each approach spider species data and ant generic data were used.

We assessed the relationship between spiders and ants based on various, biologically meaningful classifications. Ants were classified by average size in a genera; and by their taxonomic type also related to defence type: myrmicine (cuticular defence, sting) or formicine (formic acid or gland secretions) (Bettini et al., 1978; Bolton, 2003). We considered only workers. Mean worker size was taken from the literature (Somfai, 1959). Size difference between dimorphic worker classes was not small in all cases. Dimorphism was taken into consideration by calculating mean size from the worker classes. List of ant genera, their classification and mean size are given in Table 1. Spiders were divided into two groups based on their association type to ants: ant-eating "myrmecophages" and ant-mimicking "myrmecomorphs", derived from data in the literature (Cushing, 1997; Pekar, 2004; Platnick, 2010; Pekar & Jarab, 2011; Cushing, 2012; Nentwig et al., 2013), and the average size in each species was also considered; spider classification and size are given in Table 2.

In the statistical analyses we have included only species/genera where more than five individuals were found during the study. We used Spearman correlation to reveal positive or negative correlation between counts of individuals of ant genera and spider species. A non-parametric approach was used because of the skewed distribution of counts (many 0 values and some high counts). Ant and spider related factors that influence the strength of correlation were analysed by linear mixed model. The model included Spearman correlation coefficient values as response variable, spider strategy, ant defence type, average ant and spider size in given genus/species as explanatory variables, and to control for the non-independence of values within genus or species, spider species and ant genera were added to the model as random factors (Faraway, 2005). Specificity of the relationship (measured as the number of significant correlations) was analysed by nominal logistic analysis.

Table 1. List of ant genera in the present study. Subfamily and grouping according to morphs are given, together with mean worker size and number of specimens caught in the study.

Genus	Abbrev. (5 character)	Subfamily	Morph	Mean size (mm)	No. of indiv.
Bothriomyrmex	*Bothr.*	Dolichoderinae	formicine	2.5	13
Tapinoma	*Tapin.*	Dolichoderinae	formicine	3.0	5550
Camponotus	*Campo.*	Formicinae	formicine	10.0	1596
Formica	*Formi.*	Formicinae	formicine	7.0	1074
Lasius	*Lasiu.*	Formicinae	formicine	3.0	1179
Plagiolepis	*Plagi.*	Formicinae	formicine	1.5	210
Leptothorax	*Lepto.*	Mirmicinae	myrmicine	2.5	12
Messor	*Messo*	Mirmicinae	myrmicine	8.5	112
Myrmecina	*Myrme.*	Mirmicinae	myrmicine	5.5	3
Myrmica	*Myrmi.*	Mirmicinae	myrmicine	5.5	185
Solenopsis	*Solen.*	Mirmicinae	myrmicine	1.5	20
Temnothorax	*Temno.*	Mirmicinae	myrmicine	2.5	33
Tetramorium	*Tetra.*	Mirmicinae	myrmicine	2.5	243

Analyses were carried out by R 2.15.2 (R Core Team, 2013).

We used co-occurrence analysis to detect possible non-random patterns in presence absence matrices, comparing them to matrices generated by randomization. Analysis was carried out by EcoSim's (build 021605) co-occurrence module (Gotelli & Entsminger, 2010). We used location by taxon presence-absence matrices, where location datasets were either trap or trap-season, and taxon was (i) only ant genera; (ii) only ant associated spider species; (iii) both ants and ant associated spiders. The co-occurrence analysis searches for checkerboard units (CU), which are 2x2 sub matrices in the original presence-absence matrix. The number of CUs for a species pair is the number of localities where only one of the species occurs, i.e. their occurrence is mutually exclusive (Stone & Roberts, 1990). For a given species-pair the negative association is represented by a large number of CUs in every possible habitat combination. The average number of CUs for all the possible species combination is the Checkerboard score (C-score), which is a measure of negative association in the community (Stone & Roberts, 1990; Gotelli, 2000; Gotelli & Entsminger, 2010). The null-model matrices are Monte-Carlo randomizations of the original matrix. The average of such randomized C-score values represent the case without biological interactions, higher observed C-score values than that indicate negative, while lower observed values indicate positive associations between the species.

Results

Quantitative results

During the whole sampling period we emptied the 40 traps 40 times. In total 10,230 ant specimens and 751 ant associated spiders were found. The total number of ant genera was 13 (Table 1), the ant associated spiders were represented by 11 species (Table 2). Most ant associated spiders were relatively rare, the majority representing the ant-eating strategy. A single ant-eating species, *Z. rubidum*, made up nearly 90% of all ant associated spiders, and it meant a very high, 16% dominance among all spiders.

Table 2. List of ant associated (AA) spider species on Sas Hill. Number of individuals refers to total catch during the period. Catches of AA species also expressed as % of all AA (ΣAA). As a reference total number of spider individuals (including non AA) and total number of AA spiders caught are given. The mean size of each species is also given.

Species name	Family	Association type	No. of indiv.	% of ΣAA	Mean size (mm)
Callilepis schuszteri Simon	Gnaphosidae	ant-eater	15	2.0	5.2
Euryopis quinqueguttata Koch	Theridiidae	ant-eater	2	0.3	2.3
Harpactea hombergi (Scopoli)	Dysderidae	ant-mimic	1	0.1	4.8
Micaria dives (Lucas)	Gnaphosidae	ant-mimic	6	0.8	3.1
Micaria formicaria (Sundevall)	Gnaphosidae	ant-mimic	2	0.3	6.6
Micaria pulicaria (Sundevall)	Gnaphosidae	ant-mimic	1	0.1	4.3
Micaria silesiaca Koch	Gnaphosidae	ant-mimic	1	0.1	4.3
Phrurolithus festivus (Koch)	Corinnidae	ant-mimic	8	1.1	2.7
Phrurolithus szilyi Herman	Corinnidae	ant-mimic	42	5.6	2.3
Synageles hilarulus (Koch)	Salticidae	ant-mimic	2	0.3	3.0
Zodarion rubidum Simon	Zodariidae	ant-eater	671	89.3	3.5
All spiders			4,051		
All ant associated spiders (ΣAA)			751		
ΣAA as % of all spiders			18.5		

Correlation analysis

Spearman correlation analyses were performed on the trap-season dataset. There was a strong correlation between the overall number of ants and ant associated spiders ($\rho = 0.65$, $P < 0.0001$). Correlation was also calculated at the functional grouping levels. Spider ant association types showed no correlation with myrmicine ants (ant-eating spiders: $\rho = 0.025$, $P = 0.68$; ant-mimicking spiders: $\rho = 0.024$, $P = 0.69$), but correlation with formicine ants was significant and of similar strength for both spider groups (ant-eating spiders: $\rho = 0.55$, $P < 0.0001$, ant-mimicking spiders: $\rho = 0.54$, $P < 0.0001$).

Correlation analysis between individual spider species and ant genera was also performed (Table 3). Analysing the pattern of significant correlations, it is clear that the association of spiders is broader than ant genera, because all spider species were significantly positively associated with more than one ant genus (Table 3).

Analysing the number of significant correlations of the spider species in a nominal logistic model including spider strategy, ant type, average ant and spider size as explanatory variables, ant size proved to be marginally significant (Wald test: $\chi^2 = 4.052$, df $= 1$, $P < 0.04$), the spider association became more frequent with increasing ant size. Ant type proved to be highly significant (Wald test: $\chi^2 = 13.70$ df $= 1$, $P < 0.0002$), with much more significant associations of spiders with formicine ants.

We also wanted to know how the strength of associations was dependent on spider and ant strategies and average spider and ant size. We tested a linear mixed model on Spearman correlation coefficients, which had normal distribution (Kolmogorov-Smirnov test, d $= 0.131$, NS). The model included spider strategy, ant defence type and ant and spider size as explanatory variables, and spider species and ant genus as random factors. Spider size was marginally significant, with smaller spiders correlating more with ants (F $= 21.51$, df $= 1$, $P = 0.044$); spider strategy was also marginally significant, with ant-eaters more strongly associated with ants (F $= 22.83$, df $= 1$, $P = 0.041$). The most important factor proved to be ant defence type showing a much higher correlation of spiders to the formicine group than to myrmicine (F $= 12.92$, df $= 1$, $P = 0.005$).

Co-occurrence analysis

The co-occurrence analysis revealed positive association in the ant-spider assemblage. We made simulations on data of "just spider", "just ant" and "ant+spider" assemblages. Observed C-scores were consistently lower than simulated ones, as measured by standard effect size (S.E.S.) in the spider-ant assemblage, meaning that on average the mixed assemblage is more associated than pure taxa assemblages (Table 5). Considering specific species pairs, the number of CUs is a measure of negative association. Higher number of CUs was found between myrmicine ants and ant associated spiders. In *Z. rubidum* we found no CU with any of the ant genera. The CU pattern of spider-ant species pairs is given in Table 5.

Discussion

The main purpose of the present study was to reveal if known ant associated spiders respond to the distribution of ants in an ecologically measurable way. The results certainly support the hypothesis, that non-random co-occurrence patterns exist in the field between ants and ant associated spiders. Although associations were rather broad, they were influenced by spider and ant characteristics, from which ant defence type seemed to be the most important. For our purposes the Sas Hill in Budapest proved to be a very good location where we could sample 11 ant associated species. This is important, because most ant associated species are relatively rare (Cushing, 1997; Pekar, 2004b; Pekar, 2004a; Nelson & Jackson, 2009; Cushing, 2012; Nelson & Jackson, 2012), and to study their ecology and relations to other taxa is therefore not easy.

Measuring the association pattern indicated by ants and ant associated spiders first of all gave us the result that associations are not at the lowest taxonomic resolution of the present study (spider species and ant genera), but at the higher

Table 3. Spearman correlation coefficients (ϱ) of ant associated spiders and ant genera in the trap-season dataset. Row header contains ant genera, (abbreviated names, c.f. Table 2). ■ marked ants are myrmicine, unmarked ones are formicine ants. Spiders marked with ♣ are ant-eaters, unmarked ones are ant-mimics. The number of ● symbols marks the strength of the correlation (denoted by: ● – 0.1-0.19, ●● – 0.2-0.29, ●●● – 0.3-0.39, ●●●● >0.4). All marked correlations were significant at $P<0.05$.

Spider/Ant	*Micaria dives*	*Phrurolithus festivus*	*Phrurolithus szilyi*	*Callilepis schuszteri* ♣	*Zodarion rubidum* ♣
Lepto. ■	-0.04	0.1	0	0.21 ●●	0.17 ●
Messo. ■	-0.04	0.03	0.13	0.02	-0.06
Myrmi. ■	0.01	0.07	-0.11	-0.1	0
Solen. ■	0.1	-0.05	0.14	-0.05	0.01
Temno. ■	0.02	0.09	0.01	0	0.13
Tetra. ■	0.07	-0.01	0.01	-0.04	0
Bothr.	-0.04	0.11	0.02	-0.04	-0.01
Campo.	0.28 ●●	0.15 ●	0.24 ●●	0.23 ●●	0.37 ●●●
Formi.	0.17 ●●	0.15 ●	0.20 ●●	0.23 ●●	0.53 ●●●●
Lasiu.	0.11	0.11	0.16 ●	0.01	0.54 ●●●●
Plagi.	0.21 ●●	0.09	0.18 ●	0.14 ●	0.29 ●●
Tapin.	0.19 ●	0.13	0.32 ●●●	0.09	0.34 ●●●

Table 4. Observed C-scores, standard effect size (S. E. S.) and significance level of co-occurrence analysis on trap and trap season datasets.

Composition	Trap S.E.S.	Trap P	Trap (C-score)	Trap-Season S.E.S.	Trap-Season P	Trap-Season (C-score)
Only spider	-1.8	0.02	3.42	-1.1	0.13	15.53
Only ant	-2.1	0.01	2.31	-3.2	0.0002	11.82
Spider + ant	-3.2	0.0002	1.67	-3.5	0.0001	10.04

level of functional and morphological groups. At these higher levels the two statistical methods, measuring positive and negative associations, gave congruent results.

Our results are also in agreement with observations about the moderately narrow diet of ant-eating spiders. These spiders are specialised on consuming not a single ant species, but rather a broader spectrum of species, such as genera or subfamily (Pekar, 2004; Pekar et al., 2012). Based on spider response to olfactory cues produced only in a narrow range of genera, in a recent study Cardenas et al. (2012) argue that *Z. rubidum* - previously thought as an "ant generalist" - in fact, preys mostly on the genera *Lasius* and *Formica*. Our field results support this notion.

It was proved that spider and ant size are marginally significant factors in the association, most probably for different reasons in ant-mimicking and ant-eating spiders. On one hand, preying on ants is risky because of the defences, which favours a higher spider/ant size ratio, but for the same reason large ant-eaters can be less associated with ants because they may have a broader diet. On the other hand, ants below a certain size might not be preferred, because preying on them results in lower profit (less nutrition in preferred body parts) compared to cost (Pekar, 2004b; Pekar et al., 2010; Cushing, 2012). In *Z. rubidum* we know from case studies that the spider shows preference for similar sized or larger ants (Pekar, 2004b). Probably for other ant-eaters, size ratio with prey plays similar role. In ant mimics size plays important role because appropriate size enchases the accuracy of the

mimic, making it more effective.

The strongest pattern found was, that both ant-mimicking and ant eating spiders showed positive association with formicine ants and neutral or negative association with myrmicine ants, as it was confirmed by both statistical approaches. In ant-mimicking spiders the reason could be that the numerical dominance of formicine ants makes these ants a better model for Batesian-mimicry (Schowalter, 2006). The strength of the association was the strongest in ant-eating spiders, where the reason for co-occurrence pattern could be the difference how they are able to cope with different defences: thick cuticle, propodeal spikes and sting with neurotoxins vs. formic acid (Blum, 1992; Bolton, 2003).

The aim of the present study was to reveal spider-ant association patterns from a field study. This is an alternative and complementary approach to laboratory studies, where preference is tested under highly artificial circumstances, and relatively little can be said about their realisation in the field. The specificity of spider-ant association proved to be relatively broad in the field, with ant associated spiders correlating with more than one ant genera. In the present study we uncovered a non-random pattern of co-occurrence, where - possibly for different reasons for ant-mimicking and ant-eating spiders - the most substantial pattern was a stronger association with fomicine than with mymicine ants.

Acknowledgements

The authors are grateful to Erika Botos, Kinga Fetykó, Éva Szita, Gábor Lőrinczi and Zsolt Lang for their contribution in this research. We thank Stano Pekar for detailed comments on a previous version of the manuscript. We are indebted to Sándor Csősz for reviewing the manuscript before submission and giving us insightful advises. We thank four anonymous referees for their comments and criticisms on the manuscript. The project was supported by OTKA grant K81971, by the colleagues of Sas Hill Nature Reserve and by Pál Kézdy from the directorate of Duna-Ipoly National Park.

Table 5. The number of checkerboard units (CU) of ant associated spiders and ant genera in the trap-season dataset. Row header contains ant genera (names abbreviated, c.f. table 2). ■ marked ants are Myrmicine, unmarked ones are Formicine ants. Spiders marked with ♣ are ant-eaters, unmarked ones are ant-mimics. Numbers show the number of CUs observed for the species pair. the number of CUs is visually represented by the number of ● symbols (no symbol – <4, ● – 4-5, ●● – 6-7, ●●● – 8-9).

Spider/Ant	*Micaria dives*	*Phrurolithus szilyi*	*Phrurolithus festivus*	*Callilepis schuszteri* ♣	*Zodarion rubidum* ♣
Lepto.■	6 ●●	0	2	6 ●●	0
Messo.■	3	2	8 ●●●	0	0
Myrmi. ■	5 ●	5 ●	2	6 ●●	0
Solen. ■	4 ●	4 ●	9 ●●●	4 ●	0
Temno. ■	8 ●●●	2	0	6 ●●	0
Tetra. ■	0	0	0	0	0
Bothr.	3	6 ●●	8 ●●●	6 ●●	0
Campo.	0	0	0	0	0
Formi.	0	0	0	0	0
Lasiu.	0	0	0	0	0
Plagi.	0	0	0	0	0
Tapin.	0	0	0	0	0

References

Balogh, J. I. (1935). A Sashegy Pókfaunája. Faunisztikai, Rendszertani és Környezettani Tanulmány [Spider fauna of the Sas-hegy. A faunistical, taxonomical and environmental study]. Budapest: Sárkány-Nyomda Rt., 60 p

Bettini, S., S. Blum & H. R. Hermann, Jr. (1978). Venoms and Venom Apparatuses of the Formicidae: Dolichoderinae and Aneuretinae. In Arthropod Venoms. (pp. 871-894), Springer Berlin Heidelberg.

Blum, M. S. (1992). Ant Venoms: Chemical and Pharmacological Properties. Toxin Reviews, 11: 115-164.

Bolton, B. (2003). Synopsis and classification of Formicidae. Memoirs of the American Entomological Institute, 71: 1-370.

Cardenas, M., P. Jiros & S. Pekar. (2012). Selective olfactory attention of a specialised predator to intraspecific chemical signals of its prey. Naturwissenschaften, 99: 597-605. doi: 10.1007/s00114-012-0938-9

Ceccarelli, F. S. (2008). Behavioral mimicry in Myrmarachne species (Araneae, Salticidae) from North Queensland, Australia. Journal of Arachnology, 36: 344-351. doi: 10.1636/CSt07-114.1

Csősz, S., B. Markó & L. Gallé. (2011). The myrmecofauna (Hymenoptera: Formicidae) of Hungary: an updated checklist. North-Western Journal of Zoology, 7: 55.

Cushing, P. E. (1997). Myrmecomorphy and myrmecophily in spiders: A review. Florida Entomologist, 80: 165-193. doi: 10.2307/3495552

Cushing, P. E. (2012). Spider-Ant Associations: An Updated Review of Myrmecomorphy, Myrmecophily, and Myrmecophagy in Spiders. Psyche, 2012: 23. doi: 10.1155/2012/151989

Czechowski, W., A. Radchenko, W. Czechowska & K. Vepsäläinen. (2012). The ants of Poland with reference to the myrmecofauna of Europe. Fauna Poloniae 4. Warsaw: Natura Optima Dux Foundation, 496 p

Edmunds, M. (1978). On the association between *Myrmurachne* spp. (Salticidae) and ants. Bulletin of the British Arachnological Society, 4: 149-160.

Edwards, G. B., J. F. Carroll & W. H. Whitcomb. (1974). *Stoidis aurata* (Araneae: Salticidae), a Spider Predator of Ants. Florida Entomologist, 57: 337-346.

Faraway, J. J. (2005). Linear Models with R. London: Chapman and Hall.

Feener, D. J. (2000). Is the assembly of ant communities mediated by parasitoids? Oikos, 90: 79-88.

Gotelli, N. J. (2000). Null model analysis of species co-occurrence patterns. Ecology, 81: 2606-2621. doi: 10.1890/0012-9658(2000)081[2606:Nmaosc]2.0.Co;2

Gotelli, N. J. & G. L. Entsminger. (2010). EcoSim: Null models software for ecology. Version 7. http://garyentsminger.com/ecosim.htm. Jericho, VT 05465: Acquired Intelligence Inc. & Kesey-Bear.

Hermann, H. R. (1969). The hymenopterous poison apparatus: Evolutionary trends in three closely related subfamilies of ants. (Hymenoptera: Formicidae). Journal of the Georgia Entomological Society, 4: 123-141.

Hölldobler, B. & E. O. Wilson. (1990). The Ants. Belknap Press of Harvard University Press

Holway, D. A. (1999). Competitive mechanisms underlying the displacement of native ants by the invasive Argentine ant. Ecology, 80: 238-251.

Kádár, F. & F. Samu. (2006). A duplaedényes talajcsapdák használata Magyarországon [On the use of duble-cup pitfalls in Hungary]. Növényvédelem, 42: 305-312.

Loksa, I. (1969). Pókok I. - Araneae I. In Magyarország Állatvilága (Fauna Hungariae). (pp. 133), Budapest: Akadémiai Kiadó.

Loksa, I. (1972). Pókok II. - Araneae II. In Magyarország Állatvilága (Fauna Hungariae). (pp. 112), Budapest: Akadémiai Kiadó.

McIver, D. J. & G. Stonedahl. (1993). Myrmecomorphy: Morphological and behavioral mimicry of ants. Annual Review of Entomology, 38: 351-379.

Nelson, X. J. & R. R. Jackson. (2009). Collective Batesian mimicry of ant groups by aggregating spiders. Animal Behaviour, 78: 123-129. doi: 10.1016/j.anbehav.2009.04.005

Nelson, X. J. & R. R. Jackson. (2012). How spiders practice aggressive and Batesian mimicry. Current Zoology, 58: 620-629.

Nentwig, W., T. Blick, D. Gloor, A. Hänggi & C. Kropf. (2013). Spiders of Europe www.araneae.unibe.ch version 2.2013.

Pekar, S. (2004a). Poor display repertoire, tolerance and kleptobiosis: Results of specialization in an ant-eating spider (Araneae, Zodariidae). Journal of Insect Behavior, 17: 555-568. doi: 10.1023/B:Joir.0000042541.23748.D7

Pekar, S. (2004b). Predatory behavior of two European ant-eating spiders (Araneae, Zodariidae). Journal of Arachnology, 32: 31-41. doi: 10.1636/S02-15

Pekar, S. (2004). Predatory behavior of two European ant-eating spiders (Araneae, Zodariidae). Journal of Arachnology, 32: 31-41.

Pekar, S., J. A. Coddington & T. A. Blackledge. (2012). Evolution of stenophagy in spiders (Araneae): evidence based on the comparative analysis of spider diets. Evolution, 66: 776-806. doi: 10.1111/j.1558-5646.2011.01471.x

Pekar, S. & M. Jarab. (2011). Assessment of color and behavioral resemblance to models by inaccurate myrmecomorphic spiders (Araneae). Invertebrate Biology, 130: 83-90. doi: 10.1111/j.1744-7410.2010.00217.x

Pekar, S., D. Mayntz, T. Ribeiro & M.E. Herberstein. (2010). Specialist ant-eating spiders selectively feed on different body parts to balance nutrient intake. Animal Behaviour, 79: 1301-1306.

Platnick, N.I. (2010). The World Spider Catalog, Version 11.0 http://research.amnh.org/entomology/spiders/catalog/. New York: The American Museum of Natural History

Platnick, N.I. (2013). The World Spider Catalog, Version 13.5 http://research.amnh.org/entomology/spiders/catalog/. New York: The American Museum of Natural History

R Core Team. (2013). R: A language and environment for statistical computing. http://www.R-project.org/. Vienna, Austria: R Foundation for Statistical Computing.

Rákóczi, A. M. & F. Samu. (2012). Természetvédelmi célú orgonairtás rövidtávú hatása pókegyüttesekre [The short term effect of Syringa eradication conservation management on spider assemblages]. Rosalia, 8: 141-149.

Reiskind, J. (1977). Ant-mimicry in panamanian clubionid and salticid spiders (Araneae - Cubionidae, Salticidae). Biotropica, 9: 1-8. doi: 10.2307/2387854

Renault, C. K., L. M. Buffa & M. A. Delfino. (2005). An aphid-ant interaction: effects on different trophic levels. Ecological Research, 20: 71-74. doi: 10.1007/s11284-004-0015-8

Roberts, M. J. (1995). Spiders of Britain and Northern Europe. London: HarperCollins

Samu, F. & C. Szinetár. (2000). Rare species indicate ecological integrity: an example of an urban nature reserve island. In (P. Crabbé, Ed. Implementing ecological integrity. (pp. 177-184), Kluwer Academic Publishers.

Sanders, D. & F.J.F. van Veen. (2012). Indirect commensalism promotes persistence of secondary consumer species. Biology Letters, 8: 960-963. doi: 10.1098/rsbl.2012.0572

Schowalter, T.D. (2006). Insect Ecology: An Ecosystem Approach. Elsevier Science

Shattuck, S. O. (1992). Higher classification of the ant subfamilies Aneuretinae, Dolichoderinae and Formicinae (Hymenoptera: Formicidae). Systematic Entomology, 17: 199-206.

Somfai, E. (1959). Hangya alkatúak Formicoidea. In Magyarország Állatvilága (Fauna Hungariae). (pp. 79), Budapest: Akadémiai Kiadó.

Stone, L. & A. Roberts. (1990). The checkerboard score and species distributions. Oecologia, 85: 74-79. doi: 10.1007/Bf00317345

Szinetár, C., A. M. Rákóczi, K. Bleicher, E. Botos, P. Kovács & F. Samu. (2012). A Sas-hegy pókfaunája II. A Sas-hegy faunakutatásának 80 éve a hegyről kimutatott pókfajok kommentált listája [Spider fauna of Mt Sas-hegy II. 80 years of fauna research on Mt Sas-hegy, with the annotated list of spiders]. Rosalia, 8: 333-362.

Wilson, E. O. (1976). Organization of colony defense in ant Pheidole dentata Mayr (Hymenoptera-Formicidae). Behavioral Ecology and Sociobiology, 1: 63-81. doi: 10.1007/bf00299953

Yanoviak, S.P. & M. Kaspari. (2000). Community structure and the habitat templet: ants in the tropical forest canopy and litter. Oikos, 89: 259-266. doi: 10.1034/j.1600-0706.2000.890206.x

19

Composition and Diversity of Ant Species into Leaf Litter of Two Fragments of a Semi-Deciduous Seasonal Forest in the Atlantic Forest Biome in Barra do Choça, Bahia, Brazil

FREITAS, JMS[1,2,3], DELABIE, JHC[1,2] & LACAU, S[1,2,3]

1 - Universidade Estadual de Santa Cruz, Ilhéus-BA, Brazil.
2 - Laboratório de Mirmecologia, CEPLAC/CEPEC/SECEN, Ilhéus-BA, Brazil.
3 - Universidade Estadual do Sudoeste da Bahia, Itapetinga-BA, Brazil.

Keywords

Formicidae, Planalto da Conquista, Atlantic Forest, Tropical forest

Corresponding author

Juliana Martins da Silva Freitas
Univ. Estadual do Sudoeste da Bahia
Laboratório de Biossistemática Animal
Itapetinga-BA, Brazil
E-mail:julliana.martins@yahoo.com.br

Abstract

We present here the results of a study of leaf litter ant diversity in remnant areas of semi-deciduous seasonal forests in the Atlantic Forest biome. Standardized collections were made in 2011, using pitfall traps and Winkler sacks in two fragments of native forest in the municipality of Barra do Choça in the micro-region of the Planalto da Conquista, in southwestern of the state of Bahia, Brazil. A total of 107 species from 37 ant genera and 9 subfamilies was collected. The observed richness was high, and the diversity indices (Shannon-Wiener) of the two fragments suggest that in spite of being strongly impacted by anthropogenic actions, they maintained high faunal diversity levels, similar to those observed in other original Atlantic Forest sites in state of Bahia. Analyses of the species accumulation curves (Jackknife 2), however, indicated that survey effort was not sufficient to capture all of the species present. The high observed numbers of unique species, the shape of the species accumulation curves, and high values of estimated richness suggest that the survey areas were quite heterogeneous. These results provide new information concerning regional biodiversity that will be useful for continuing studies on fragmentation processes in the region.

Introduction

Ants (Hymenoptera: Formicidae) form one of the most diverse and ecologically important insect groups in terms of their diverse and essential functions in terrestrial ecosystems (Wilson & Hölldobler, 1990; Alonso & Agosti, 2000). Their predominance can be attributed in part to their eusocial nature, which favors their dispersal and successful occupation of new habitats (Wilson & Hölldobler, 2005). Since the Cretaceous period, these animals have demonstrated successful radiation throughout almost all terrestrial habitats, with numerical and biomass predominance in most of them (Fernández & Ospina, 2003; Wilson & Hölldobler, 2005).

Ant diversity in forest ecosystems is particularly high in the leaf litter (Alonso & Agosti, 2000; Silva & Brandão, 2010), although community composition is influenced by numerous factors, including the nature of the surrounding plant formations, soil composition and the local microclimate (Schowalter & Sabin, 1991). Ant community structures respond directly and quickly to both quantitative and qualitative environmental changes, and have therefore been the focus of stu-dies investigating the effects of environmental disturbances on ecological communities (Veiga-Ferreira et al., 2005; Delabie et al., 2006, 2007).

Ants maintain numerous biotic associations with other organisms in their environments (Wilson & Hölldobler, 1990; Rico-Gray & Oliveira, 2007), rapidly respond to habitat alterations such as fragmentation (Peck et al., 1998; Veiga-Ferreira et al., 2005; Delabie et al., 2006) and are relatively easily collected and identified (Peck et al., 1998), making them ideal models for studying and monitoring global biodiversity and useful as bioindicators of disturbances caused by ecosystem size reductions

The Atlantic Forest biome has been a focal area for environmental conservation efforts (Dean, 2002). Studies of Atlantic Forest biodiversity have almost exclusively focused on ombrophilous forests on the coastal plains of Brazil (Ivanauskas & Rodrigues, 2000; Costa & Mantovani, 1993; Martins, 1993). However, a number of diverse ecosystems are found in this biome (Brasil, 2000), including semi-deciduous and deciduous seasonal forests in the region of southwestern of the state of Bahia, and some have been poorly studied (especially those situated more inland) (Brasil, 2000). The semi-deciduous seasonal forest exhibits high biodiversity due to the confluence between Atlantic Forest, Caatinga (dryland vegetation), and Cerrado (Neotropical savanna) biomes (Soares-Filho, 2000; Daniel & Arruda, 2005; Dean, 2002). They are highly threatened and have experienced critical levels of fragmentation due to agricultural expansion, pasture formation, and urban occupation, among other factors (Campanili & Prochnow, 2006).

One of the most highly neglected regions in terms of studies of ant fauna diversity is the Atlantic Forest in the southwestern region of the state of Bahia, principally the Planalto da Conquista. These vegetation formations are considered "Inland Atlantic Forests of Bahia" (classified as semi-deciduous seasonal forests) The present study was designed to examine the ant fauna of this region and characterize the composition and diversity of ant species in the leaf litter of two remnant forest fragments situated in the municipality of Barra do Choça, in the Planalto da Conquista, state of Bahia State, Brazil.

Materials and Methods

Collection sites

The surveys were performed in two areas of Semi-Deciduous Seasonal Forest: "Remnant 1" (14°50'00"S 40°33'13"W; 86 hectares) and "Remnant 2" (14°48'29"S 40°35'23"W; 62 hectares) (Fig. 1). Both fragments were located in the municipality of Barra do Choça, in the state of Bahia, Brazil, within the transition zone between dense ombrophilous forests and seasonal deciduous forest areas (IBGE, 1993; 1997) in formations locally known as "mata de cipó" (Soares-Filho, 2000) ; between 20 and 50% of the trees there are large deciduous species (IBGE, 1993; 1997; Soares-Filho, 2000).

These once extensive native forest formations are currently represented only by remnant fragments that have experienced intense processes of environmental degradation from agro-pastoral activities and the selective extraction of commercially valuable trees (Soares-Filho, 2000; Projeto Mata Atlântica Interiorana da Bahia, 2002; Oliveira-Filho et al., 1994). The fragments studied here are embedded within monoculture and pasture matrices, and their interiors demonstrate clear evidence of selective cutting and cattle trails. The regional climate is classified as high-elevation tropical (IBGE, 1993; 1997), with a mean annual temperature of 19.8 °C, and a mean annual rainfall rate of 734 mm.

Fig 1 - Satellite picture of the two forest remnant where the experiment was conducted. Barra do Choça, Bahia, Brazil (Source: Google Earth, 2011).

Collection methodology

Ant collections were made in January and April/2011 using 47 Winkler sacks and 47 pitfall traps (Bestelmeyer *et al.*, 2000) in each fragment (one Winkler sack was destroyed in fragment 1, Table 1) distributed at intervals of 30 m within an area of approximately 10 hectares – but always at least 100 m from the external edges of the fragment. A pitfall trap was installed at each collection point and left for two days. These traps consisted of cups of 7 cm diameter by 10 cm height containing only water and detergent. When the pitfall trap was removed, an additional sample of 1 m² of leaf litter was removed from the same site, passed through a sieve, and then processed in a Winkler extractor for 48 hours (Bestelmeyer *et al.*, 2000). This standardized methodology was adapted from the *Ants of the Leaf Litter Protocol* of Agosti & Alonso (2001).

Biological material

The biological material collected in the field was preserved in ethanol and then taken to the Laboratório de Mirmecologia (CEPLAC/CEPEC/SECEN) and Laboratório de Biossistemática Animal (UESB/DEBI) where ant specimens were sorted out from the samples, mounted and identified to species level. The nomenclature follows Bolton *et al.* (2011) and Wilson (2003). Representative materials of all of the species are deposited in the Myrmecology Laboratory Collection (CPDC) under the reference number #5729.

Data analyses

Data was recorded using Excel version 10 software (Microsoft, 2007) which was used to calculate the relative frequencies of the species and their species richness for each different area and for each type of trap. EstimateS software version 8.2 (Colwell, 1997) was used to generate species accumulation curves for each area and each type of trap in terms of the sampling effort employed (Santos, 2003). The estimated species richness was subsequently calculated for each area using the Jackknife 2 index – an index based on the numbers of species that occur only once in a sample (*singletons*) and those occurring twice (*doubletons*). To determine if there were significant differences between the occurrences of ant species between the different fragments and between the different types of traps utilized, two-way analyses of variance (ANOVA Factorial) were performed using PRIMER 5 software (Clarke & Gorley 2001). The Shannon-Wiener diversity index was used to calculate the alpha diversity of the two fragments using Bioestat 5.3 software (Ayres, 2011). This index was chosen as it gives the same weight to both rare and abundant species. The t test was used to determine if there were significant differences between the diversity index values (also using Bioestat 5.3 software).

Results and Discussion

Observed richness

A synoptic list of the species collected in the present study, and their occurrences as a function of the collection areas and types of trap utilized, is presented in the Appendix. A total of 107 ant species belonging to 36 genera and 9 subfamilies were observed, with 83 species found in fragment 1, and 67 in fragment 2 (Table 1). These results indicated that the fragments analyzed retained relatively high faunal richnesses – even greater than reports for other areas of semideciduous forests in the Atlantic Forest biome (Mentoni *et al.*, 2011; Dias *et al.*, 2008, Castillho *et al.*, 2011).

The only other study that has examined ant diversity in remnant seasonal semi-deciduous forests in the region around the Planalto da Conquista was undertaken in 2011 by Sofia Campiolo, Ivan Nascimento, and Jacques Delabie (personal communication, May 4, 2011). These investigators undertook collections during the dry season using Winkler sacks in five areas relatively close to the present research site (in the municipalities of Barra do Choça, Itambé, and Vitória da Conquista). Their collection efforts were very similar to those of the present study, and they found between 47 and 86 species belonging to between 23 and 33 genera in the five fragments examined – results that are reasonably close to those of the present study (Table 1).

In both types of trap, fragment 1 demonstrated greater taxonomic richness than fragment 2 in terms of the species, genus, and subfamily levels. This result is somewhat surprising, as this fragment was 28% smaller than the other. In similar studies, and in accordance with the theory of island biogeography (MacArthur and Wilson, 1967), positive correlations have been found between species richness and fragment sizes (Morini *et al.*, 2007). One explanation for the observed discrepancy reported here could be that, in spite of the fact that the two fragments appeared to be phytosociologically similar (Avaldo de Oliveira Soares Filho, personal communication, January 10, 2011), the first fragment had more ecological niches available for ants. As such, and to better interpret the results, it would be interesting to quantify other diverse parameters in these forests fragments in future studies, such as the richness and diversity of their vegetation, their spatial structuring in terms of microhabitats, and available trophic resources.

Table 1 - Summary of the results found in two forest remnants at Barra do Choça, Bahia, Brazil. W: Winkler sack, P: pitfall trap.

	Remnant 1		Remnant 2	
	W	**P**	**W**	**P**
Number of samples	46	47	47	47
Number of species occurrences	349	201	382	231
Number of subfamilies	9	7	7	7
Number of genus	31	28	22	22
Number of species	66	54	50	44
Number of estimated species (Jackknife2)	109.6	94.9	66.6	73.3
Singletons	25	26	11	19
Doubletons	5	10	5	8
Shannon-Wiener	3.68	3.44	3.45	3.05

Table 2 – Variations fonts of ANOVA test (two-way analyzes of variance - Factorial ANOVA) for comparisons of samples collected in two forest remnants at Barra do Choça, Bahia, Brazil.

	SS	Degree of Freedom	MS	F	p
Intercept	6287.75	1	6287.75	140.653	0
Place/trap	209.446	1	209.446	4.6852	0.03156
Remnant	191.356	1	191.356	4.2805	0.03978
Remnant/trap	7.813	1	7.813	0.1748	0.67634
Error	9343.16	209	44.704		

Winkler sacks collected larger numbers of species in both fragments than did pitfall traps (Table 1), although these differences were not statistically significant (F = 4.69 and P = 0.032, Table 2). Most published studies have shown that Winkler sacks can collect greater numbers of species than most other techniques (Sabu et al., 2011; Vargas et al., 2009), although this assertion was recently challenged by Souza et al., (2012) to the profit of pitfall traps. Nevertheless, both techniques complement each other for maximum sampling efficiency (Delabie et al., 2000). Winkler sacks are most efficient for collecting smaller species with cryptic behavior and low activity levels (Agosti et al, 2000), while pitfall traps are better for collecting large, active foraging ants that are not easily collected through Winkler method, such as species of the genera *Camponotus, Pachycondyla* or *Odontomachus.*

Winkler sacks also harvested higher mean numbers of species per sampling point than pitfall traps (Table 3). These results are similar to those previously reported for native forest sites in the Atlantic Forest biome (Marinho et al., 2002; Suguituru et al., 2013). It is worth noting that there were differences between the two fragments studied here in terms of their mean species richness per sampling point, with this value being lower in fragment 2 (although these differences were not statistically significant) (Table 3).

Table 3 -Mean number of ant species (mean occurrence) observed in forest remnants in Barra do Choça, Bahia, Brazil. Values in the main diagonal of the matrix are means ± 1 SE. Differences between the pairs of ant species are based on Fisher post-hocTest. ns = not significant; * = $P < 0.05$; MW1 = species collected with Winkler sacks in first forest remnant; MW2 = species collected with Winkler sacks in second forest remnant; MP1 = species collected with pitfall traps in first forest remnant; MP2 = species collected with pitfall traps in second forest remnant;

		MW	MW	MP	MP
		1	2	1	2
MW	1	5.33±0.76	NS	NS	NS
MW	2		7.64±1.13	*	NS
MP	1			3.72±0.65	NS
MP	2				5.25±1.18

Richness estimates

Richness accumulation curves for each type of trap for each area are shown in fig 2A and 2B. Analyses of the accumulated richness curves indicated complete sampling of the ant fauna was not achieved in either type of trap, as none of the curves approached the asymptote. Thus, projected richness values were always greater than observed values (Table 1), which is the usual situation in biodiversity studies in the tropics (Martins & Santos, 1999; Feitosa & Ribeiro, 2005; Baccaro et al., 2011; Braga et al., 2010; Leponce et al., 2004; Delabie et al., 2007), since many rare species continue to be encountered even after extremely intense sampling efforts

(Santos, 2003).

The high estimated richness values could be explained by the study areas possibly having heterogeneous species distributions, because, according to Chao (1987), the greater the heterogeneity of the species' spatial distributions, the greater will be the observed divergences between observed and expected richness.

The large number of unique species encountered (Table 1) corroborated the results of the observed richness curves and diversity estimators. Unique species may be present in the area only as foragers ("tourist" species, according to Belshaw & Bolton, 1993), as rare species, as generalist species occasionally feeding in the locality, or as specialist that feed exclusively on plants occurring only in a single site in the study area. Also, the surveys may have been undertaken using inadequate methodologies (particularly an issue with small populations) (Novotny & Basset, 2000). The proportions of unique species in the present work were always greater than 30% of the total numbers of species collected – reaching up to almost 50% in the case of pitfall traps in the first fragment. Such high proportions of unique species are consistent with other biodiversity studies of arthropods in tropical regions (Coddington et al., 2009) and with other ant studies undertaken in Atlantic Forest areas (Pacheco et al., 2009).

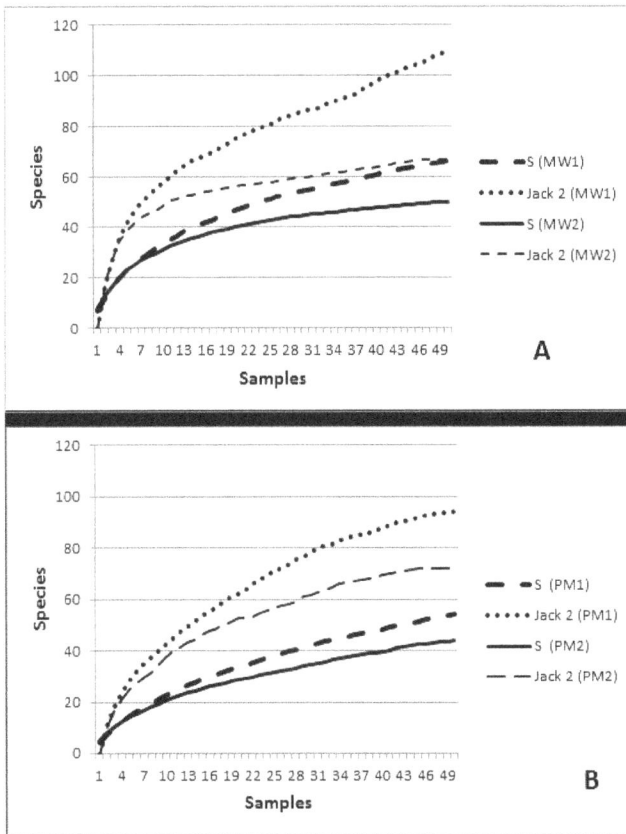

Fig 2 - Species accumulation curves for ant species collected in Winkler sacks (A) and pitfall traps (B) in two forest remnant in Barra do Choça, Bahia, Brazil. MW1: first forest remnant, MW2: second forest remnant, MP1: first forest remnant, MP2: second forest remnant, S: observed richness, Jack 2: richness index estimated by Jackknife 2.

Analyses of faunal composition

The richest subfamilies at the species level in fragment 1 were: Myrmicinae (59% of the total number of species), Formicinae (15.7%), and Ponerinae (13.3%). The same tendency was observed at the genus level: Myrmicinae (52.6% of the total number of genera), Formicinae (13.2%), and Ponerinae (10.5%) (Appendix). The richest subfamilies at the species level in fragment 2 were: Myrmicinae (58% of the total number of species encountered), Ponerinae (14.5%), and Formicinae (14.5%), while at the generic level, the richest subfamilies were: Myrmicinae (44.4% of all of the genera encountered), Formicinae (18.5%), and Ponerinae (11.1%) (Appendix).

The observation that the subfamily Myrmicinae was the richest in both fragments at both the genus and species levels, while the subfamilies Formicinae and Ponerinae demonstrated similar species richness, had been reported in other surveys of forested areas in the Neotropical region (Miranda *et al.*, 2012).

A particularly interesting find in fragment 1 was the capture of two specimens of *Ochetomyrmex* (*Ochetomyrmex* sp_LBSA_14010266) (very similar to *Ochetomyrmex semipolitus* but possibly distinct) – a new record for this genus in the state of Bahia, and extending the geographical distribution of this species by more than 2,380 km to the east (referring to the map given by Fernandez, 2003). A new species of *Oxyepoecus* Santschi, 1923 was also encountered (six specimens in fragment 2), which is now being described (Sebastien Lacau, personal communication, June 4, 2012). Another important collection was the capture of three specimens of *Monomorium delabiei* Fernandez, 2007 in fragment 2 – a species up to now only known from the holotype, described from Guaratinga, Bahia, Brazil (Fernandez, 2007).

Differences existed in the composition of the species encountered in the two study areas, with only 38% of the species common to both areas – indicating considerable differences in the compositions of their respective species communities (Appendix). There was also a relationship between the global richness of each remnant at the species level and its degree of exclusivity, with fragment 1 being the most species rich (83 species) and having the most exclusive faunal composition (with 50.6% of the species and 36.8% of the genera being exclusive to that fragment), while fragment site 2 was the least rich (67 species) and had the least exclusive faunal composition (with 34.3% of the species and 7.6% of the genera being exclusive to that fragment).

Relative frequency and dominance

To determine if the ant communities are organized into defined structural patterns, we examined the species relative frequencies in the two areas according to the types of traps in which they were caught. In the case of Winkler sacks, it was observed that 10 species were responsible for half of the oc-currences in fragment 1 (fig 3A), and eight species for half of the occurrences in fragment 2 (fig 3B). Four of those species (*Hypoponera sp.* 3, *Nylanderia sp.* 2, *Gnamptogenys striatula* and *Solenopsis sp.* 6) only were common to both areas. In the case of the pitfall traps, eight species were responsible for half of the occurrences in fragment 1 (fig 3C) and five species in fragment 2 (fig 3D), with three species (*Pheidole radoszkowskii*, *Gnamptogenys striatula*, and *Nylanderia* sp. 2) being encountered in both fragments.

The overall results demonstrated that the three most frequent species (*Hypoponera sp.* 3, *Octostruma sp.* 1, and *Strumigenys sp.* 4) were exclusively collected with Winkler sacks, suggesting that they have cryptic lifestyles, and nest and forage below the leaf litter surface.

The same types of frequency analyses were performed to determine the identities and relative frequencies of the genera responsible for 50% of the total occurrences in the collections. In the case of the Winkler sacks, five genera (*Pheidole, Solenopsis, Hypoponera, Nylanderia*, and *Strumigenys*) were responsible for 50% of the occurrences in fragment 1 (fig 4A), while four genera (*Solenopsis, Hypoponera, Nylanderia* and *Pheidole)* were responsible for this same percentage in fragment 2 (fig 4B). Four genera (*Pheidole, Solenopsis, Hypoponera*, and *Nylanderia*) were encountered in both areas.

In the case of the pitfall traps, three genera (*Pheidole, Gnamptogenys*, and *Ectatomma*) were found to be responsible for 50% of the occurrences in fragment 1 (fig 4C), and four genera in fragment 2 (*Pheidole, Pachycondyla, Linepithema*, and *Solenopsis*) (fig 4D). Only one genus (*Pheidole*) was encountered in both fragments.

These results emphasize the notable dominance of the genera *Pheidole, Solenopsis*, and *Nylanderia*. The genus *Pheidole* has been observed to be the most dominant in many studies of ant diversity, with two of its species consistently appearing among the most abundant arthropod representatives in the leaf litter. *Pheidole* is the most diversified genus in the family Formicidae (Wilson, 2003) and its species are encountered in all soil microhabitats, have wide ranges of feeding habits (most are omnivores), and demonstrate great efficiency in recruiting workers to exploit many trophic resources (Fernández, 2003). Some *Pheidole* species are rather aggressive in their relationships with competitors, are opportunists and can colonize a wide diversity of environments (Wilson, 2003). The observed dominance of this genus in the present study therefore corroborates the recognition of this genus as the most abundant and diversified in the Neotropical region (Majer & Delabie, 1994; Marinho, *et al.* 2002; Vasconcelos, 1999).

In the same sense, the high abundance of *Solenopsis* species observed in the present study confirms previous observations in the literature. Some species of this genus are common throughout the world (Fernandéz, 2003), principally in the leaf litter, with many of them being generalists in terms of their habitats and diets (Fowler *et al.,* 1991). These species

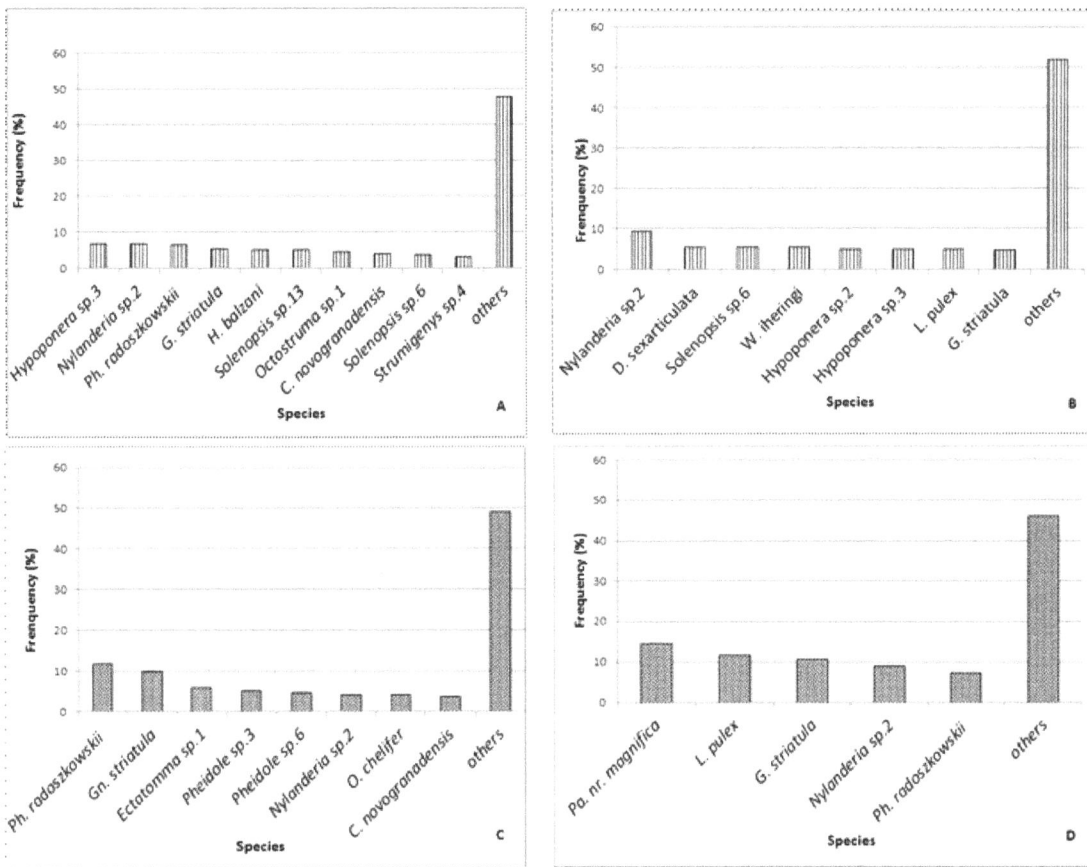

Fig 3 - Relative frequency of the species more collected in two forest remnant at Barra do Choça, Bahia, Brazil. (A) Samples collected with Winkler sacks in the first forest remnant; (B) Samples collected with Winkler sacks in the second forest remnant; (C) Samples collected with pitfall traps in the first forest remnant; (D) Samples collected with pitfall traps in the second forest remnant;

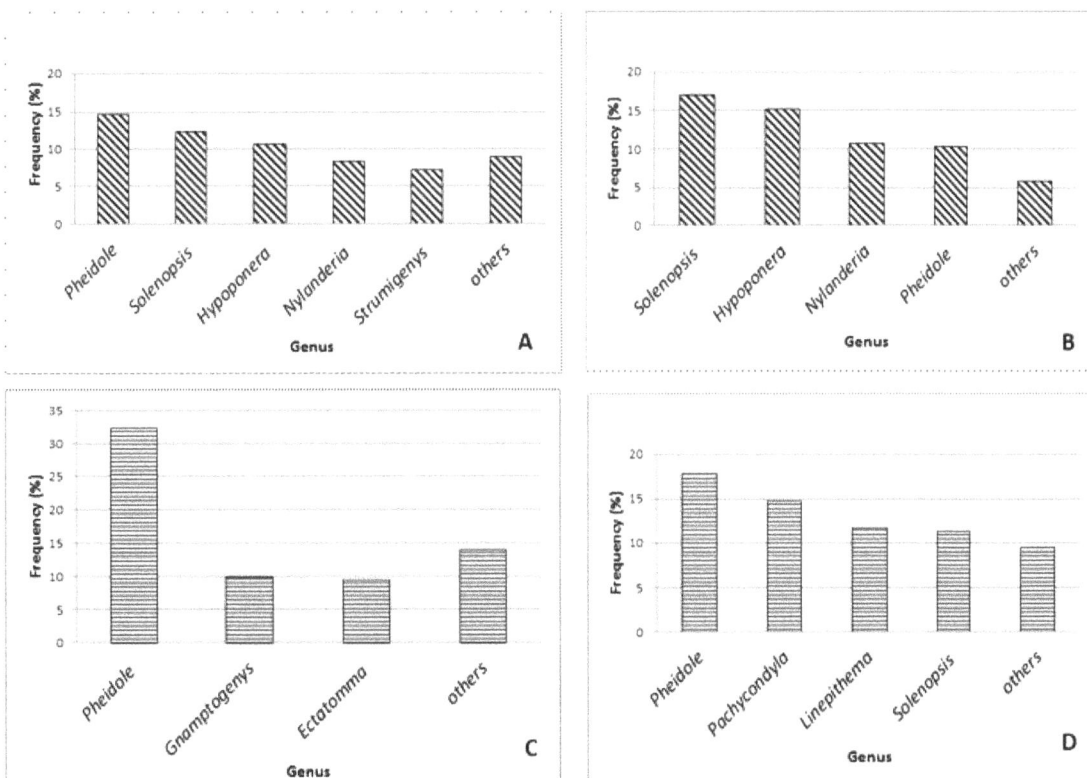

Fig 4 - Relative frequency of the most frequently collected genera in two forest remnants at Barra do Choça, Bahia, Brazil. (A) Samples collected with Winkler sacks in the first forest remnant; (B) Samples collected with Winkler sacks in the second forest remnant; (C) Samples collected with pitfall traps in the first forest remnant; (D) Samples collected with pitfall traps in the second forest remnant.

are also encountered with relatively high frequencies in agricultural areas (Dias *et al.*, 2008).

The genus *Nylanderia* was also very frequent in the present study, and previous publications have shown that the relatively small individuals that are characteristic of this genus are quite abundant (Mentone, 2011), being terricolous or arborous, or occupying the leaf litter, in both natural and disturbed environments (Fernández, 2003).

Diversity indices

The alpha diversity estimates for the fragments, as calculated by the Shannon and Wiener index, can be found in Table 1 and are comparable to other biodiversity studies of ants in the Neotropical region (Lopes *et al.*, 2010; Lutinski *et al.*, 2008). The greatest alpha diversity was observed in fragment 1, corroborating the hypothesis that fragment 1 had the best phytosociological quality and favored the occurrence of a wider diversity of species. The situation in fragment 2 apparently represents a simplification of the original community structure. The differences between fragments were not, however, statistically significant (t = 1.3291; *P* = 0.1576).

Concluding comments

The results obtained in the present study are totally original, and no similar research has previously been undertaken in the Planalto da Conquista region. It was observed that while the two fragments analyzed had both been subjected to anthropogenic modifications, they still maintained high natural faunal diversities typical of inland Atlantic Forest sites in state of Bahia. The diversity and species richness observed in this study were actually greater than those reported in the literature for other semi-deciduous forests of the Atlantic forest biome (Mentoni *et al.*, 2011; Dias *et al.*, 2008, Castilho *et al.*, 2011).

When the two fragments were compared, it could be seen that the numbers of species, genera, and sub-families were greater in fragment 1. This result suggests that although the two fragments were superficially similar, the first was better preserved in terms of the ecological niches available for the ant fauna and therefore better reflected the original community structures of the Formicidae in this ecosystem; the second fragment represented a greater simplification of the more complex original community.

The results presented here will hopefully be useful to future conservation plans for remnant forest areas in the Planalto da Conquista, as ants can be easily used as biological indicators of the degradation (or preservation) of areas subjected to anthropogenic impacts (or management). Additionally, it is hoped that this study will serve as a baseline for further investigations about regional biodiversity, as this region is desperately lacking this kind of information.

Acknowledgments

We would like to thank Drs. Sofia Campiolo and Ivan Nascimento allowing access to the unpublished ant data of the Planalto da Conquista experiment. We convey our thanks to the graduate and post-graduate students who assisted in data collection of this study; and to our friend Dr. Paulo Sávio for the help in the statistical analysis. Also we thank the granting bodies of CNPq, CAPES, 019/2013 FAPESB and the PRONEX program, project SECTI-FAPESB/ CNPq PNX 001/2009.

References

Ayres, M. (2011) Bioestat 5.3: aplicações estatísticas nas áreas das ciências biológicas e médicas. 3.ed. Sonopress, Belém. Retrieved from: www.mamiraua.org.br/pt-br

Agosti, D., Majer, J. D., Alonso, L. E. & Schultz, T.R. (2000). Ants: standard methods for measuring and monitoring biodiversity. Washington D.C. Smithsonian Institution Press.

Agosti, D. & Alonso, L. E. (2000) The ALL Protocol: a standard protocol for the collection of ground-dwelling ants. In: Agosti, D., Majer, J.D.; Alonso, L. E. & Schultz, T.R. (Eds.). Ants: standard methods for measuring and monitoring biodiversity (pp. 204–206). Washington, D. C.: Smithsonian Institution Press.

Alonso, L.E., & D. Agosti. (2000) Biodiversity studies, monitoring, and ants: an overview. In: Agosti, D. Majer, J. D., Alonso, L. E. & Schultz, T. R. Ants: standard methods for measuring and monitoring biodiversity (p.1-8). Washington, D. C.: Smithsonian Institution Press.

Baccaro, F. B., Ketelhut, S. M. & Morais, J. W. (2011) Efeitos da distância entre iscas nas estimativas de abundância e riqueza de formigas em uma floresta de terra-firme na Amazônia Central. Acta Amazonica, 41: 115-122.

Belshaw, R. & Bolton, B. (1993) The effect of forest disturbance on leaf litter ant fauna in Ghana. Biodiversity and Conservation, 2: 656-666.

Bestelmeyer, B.T., Agosti, D., Alonso, L.E., Brandão C.R.F., Brown, W.L. Jr, Delabie, J.H.C. & Silvestre, R. (2000) Field techniques for the study of ground-dwelling ants: an overview, description, and evaluation. In: Agosti, D., Majer, J.D.; Alonso, L. E. & Schultz, T.R. (Eds.). Ants: standard methods for measuring and monitoring biodiversity (p.22-144). Washington D.C. Smithsonian Institution Press.

Bolton, B. & Alpert G. D. (2011) Barry Bolton's synopsis of the formicidae and catalogue of ants of the world. Retrieved from: http://gap.entclub.org/ (accessed date: 1 March, 2011)

Braga, D. L, Louzada, J. N. C., Zanetti, R. & Delabie, J. C. H. (2010) Avaliação rápida da diversidade de formigas em sistemas de uso do solo no sul da Bahia. Londrina: Neotropical Entomology, 39: 464-469.

Brasil. Ministério do Meio Ambiente, Secretaria de Biodiversidade e Florestas (2000). Avaliação e ações prioritárias para a conservação da biodiversidade da Mata Atlântica e Campos Sulinos. Brasília. Retrieved from: www.rbma.org.br/anuario/pdf/areasprioritarias.pdf (accessed date: 6 March, 2011).

Campanili, M. & Prochnow, M. (2006). Mata Atlântica – uma rede pela floresta, Brasília: RMA. Retrieved from: www.apremavi.org.br (accessed date: 8 June, 2010)

Castilho, G. A., Noll, F. B., Silva, R. E. & Santos, E. F. (2011). Diversidade de Formicidae (Hymenoptera) em um fragmento de floresta estacional semidecídua no noroeste do estado de São Paulo, Brasil. Revista Brasileira de Biologia, 9: 224-230.

Chao A. (1987). Estimating the population size for capture-recapture data with unequal catchability. Biometrics, 43: 783-791.

Clarke, K. R. & Gorley, R. N. (2001) Software PRIMER v5. Plymouth, PRIMER-E. Retrieved from: http://www.primer-e.com/index.htm (accessed date: 8 June, 2010)

Coddington, J.A., Agnarsson, I., Miller, J.A., Kuntner, M. & Hormiga, G. (2009). Undersampling bias: the null hypothesis for singleton species in tropical arthropod surveys. Journal of Animal Ecology, 78: 573-84.

Colwell, R.K. (1997) EstimateS: Statistical estimation of species richness and shared species from samples. Version 7.5. User's guide and applications published. Retrieved from: http://viceroy.eeb.uconn.edu/estimates. (accessed date: 10 august, 2010)

Costa, L. G. S. & Mantovani, W. (1993) Flora arbustivo-arbórea de trecho de mata mesófila semidecídua na Estação Ecológica de Ibicatu, Piracaba - SP. Hoehnea, 22: 47-59.

Daniel, O. & Arruda, L. (2005) Fitossociologia de um fragmento de floresta estacional semidecidual aluvial as margens do Rio Dourados, MS. Scientia Forestalis (IPEF), 68: 69-86.

Dean, W. (2002) A ferro e fogo: A história e a devastação da mata atlântica brasileira. São Paulo: Companhia das Letras. 484 p.

Delabie, J.H.C., Fisher, B. L., Majer, J. D. & Wright, I. W. (2000) Sampling effort and choice of methods. In: Agosti, D., Majer, J., Alonso, L.E. & Schultz, T.R. (eds.), Ants: standard methods for measuring and monitoring biodiversity (pp. 145-154). Washington: Smithsonian Institution Press.

Delabie, J. H. C., Jahyny, B., Nascimento, I. C., Mariano, C. S. F., Lacau, S., Campiolo, S., Philpott, S. M. & Leponce, M. (2007) Contribution of cocoa plantations to the conservation of native ants (Insecta: Hymenoptera: Formicidae) with a special emphasis on the Atlantic Forest fauna of southern Bahia, Brazil. Biodiversity and Conservation, 16: 2359-2384.

Delabie, J. H. C., Paim, V. R. L. M., Nascimento, I. C., Campiolo, S. & Mariano, C. S. F. (2006) As formigas como indicadores biológicos do impacto humano em manguezais da costa sudeste da Bahia. Neotropical Entomology, 35: 602-615.

Dias, N.S., Zanetti, R., Santos, M.S., Louzada, J. & Delabie, J.H.C. (2008) Interação de fragmentos florestais com agro-ecossistemas adjacentes de café e pastagem: respostas das comunidades de formigas (Hymenoptera, Formicidae). Iheringia, 98 (1): 136-142.

Feitosa, R. S. M. & Ribeiro, A. S. (2005) Mirmecofauna (Hymenoptera: Formicidae) de serapilheira de uma área de Floresta Atlântica no Parque Estadual da Cantareira – São Paulo, Brasil. Biotemas, 18(1): 51-71.

Fernández, F. (2003). Introducion a las hormigas de la region Neotropical. Instituto de Investigacion de Recursos Biologicos Alexander von Humboldt, Bogota, Colombia.

Fernández, F. & Ospina, M. (2003) Sinopsis de las hormigas de la región Neotropical. In: Fernández, F. (Eds). Introductión a las hormigas de la región Neotropical (pp. 337-349). Instituto de Investigacion de Recursos Biologicos Alexander von Humboldt, Bogota, Colombia.

Fernández, F. (2007) Two new South American species of *Monomorium* Mayr with taxonomic notes on the genus (pp. 128-145). In Snelling, R. R., Fisher, B. L. & Ward, P. S. (Eds) Advances In Ant Systematics (Hymenoptera: Formicidae): Homage to E. O. Wilson – 50 Years of Contributions. Memoirs of the American Entomological Institute, 80.

Fowler, H. G., Forti, L. C., Brandão, C. R. F., Delabie, J. H. C & Vasconcelos, H. L. (1991) Ecologia nutricional de formigas. In: Pazzini, A. R. & Parra , J. R. P. (Eds) Ecologia nutricional de insetos e suas implicações no manejo de pragas (pp.131-209). São Paulo: Manole.

Instituto Brasileiro de Geografia e Estatística – IBGE (1993). Mapa de vegetação do Brasil. Rio de Janeiro, Fundação Instituto Brasileiro de Geografia e Estatística. Retrieved from: www.ibge.gov.br (accessed date: 10 august, 2011)

Instituto Brasileiro de Geografia e Estatística – IBGE (1997) Recursos naturais e meio ambiente: uma visão do Brasil. 2ª edição. Rio de Janeiro. Retrieved from: www.ibge.gov.br (accessed date: 10 august, 2011)

Ivanauskas, N. M. & Rodrigues, R. R. (2000) Florística e fitossociologia de remanescentes de floresta estacional decidual em Piracicaba, São Paulo, Brasil. Revista Brasileira de Botânica, 23: 291-304.

Leponce, M.; Theunis, L.; Delabie, J.H.C & Roisin, Y. (2004) Scale dependence of diversity measures in a leaf-litter ant assemblage. Ecography, 27: 253-257.

Lopes, D. T., Lopes, J., Nascimento, I. C. & Delabie, J. H. C. (2010) Diversidade de formigas epigéicas (Hymenoptera,

Formicidae) em três ambientes no Parque Estadual Mata dos Godoy, Londrina, Paraná. Iheringia Ser. Zool., 100: 84-90.

Lutinski, J. A., Garcia, F. R. G., Lutinski, C. J. & Iop, S. (2008) Diversidade de formigas na Floresta Nacional de Chapecó, Santa Catarina, Brasil. Ciência Rural, 38: 1810-1816.

Magurran, A.E. (2004) Measuring Biological Diversity. Blackwell Publishing.

Majer, J.D. & Delabie, J.H.C. (1994) Comparision of the ant communities of annually inundated and terra firme forests at Trombetas in Brazilian Amazon. Insectes Socieux, 41: 343-359.

Marinho, C.G.S., Zanetti, R., Delabie, J.H.C., Schlindwein, M.N. & Ramos, L.S. (2002) Diversidade de formigas (Hymenoptera: Formicidae) da serapilheira em eucaliptais (Myrtaceae) e áreas de cerrado de Minas Gerais. Neotropical Entomology, 31: 187-195.

Martins, F.R. & Santos, F.A.M. (1999) Técnicas usuais de estimativa da biodiversidade. Revista Holos 1 (edição especial): 236-267.

Martins, F.R. (1993) Estrutura de uma floresta mesófila. Campinas: Ed. Universidade Estadual de Campinas, 246 p.

Mentone, T.O., Diniz, E. A., Munhae, C. B., Bueno, O. C. & Morini, M. S. C. (2011) Composição da fauna de formigas (Hymenoptera: Formicidae) de serapilheira em florestas semidecídua e de *Eucalyptus* spp., na região sudeste do Brasil. Biota Neotropica, 11: 237-246.

MacArthur, R. H. & Wilson, E. O. (1967). The Theory of Island Biogeography. Princeton, N.J.: Princeton University Press.

Miranda, P. N., Oliveira, M. A., Baccaro, F. B., Morato, E. F. & Delabie, J. H. C. (2012) Check list of ground-dwelling ants (Hymenoptera: Formicidae) of the eastern Acre, Amazon, Brazil. CheckList, 8: 722-730.

Morini, M. S. C., Munhae, C. B., Leung, R., Candiani, D. F. & Voltolini, J. C. (2007) Comunidades de formigas (Hymenoptera, Formicidae) em fragmentos de Mata Atlântica situados em áreas urbanizadas. Iheringia Sér Zool., 97: 246-252.

Novotny, V. & Basset, Y. (2000) Rare species in communities of tropical insect herbivores: pondering the mystery of singletons. Oikos, 89: 564-572.

Oliveira-Filho, A. T.; Scolforo, J. R. & Mello, J. M. (1994) Composição florística e estrutura comunitária de um remanescente de floresta semidecídua Montana em Lavras (MG). Revista Brasileira de Botânica, 17: 159-174.

Pacheco, R., Silva, R.R., Morini, M.S. DE C. & Brandão, C.R.F. (2009) A comparison of the leaf-litter ant fauna in a secondary Atlantic Forest with an adjacent pine plantation in Southeastern Brazil. Neotropical Entomology, 38: 55-65

Peck, S. L., Mcquaid B. & Campbell, C. L. (1998) Using ant species (Hymenoptera: Formicidae) as a biological indicator of agroecosystem condition. Enviromental Entomology, 27: 1102-1110.

Projeto Mata Atlântica Interiorana da Bahia (2002). O Planalto da Conquista. Retrieved from: http://www.uesc.br/biota/ (accessed date: 10 august, 2011)

Rico-Gray, V. & Oliveira, P. S. (2007) The ecology and evolution of ant-plant interactions. The University of Chicago Press: 320 p.

Sabu T. K., Shiju R. T, Vinod K. V. & Nithya S. (2011) A comparison of the pitfall trap, Winkler extractor and Berlese funnel for sampling ground-dwelling arthropods in tropical montane cloud forests. Journal of Insect Science, 11(28). Retrieved from: insectscience.org/11.28 (accessed date: 10 august, 2011)

Santos, A. J. (2003) Estimativas de riqueza em espécies. In: Cullen L. JR.; Valladares C. P.; Rudran R. (Eds). Métodos de estudos em biologia da conservação e manejo da vida silvestre (pp.19-41). Fundação O Boticário de Proteção à Natureza. Curitiba: UFPR.

Schowalter, T.D. & Sabin T.E. (1991) Serrapilheira microarthropod responses to the canopy herbivory, season and decomposition in serrapilheira bags in a regenerating conifer ecosystem in Western Oregon. Biology and Fertility of Soils, 11: 93-96.

Silva, R.R. & Brandão, C.R.F. (2010) Morphological patterns and community organization in leaf-litter ant assemblages. Ecological Monographs, 80: 107–124

Soares-Filho, A.O. (2000) Estudo fitossociológico de duas florestas em região ecotonal no planalto de Vitória da Conquista, Bahia, Brasil. Dissertação de Mestrado — Pontifícia Universidade de São Paulo.

Souza, J. L. P., Baccaro, F. B., Landeiro, V. L., Franklin, E. & Magnusson, W.E. (2012) Trade-offs between complementarity and redundancy in the use of different sampling techniques for ground-dwelling ant assemblages. Applied Soil Ecology, 56: 63-73.

Suguituru, S. S., Souza, D. R., Munhae, C. B., Pacheco, R. & Morini, M. S. C. (2013) Ant species richness and diversity (Hymenoptera: Formicidae) in Atlantic Forest remnants in the Upper Tietê River Basin. Biota Neotropica, 13: 141-152.

Vargas, A. B., Queiroz, J. M., Mayhé-Nunes, A. J.; Souza, G. O. & Ramos, E. F. (2009) Teste da regra de equivalência energética para formigas de serapilheira: efeitos de diferentes métodos de estimativa de abundância em floresta ombrófila. Neotropical Entomology, 38: 867-870.

Vasconcelos, H.L. (1999) Effects of forest disturbance on the structure of ground-foraging ant communities in Central Amazonia. Biodiversity and Conservation, 8: 409-420.

Veiga-Ferreira, S., Mayhé-Nunes, A. J. & Queiroz, J. M. (2005). Formigas de Serapilheira na Reserva Biológica do Tinguá, Estado do Rio de Janeiro, Brasil (Hymenoptera: Formicidae). Revista da Universidade Rural, Sér. Ciênc. Vida.. Seropédica, 25: 49-54.

Wilson, E. O. & Hölldobler, B. (2005) The rise of the ants: A phylogenetic and ecological explanation. PNAS, 102: 7411 – 7414.

Wilson, E.O. (2003) The genus *Pheidole* in the New World: A Dominant, Hyperdiverse Ant Genus. Harvard University Press.

Appendix - List of ant species and their frequency of occurrence collected by two sampling methods and in two forest remnants in Barra do Choça, Bahia, Brazil: MW1 and MP1 represent the species collected with Winkler sacks and pitfall traps in first forest remnant; MW2 and MP2 represent the species collected with Winkler sacks and pitfall traps in the second forest remnant.

Subfamilies	Species	MW1	MP1	MW2	MP2
Dolichoderinae	*Azteca sp. 1*		1	7	
	Azteca sp. 2	1			
	Dolichoderus attelaboides (Fabricius, 1775)	1	1		
	Linepithema pulex Wild, 2007			20	27
Ecitoninae	*Eciton burchelli* Westwood, 1842		1		
	Labidus praedator (Smith, 1858)		1		
Ectatomminae	*Ectatomma edentatum* Roger 1863	1	7		
	Ectatomma sp. 1		12	1	
	Ectatomma sp. 3				10
	Gnamptogenys striatula Mayr, 1884	19	20	19	25
Formicinae	*Acropyga guianensis* Weber, 1944				1
	Acropyga sp. 1		1		
	Brachymyrmex patagonicus Mayr, 1868	4		6	
	Brachymyrmex sp. 1			7	
	Brachymyrmex sp. 3		2		6
	Brachymyrmex sp. 4	8			
	Brachymyrmex sp. 5	6	1		
	Camponotus cingulatus Mayr, 1862		7		4
	Camponotus novogranadensis Mayr, 1870	14	8	5	5
	Camponotus renggeri Emery, 1894	1			
	Camponotus sp. 3	2			
	Camponotus (Tanaemyrmex) sp. 5		1		
	Nylanderia sp. 1	5	1	2	1
	Nylanderia sp. 2	24	9	37	21
	Nylanderia sp. 4			2	
	Nylanderia sp. 5			1	
	Paratrechina longicornis (Latreille, 1802)		1		
Heteroponerinae	*Heteroponera mayri* Kempf, 1962		3		6
Myrmicinae	*Acanthognathus sp 1*	1			
	Acromyrmex aspersus (Smith, 1858)	1			
	Acromyrmex sp. 1		1		1
	Apterostigma pilosum Mayr, 1865	3			

	Basiceros disciger (Mayr, 1887)	1			
	Carebara sp. 1	6		4	1
	Carebara sp. 2	4			
	Crematogaster distans Mayr, 1870	2	1		
	Crematogaster sp. 2	1		5	1
	Crematogaster sp. 3	1			
	Cyphomyrmex transversus Emery, 1894	7	2	5	1
	Cyphomyrmex strigatus gp. sp. 1	1			
	Hylomyrma balzani (Emery, 1894)	18	7	19	3
	Megalomyrmex drifti Kempf, 1961		1		
	Megalomyrmex goeldii Forel, 1912	7	8		
	Monomorium delabiei Fernández, 2007			1	
	Myrmicocrypta sp. 1		1		
	Myrmicocrypta sp. 2	4			
	Ochetomyrmex sp. 1 (LBSA 40 10 266)	1	1		
	Octostruma sp. 1	16		3	
	Oxyepoecus myops Albuquerque & Brandão, 2009			5	
Myrmicinae (cont.)	*Oxyepoecus sp. 2* (LBSA_ 1 40 10 26 4)			4	2
	Pheidole radoszkowskii Mayr, 1884	23	24	8	17
	Pheidole tristis gp. sp. 1		1		
	Pheidole sp. 2			1	
	Pheidole fallax gp. sp. 3	1	1 1	6	9
	Pheidole tristis gp. sp. 4	5	1	2	
	Pheidole sp. 5	6	4	1	
	Pheidole flavens gp. sp. 6		10	7	6
	Pheidole tristis gp. sp. 7		2	4	2
	Pheidole sp. 8	1	1	1	1
	Pheidole diligens gp. sp. 9	1	1		
	Pheidole diligens gp. sp. 10	1	2	1	2
	Pheidole fallax gp. sp. 11	1	3		
	Pheidole sp. 12	5	1		
	Pheidole tristis gp. sp. 13	4	1		
	Pheidole diligens gp. sp. 15	1	1		2
	Pheidole sp. 17	1	1		
	Pheidole sp. 18			7	
	Pheidole sp. 21			1	2
	Pheidole transversostriata Mayr, 1887	1	1		
	Procryptocerus hylaeus Kempf, 1951		1		
	Solenopsis sp. 3	2			2
	Solenopsis sp. 4			17	
	Solenopsis sp. 6	1 3	2	22	1 3
	Solenopsis sp. 7			1	

Myrmicinae (cont.)	*Solenopsis sp. 8*				3
	Solenopsis sp. 9	8	6	14	3
	Solenopsis sp. 10				1
	Solenopsis sp. 11				1
	Solenopsis sp. 12	2	1	4	1
	Solenopsis sp. 13	18	2	5	1
	Solenopsis sp. 14				1
	Strumigenys appretiata (Borgmeier, 1954)			2	1
	Strumigenys denticulata Mayr, 1887	2			
	Strumigenys sp. 1			2	
	Strumigenys sp. 2	9	2		
	Strumigenys sp. 3	5			1
	Strumigenys sp. 4	11		8	
	Trachymyrmex sp. 1	1			
	Wasmannia auropunctata (Roger, 1863)	4	2	6	2
	Wasmannia iheringi Forel, 1908	8	2	22	2
	Wasmannia sp. 3	4			
Ponerinae	*Anochetus simoni* Emery, 1890	3			
	Hypoponera foreli (Mayr, 1887)	5		5	2
	Hypoponera sp. 1	3			
	Hypoponera sp. 2	5	2	20	1
	Hypoponera sp. 3	24		20	
	Hypoponera sp. 5			1	
	Hypoponera sp. 6			9	
	Hypoponera sp. 7			3	
	Odontomachus chelifer (Latreille, 1802)		9		8
	Odontomachus sp. 2	3	3		1
	Odontomachus sp. 3	1			
	Pachycondyla crenata (Roger, 1861)	1			
	Pachycondyla moesta Mayr, 1870	1		1	
	Pachycondyla nr. magnifica		7		34
Proceratiinae	*Discothyrea sexarticulata* Borgmeier, 1954	1		22	1
Pseudomyrmecinae	*Pseudomyrmex tenuis* (Fabricius, 1804)	1	1		1

20

Size variation in eggs laid by normal-sized and miniature queens of *Plebeia remota* (Holmberg) (Hymenoptera: Apidae: Meliponini)

MF Ribeiro[1,2], PS Santos Filho[1]

1 - Universidade de São Paulo, São Paulo, SP, Brazil

2 - Empresa Brasileira de Pesquisa Agropecuária (EMBRAPA), Petrolina, PE, Brazil

Keywords

Queen body size, egg morphometry, stingless bees.

Corresponding author

Márcia de Fátima Ribeiro
EMBRAPA SEMIÁRIDO, BR 428, Km 152
Zona Rural, C.P. 23, 56302-970
Petrolina - PE, Brazil
E-Mail: marcia.ribeiro@embrapa.br

ABSTRACT

Miniature stingless bee queens have been studied concerning frequency distribution, production and egg laying performance. This study aimed to investigate size variation in eggs laid by *Plebeia remota* (Holmberg) queens and whether it is due to differences in queen size or colony conditions. A sample of 10 queens (8 of typical size and 2 miniature) was measured morphometrically (head width, interorbital distance, and intertegular distance) as well the eggs they laid (length, width and volume). Initially, eggs were analyzed when laid by queens in their own colonies. Significant differences were found for length, width and volume of eggs considering the total group of queens or both queen morphotypes. However, no significant correlations were found between queen size and egg size. Afterwards, two experiments were performed to evaluate the influence of colony conditions on egg size. Firstly, we shifted the queens from their original colonies (i.e., a typical queen was placed into a miniature queen colony, and vice-versa). Secondly, they were put into another colony (both types of queens, one each time, were placed on a third colony, a 'host colony'). In all situations, both queen morphotypes laid eggs of similar or different sizes than before, often with significant differences. The results indicate that variation in egg size is due to conditions imposed to queens in the colony (e.g. queen feeding status, number of cells available to be oviposited), and not due to variation in queen body size.

Introduction

Variation in queen body size has been registered for several species of stingless bees. Dwarf or miniature queens have been described as small sized queens that may be eventually present in the nests. However, details related to proportion of occurrence and behavior when heading colonies, have been investigated practically only for *Schwarziana quadripunctata* (Lepeletier) and *Plebeia remota* (Holmberg) (Ribeiro, 2002; Ribeiro & Alves, 2001; Ribeiro et al, 2003; Wenseleers et al., 2005; Ribeiro et al, 2006a, b). These studies reported that miniature mated queens occur naturally at a low frequency in the population and, at least for *P. remota*, they can be as efficient in laying eggs as typical-sized queens. Nevertheless, even when miniature queens lay the same amount of eggs as normal-sized queens, it is unknown whether the eggs of both queen morphotypes differ in size, and whether this variation is related to body size. In case the eggs produced by small queens are also small, this could result in smaller individuals, or when hatching out, the larvae would need larger amounts of food to develop into normal-sized adults. As suggested for honeybees (Henderson, 1992), the amount of food available to the larvae is, in fact,

more important to determine adult size rather than egg size. However, to date, the question remains unclear and it is not possible to determine the implications of small and large eggs to the colony.

Actually, egg size in bees has been poorly studied. In honeybees, egg size varies due to several factors: between castes, i.e. queen and workers (Woyke & Wongsiri, 1992; Woyke, 1994; Gençer & Woyke, 2006); according to seasons (Henderson, 1992); and due to changes in the metabolic process (Woyke, 1998). In stingless bees, eggs differ in size according to the species (Velthuis & Sommeijer, 1991), but there is little information on intra-specific variation. Eggs can also differ in morphology due to their different functions: trophic eggs (laid by workers) or reproductive eggs (laid by queens) (Koedam et al., 1996; 2001). Variation in egg size produced by a single queen was studied only for *Scaptotrigona* aff. *depilis* Moure and *S. quadripunctata* (Lacerda, unpublished data; Lacerda & Simões, 2006a, b; Ribeiro et al., unpublished data).

In this context, this study aimed (i) to investigate the size of eggs laid by queens of *P. remota*; (ii) to check for the relationship between the variation in egg size and queen body size, and (iii) the verify influence of the environment (colony) on egg size variation. The hypothesis is that small queens do lay smaller eggs than normal-sized queens, similarly to that found for *S. quadripunctata* (Ribeiro et al., unpublished data).

Material and Methods

Queen measurements

Ten *P. remota* queens of different sizes were collected with an insect aspirator, directly from colonies kept at the Bees Laboratory, University of São Paulo (USP), São Paulo. Their maximum width, average interorbital distance, and intertegular distance were measured under a stereomicroscope with a micrometer eyepiece (for details on the measuring method see Ribeiro & Alves, 2001). The queens were separated into two morphotypes: normal-sized (or typical) and miniature queens based on a previous study (Ribeiro et al., 2006a). A queen was classified as miniature using the equation HEAD < 2.76 – 0.378.IOD – 0.416 ITEG, where HEAD: head width; IOD: interorbital distance, and ITEG: intertegular distance. Although the group of queens used in this study does not represent the total range of sizes, they were the only ones available at the laboratory when the experiments were performed, and they were classified into both morphotypes using the formula described above without any restriction.

Egg measurement: non-experimental and experimental situation

Eggs were collected from the periphery of the upper combs, in which the cells have been oviposited recently. Once the hatching of the larvae is virtually identified by the horizontal position of the egg (Sommeijer et al., 1984), it was possible to collect eggs even without knowing exactly when the queen had laid them. In this way, we collected only eggs in the upright position, which therefore did not go through embryonic development. After opening the brood cell with a warmed entomological pin, the egg was collected using another pin, curved at the extremity. The egg was then immediately placed on aluminum foil, with a little amount of larval food, to prevent its dehydration. After that, the egg was measured under a stereomicroscope with a micrometer eyepiece, for length and width. Egg volume was then calculated considering the egg as a prolate spheroid, and using the formula: $V= 4/3.\pi.L/2.(L/2)^2$, where V= volume, L= length and W= width of the egg.

To analyze the possible effects of colony conditions, some experiments were performed considering other situations for eggs collection. Thus, in the first situation, non-experimental, queens were in their own colonies (Qown). In the second situation, we shifted the position of queens (Qexch), i.e., a miniature queen was placed into the colony of a typical sized queen, and vice-versa. In the third situation, queens (miniature and typical sized), one at time, were placed into a third colony, a 'host colony' (Qhost). The use of this third colony ensured that both queens (normal-sized and miniature) were subjected to the same colony conditions, which could be eventually different from their own colony, or the colony to which they were shifted to. The queen of the host colony was simultaneously placed in the colony from where the queen was removed to be tested. In order to provide time for adaptation of the queen time to a new colony, an interval of three days was allowed before a new egg collection. This method also assured that the sampled eggs were originated from the newly inserted queen and not from the former one; eggs were collected in the same way already described. In some cases, a new sample of eggs was obtained after the return of queens to their own colonies. In this way, in tables 3, 4 and 5, the numbers 1 and 2 after the code indicate the first and second times the queen was subjected to that situation. For example, Qown 2 means that the queen returned to its own colony, after a shift; Qexch 2 means the second time the queen was shifted, and so on. The table also mentions, in parentheses, in which colony the queen was at the moment of egg collection.

Statistical analysis

Statistical tests (Zar, 1999) were applied (1) to check for normality of the data (Kolmogorov-Smirnov); (2) to test the differences between the eggs laid by different queens (Kruskal-Wallis); (3) to compare the eggs laid by both morphotypes of queens (Mann-Whitney U test); and (4) to check for correlation between the queen size and the egg size (Spearman correlation). The software used to perform these tests was SPSS.

Results

Measurements of queens and eggs in own colony (Qown)

From the group of ten queens, eight were classified as normal-sized (numbered from one to eight) and two, as miniature (numbered from nine to ten). Their body measurements are presented in Table 1.

Considering all situations, a total of 642 eggs were collected, and for each individual queen, in each situation, up to 30 eggs were collected. Because data showed no normal distribution, non-parametric tests were applied. Table 1 lists the mean values (and SD) of egg measurements (length, width and volume) in Qown 1. Significant differences were found for all variables (Kruskal-Wallis, $P= 0.000$, N= 275 eggs, for length, width and volume). When comparing the two sets of queens, all variables analyzed for their eggs were also significant (Mann-Whitney, $P= 0.000$, N= 275 eggs, for length, width and volume).

Queen body size (head width, interorbital distance and intertegular distance) presented negative non-significant correlations with all variables of eggs (egg length, egg width and egg volume; Table 2).

Table 1. Morphometric measures of body and eggs (averages ± SD) fromnormal and miniature queens of *Plebeia remota* in their own colonies, in the first analyzed situation (Qown 1). (N = number of eggs).

Queens (morphotypes)	Head width (mm)	Interorbital distance (mm)	Intertegular distance (mm)	Egg Length (mm) X ± SD	Egg Width (mm) X ± SD	Egg volume (mm³) X ± SD
1 (Normal)	1.93	1.48	1.78	1.20 ± 0.05 (N=28)	0.45 ± 0.04 (N= 28)	1.13 ± 0.12 (N= 28)
2 (Normal)	1.93	1.48	1.78	1.24 ± 0.04 (N= 30)	0.46 ± 0.04 (N= 30)	1.20 ± 0.13 (N= 30)
3 (Normal)	1.85	1.41	1.70	1.21 ± 0.05 (N= 30)	0.47 ± 0.04 (N= 30)	1.18 ± 0.14 (N= 30)
4 (Normal)	1.85	1.41	1.70	1.20 ± 0.04 (N= 30)	0.46 ± 0.03 (N= 30)	1.16 ± 0.11 (N= 30)
5 (Normal)	1.78	1.33	1.63	1.24 ± 0.05 (N= 16)	0.52 ± 0.04 (N= 16)	1.35 ± 0.13 (N= 16)
6 (Normal)	1.78	1.41	1.63	1.25 ± 0.04 (N= 30)	0.47 ± 0.05 (N= 30)	1.24 ± 0.14 (N= 30)
7 (Normal)	1.78	1.33	1.56	1.20 ± 0.04 (N= 30)	0.44 ± 0.04 (N= 30)	1.10 ± 0.12 (N= 30)
8 (Normal)	1.70	1.33	1.48	1.22 ± 0.03 (N= 30)	0.45 ± 0.03 (N= 30)	1.14 ± 0.10 (N= 30)
9 (Miniature)	1.63	1.26	1.41	1.17 ± 0.04 (N= 21)	0.45 ± 0.03 (N= 21)	1.09 ± 0.08 (N= 21)
10 (Miniature)	1.48	1.11	1.18	1.16 ± 0.04 (N= 30)	0.43 ± 0.03 (N= 30)	1.04 ± 0.09 (N= 30)

On the other hand, when we included Qown2 in the analysis of all queens (i.e. Qown1 + Qown2), significant differences were also found for all variables (Kruskal-Wallis, $P= 0.000$, N= 275 eggs, for length, width and volume). This was not observed when the queens were compared between morphotypes, for egg length and volume (Mann-Whitney, $P= 0.427$, $P= 0.517$, respectively, N= 365 eggs). The differences for egg width were significant (Mann-Whitney, $P= 0.050$, N= 365 eggs).

Measurements of queens and eggs in another colony (Qexch)

The results found for egg measurements under Qexch situations (1 and 2), including Qhost and Qown (situations 1 and 2), as well as the *P* values for the statistical tests are shown in Tables 3, 4 and 5.

When queens were interchanged between colonies (Qexch 1 and Qexch 2), eggs presented significant differences (Kruskal-Wallis, $P= 0.000$, for all variables, N= 277 eggs). When both morphotypes were compared, differences were

significant only for egg volume (Mann-Whitney, $P= 0.879$, $P= 0.134$, and $P= 0.000$, respectively for length, width and volume, N= 277 eggs).

It is remarkable that the lowest and highest mean values of egg length were presented by a miniature queen (number 10), respectively, 1.16 mm and 1.29 mm. In relation to egg width, the lowest mean value (0.43 mm) was once again presented by the queen 10, but the highest (0.52 mm) was exhibited by a normal-sized queen (number 5). With respect to egg volume, the lowest mean value was found for the queen 10 (1.04 mm³) and the highest by the queen 5 (1.35 mm³; Tables 2, 3 and 4).

Fig. 1 illustrates the changes observed in egg volume for the queens subjected to more than two different experimental situations. Two normal-sized queens (numbers 2 and 5) laid smaller eggs after the shift, while the others (one normal-sized: 1, and both miniatures: 9 and 10) laid larger eggs after the shift. When they returned to their own colonies,

Table 2. Spearman rank correlations (Rho) and P values for comparisons between queens' size (head width, interorbital distance and intertegular distance) and average values for eggs' measures (length, width and volume) of *Plebeia remota* queens in their own colonies (Qown1). (N = number of eggs).

Eggs measures	Head width (mm)	Interorbital distance (mm)	Intertegular distance (mm)
Average Length (mm)	Rho= - 0.127 P= 0.727 (N= 10)	Rho= - 0.211 P= 0.559 (N= 10)	Rho= - 0.166 P= 0.646 (N= 10)
Average Width (mm)	Rho= - 0.267 P= 0.456 (N= 10)	Rho= - 0.295 P= 0.408 (N= 10)	Rho= - 0.274 P= 0.444 (N= 10)
Average Volume (mm³)	Rho= - 0.268 P= 0.454 (N= 10)	Rho= - 0.334 P= 0.346 (N= 10)	Rho= - 0.296 P= 0.406 (N= 10)

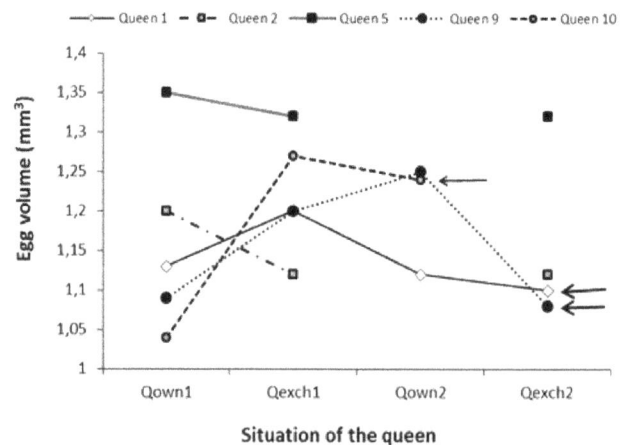

Fig 1. Average volume of eggs (mm3) laid by *Plebeia remota* queens (number 1, 2 and 5: normal-sized; 9 and 10: miniature) in the different situations (own colony or another colony). Legend: Qown 1: queen in her own colony, and analyzed for the first time; Qown 2: queen returned to her own colony, after being in another colony; Qexch 1: queen in another colony, for the first time; Qexch 2: queen in another colony, for the second time. Arrows show the host colony.

Table 3. Comparisons of morphometric measures (averages ± SD) obtained for eggs laid by the different normal-sizedqueens of *Plebeia remota* in their colonies and other colonies. The number after Qown or Qexch refers to the situation, i.e., the first or second time the queen was in that situation. (Legend: Qown = Queen in own colony; Qexch = Queen exchanged, in another colony; N = number of eggs; col. = colony).

Queens	Comparison Experimental condition	Egg Length (mm) x ± SD	Mann-Whitney test P	Egg Width (mm) x ± SD	Mann-Whitney test P	Egg Volume (mm³) x ± SD	Mann-Whitney test P
1	Qown 1 (col. 1) vs. Qexch 1 (col. 9)	1.20 ± 0.05 (N=28) 1.25 ± 0.04 (N=24)	P= 0.001**	0.45 ± 0.04 (N= 28) 0.46 ± 0.05 (N= 24)	P= 0.181	1.13 ± 0.12 (N= 28) 1.20 ± 0.14 (N= 24)	P= 0.009**
1	Qown 1 (col. 1) vs. Qown 2 (col. 1)	1.20 ± 0.05 (N=28) 1.22 ± 0.03 (N= 30)	P= 0.014*	0.45 ± 0.04 (N= 28) 0.44 ± 0.01 (N= 30)	P= 0.771	1.13 ± 0.12 (N= 28) 1.12 ± 0.04 (N= 30)	P= 0.105
1	Qexch 1 (col. 9) vs. Qown 2 (col.1)	1.25 ± 0.04 (N=24) 1.22 ± 0.03 (N= 30)	P= 0.024*	0.46 ± 0.05 (N= 24) 0.44 ± 0.01 (N= 30)	P= 0.034*	1.20 ± 0.14 (N= 24) 1.12 ± 0.04 (N= 30)	P= 0.040*
2	Qown 1 (col. 2) vs. Qexch 1 (col. 1)	1.24 ± 0.04 (N=30) 1.23 ± 0.04 (N= 30)	P= 0.196	0.46 ± 0.04 (N= 30) 0.44 ± 0.03 (N= 30)	P= 0.002**	1.20 ± 0.13 (N= 30) 1.12 ± 0.10 (N= 30)	P= 0.030*
2	Qown 1 (col. 2) vs. Qexch 2 (col. 9)	1.24 ± 0.04 (N=30) 1.24 ± 0.03 (N= 30)	P= 0.938	0.46 ± 0.04 (N= 30) 0.43 ± 0.02 (N= 30)	P= 0.002**	1.20 ± 0.13 (N= 30) 1.12 ± 0.08 (N= 30)	P= 0.049*
2	Qexch 1 (col. 1)vs. Qexch 2 (col. 9)	1.23 ± 0.04 (N=30) 1.24 ± 0.03 (N= 30)	P= 0.169	0.44 ± 0.03 (N= 30) 0.43 ± 0.02 (N= 30)	P= 0.603	1.12 ± 0.10 (N= 30) 1.12 ± 0.08 (N= 30)	P= 0.146
5	Qown 1 (col. 5) vs. Qexch 1 (col. 3)	1.24 ± 0.05 (N=16) 1.24 ± 0.05 (N= 30)	P= 0.626	0.52 ± 0.04 (N= 16) 0.51 ± 0.04 (N= 30)	P= 0.399	1.35 ± 0.13 (N= 16) 1.32 ± 0.13 (N= 30)	P= 0.388
5	Qown 1 (col. 5) vs. Qexch 2 (col. 10)	1.24 ± 0.05 (N=16) 1.24 ± 0.05 (N= 30)	P= 0.921	0.52 ± 0.04 (N= 16) 0.51 ± 0.04 (N= 30)	P= 0.258	1.35 ± 0.13 (N= 16) 1.32 ± 0.12 (N= 30)	P= 0.466
5	Qexch 1 (col. 3) Qexch 2 (col. 10)	1.24 ± 0.05 (N=30) 1.24 ± 0.05 (N= 30)	P= 0.597	0.51 ± 0.04 (N= 30) 0.51 ± 0.04 (N= 30)	P= 0.811	1.32 ± 0.13 (N= 30) 1.32 ± 0.12 (N= 30)	P= 0.852

Table 4. Comparisons of morphometric measures (averages ± SD) obtained for eggs laid by the different miniature queens of *Plebeia remota* in their colonies and other colonies. The number after Qown or Qexch refers to the situation, i.e., the first or second time the queen was in that situation. (Legend: Qown = Queen in own colony; Qexch = Queen exchanged, in another colony; N = number of eggs; col. = colony).

Queens	Comparison Experimental condition	Egg Length (mm) x ± SD	Mann-Whitney test P	Egg Width (mm) x ± SD	Mann-Whitney test P	Egg Volume (mm³) x ± SD	Mann-Whitney test P
9	Qown 1 (col. 9) vs. Qexch 1 (col. 1)	1.18 ± 0.04 (N=21) 1.22 ± 0.05 (N=20)	P= 0.001**	0.44 ± 0.03 (N= 21) 0.47 ± 0.04 (N= 20)	P= 0.027*	1.09 ± 0.08 (N= 21) 1.20 ± 0.12 (N= 20)	P= 0.001**
9	Qown 1 (col. 9) vs. Qown 2 (col. 9)	1.18 ± 0.04 (N=21) 1.28 ± 0.04 (N= 30)	P= 0.000**	0.44 ± 0.03 (N= 21) 0.47 ± 0.04 (N= 30)	P= 0.004*	1.09 ± 0.08 (N= 21) 1.25 ± 0.10 (N= 30)	P= 0.000**
9	Qexch 1 (col. 1) vs. Qown 2 (col.9)	1.22 ± 0.05 (N=20) 1.28 ± 0.04 (N= 30)	P= 0.000**	0.47 ± 0.04 (N= 20) 0.47 ± 0.04 (N= 30)	P= 0.799	1.20 ± 0.12 (N= 20) 1.25 ± 0.10 (N= 30)	P= 0.028*
10	Qown 1 (col. 10) vs. Qown 2 (col. 10)	1.16 ± 0.04 (N=30) 1.29 ± 0.07 (N=30)	P= 0.000**	0.43 ± 0.03 (N= 30) 0.46 ± 0.05 (N= 30)	P= 0.011*	1.04 ± 0.09 (N= 30) 1.24 ± 0.17 (N= 30)	P= 0.000**

queens 1 (normal-sized) and 10 (miniature) laid smaller eggs again, but the queen 10 (the smallest queen) laid even larger eggs. After Qexch2, all queens laid similar or smaller eggs than in the previous situation.

Regarding the third situation (Qhost: colonies 2 and 5), queens of both morphotypes showed similar performance (Table 5). Thus, no significant differences were detected between the eggs from normal-sized (1 and 3) and miniature queens (9 and 10), when considered both morphotypes for egg length and width, but not for egg volume (Mann-Whitney, P= 0.147, P= 0.306, and P= 0,004, respectively, N= 113 eggs). When queens were considered in pairs, i.e., queens 1 and 9 in host colony 2, the results were similar (Mann-Whitney, P= 0.658, P= 0.366, P= 0.000, respectively for length and width, and volume, N= 60 eggs). However, when queens 3 and 10 in the host colony 5 were compared, no differences were significant for all analyzed variables (Mann-Whitney, P= 0.056 and P= 0.160, and P= 0.601, respectively for length, width, and volume N= 53 eggs).

Discussion

The results for the situation in which the queens remained in their own colonies (Qown1) could suggest that normal-sized queens lay larger eggs than miniature queens (Table 1). However, analyzing in detail the results of queens in other colonies (especially Qhost, Table 5; Fig 1), it was verified that sometimes miniature queens are able to lay larger eggs than normal-sized queens. In fact, queens of different sizes laid, in all situations analyzed (i.e., in Qown 1 and 2, Qexch 1 and 2, or Qhost), eggs of similar sizes, smaller or larger than before, and often these differences were significant (Tables 3, 4 and 5; Fig. 1). Moreover, non-significant correlations between body size and egg size evidenced that the variation in egg size is not due to differences in the queen size. Likewise, in different honeybee species, Woyke et al. (2003) found no relationship between eggs' size among queens of different species and sizes.

Table 5. Comparisons of morphometric measures (averages ± SD) obtained for eggs laid by the different queens of *Plebeia remota* in their colonies and other colonies. The number after Qown or Qexch refers to the situation, i.e., the first or second time the queen was in that situation. (Legend: Qown = Queen in own colony; Qexch = Queen exchanged, in another colony; Qhost= Queen in host colony; N = number of eggs; col. = colony).

Queens (morphotypes)	Comparison Experimental condition	Egg Length (mm) x ± SD	Mann-Whitney test P	Egg Width (mm) x ± SD	Mann-Whitney test P	Egg Volume (mm³) x ± SD	Mann-Whitney test P
1 (Normal)	Qown 1 (col. 1) vs. Qexch 2 (col. 2) = Q host	1.20 ± 0.05 (N=28) 1.20 ± 0.04 (N= 30)	P= 0.974	0.45 ± 0.04 (N= 28) 0.44 ± 0.03 (N= 30)	P= 0.355	1.13 ± 0.12 (N= 28) 1.10 ± 0.10 (N= 30)	P= 0.627
1 (Normal)	Qexch 1 (col. 9) vs. Qexch 2 (col. 2) = Q host	1.25 ± 0.04 (N=24) 1.20 ± 0.04 (N= 30)	P= 0.000**	0.46 ± 0.05 (N= 24) 0.44 ± 0.03 (N= 30)	P= 0.026*	1.20 ± 0.14 (N= 24) 1.10 ± 0.10 (N= 30)	P= 0.001**
1 (Normal)	Qown 2 (col.1) vs. Qexch 2 (col. 2) = Q host	1.22 ± 0.03 (N= 30) 1.20 ± 0.04 (N= 30)	P= 0.007*	0.44 ± 0.01 (N= 30) 0.44 ± 0.03 (N= 30)	P= 0.307	1.12 ± 0.04 (N= 30) 1.10 ± 0.10 (N= 30)	P= 0.015*
9 (Miniature)	Qown 1 (col. 9) vs. Qexch 2 (col. 2) = Q host	1.18 ± 0.04 (N=21) 1.20 ± 0.05 (N= 30)	P= 0.112	0.44 ± 0.03 (N= 21) 0.43 ± 0.02 (N= 30)	P= 0.102	1.09 ± 0.08 (N= 21) 1.08 ± 0.09 (N= 30)	P= 0.862
9 (Miniature)	Qexch 1 (col. 1) vs. Qexch 2 (col. 2) = Q host	1.22 ± 0.05 (N=20) 1.20 ± 0.05 (N= 30)	P= 0.029*	0.47 ± 0.04 (N= 20) 0.43 ± 0.02 (N= 30)	P= 0.001**	1.20 ± 0.12 (N= 20) 1.08 ± 0.09 (N= 30)	P= 0.001**
9 (Miniature)	Qown 2 (col.9) vs. Qexch 2 (col. 2) = Q host	1.28 ± 0.04 (N= 30) 1.20 ± 0.05 (N= 30)	P= 0.000**	0.47 ± 0.04 (N= 30) 0.43 ± 0.02 (N= 30)	P= 0.000**	1.25 ± 0.10 (N= 30) 1.08 ± 0.09 (N= 30)	P= 0.000**
3 (Normal)	Qown 1 (col. 3) vs. Qexch 1 (col. 5) = Q host	1.21 ± 0.05 (N=30) 1.25 ± 0.06 (N=23)	P= 0.012*	0.46 ± 0.05 (N= 30) 0.49 ± 0.04 (N= 23)	P= 0.008**	1.18 ± 0.14 (N= 30) 1.29 ± 0.15 (N= 23)	P= 0.001**
10 (Miniature)	Qown 1 (col. 10) vs. Qexch 1 (col. 5) = Q host	1.16 ± 0.04 (N=30) 1.28 ± 0.05 (N=30)	P= 0.000**	0.43 ± 0.03 (N= 30) 0.48 ± 0.05 (N= 30)	P= 0.000**	1.04 ± 0.09 (N= 30) 1.27 ± 0.15 (N= 30)	P= 0.000**
10 (Miniature)	Qown 2 (col. 10) vs Qexch 1 (col. 5) = Q host	1.29 ± 0.07 (N=30) 1.28 ± 0.05 (N=30)	P= 0.477	0.46 ± 0.05 (N= 30) 0.48 ± 0.05 (N= 30)	P= 0.058	1.24 ± 0.17 (N= 30) 1.27 ± 0.15 (N= 30)	P= 0.329

Thus, our results indicate that it is not the size of the queens that influences the size of the eggs, but rather the conditions imposed to the queens. Probably the condition of each colony was different and influenced the egg size, whether positively (larger eggs) or negatively (smaller eggs). On the other hand, when queens laid similar-sized eggs in both situations, all the colonies probably provided similar conditions.

A possible explanation for the differences in egg size is the development of the embryos. Thus, changes in egg size would imply in embryonic development. The exact moment of egg laying by the queen was not observed in this work. Nevertheless, eggs were always collected in very new combs, i.e., with cells recently operculated and, therefore, the possibility of collecting eggs at an advanced stage of development is excluded. Moreover, as already mentioned, when eggs hatch they did not stay in a vertical position, but lay down on the larval food and only eggs in vertical position were collected.

Egg size variation may be caused by nutritional conditions to which queens were exposed. Therefore, in colonies with abundant food reserve, queens probably received more food, and could produce larger eggs. Simultaneously, queens placed in colonies with fewer reserves, and supposedly lower food intake, would produce smaller eggs. Opposite results, however, were found for *S.* aff. *depilis* (Lacerda, unpublished data; Lacerda & Simões, 2006b). Queens of this species in 'weak' colonies produced larger eggs than in 'strong' ones. However, the authors suggested that these eggs were haploid eggs, which are larger than diploid eggs. In *P. remota*, most haploid eggs are produced by the queen (Tóth et al., 2002).

Nevertheless, the variation in egg size found herein was not due to male production, since adult males were not registered in the studied colonies after the experimental period.

So, if the variation in egg size is associated to queen feeding status, this would be imposed by colony condition, and in some way, by the workers. Queens get food through three different ways: by eating larval food from the open cell just before laying the egg, by eating trophic eggs laid by the workers or by trophallaxis with workers. This food ingestion could be important for the production and development of eggs inside the queen ovarioles. In this way, the workers, besides controlling the number of cells available for egg laying (Ribeiro, 2002), indirectly are also responsible for the size of the eggs laid by queens.

On the other hand, as suggested by J. Woyke (personal communication, March 8, 2008) egg size may still be related to the cell production rate of the colony. Analyzing the relationship between the number of cells available for oviposition in each colony with egg size, negative correlations were recorded for egg length and width (Spearman, rho= -0.040, rho= -0.144, respectively, $P= 0.01$, N= 642 eggs). The longer the queen had to wait to lay eggs, the larger these eggs could become, probably by accumulation of vitellogenin. Certainly, before the end of egg development, during vitellogenesis, a deposit of vitello can be established thus increasing egg volume up to the moment of ovulation (i.e. the expulsion of the egg through the oviducts) (Cruz-Landim, 2009). In this work, when the queen was transferred to a smaller colony, this would result in a delay in the oviposition and consequently could result in an increased size of the eggs ready to be laid. In an opposite situation, one could expect a contrary result.

Therefore, the results of the present study indicate that egg size is influenced by colony conditions (and consequently, food supply for the queens and/or rate of cell construction) imposed to the queen, rather than by body size.

Acknowledgments

We are grateful to Dr. J. Woyke, Dr. Vera L. Imperatriz-Fonseca, Dr. Dirk Koedam, Dr. Beatriz A. J. Paranhos and two anonymous referees for the critical reading of the manuscript; Fundação de Amparo à Pesquisa do Estado de São Paulo (FAPESP, proc. n. 98/01679-5), Conselho Nacional de Desenvolvimento Científico e Tecnológico (CNPq) and Fundação Cearense de Apoio ao Desenvolvimento Científico e Tecnológico (FUNCAP, proc. n. 35.0781/2004-4) for financial support; and Erica M. T. de Alencar, for language advice.

References

Cruz-Landim, C. (2009). Aparelho reprodutor e gametogênese. In C. Cruz-Landim (Ed.), Abelhas - Morfologia e função de sistemas (pp. 15-57). São Paulo: Editora UNESP.

Gençer, H.V. & Woyke, J. (2006). Eggs from *Apis mellifera caucasica* laying workers are larger than from queens. J. Apicult. Res. 45(4): 173-179. doi: 10.3896/IBRA.1.45.4.02

Henderson, C.E. (1992). Variability in the size of emerging drones and of drone and worker eggs in honey bee (*Apis mellifera* L.) colonies. J. Apicult. Res., 31(3/4): 114-118. doi:

Katzav-Gozansky, T., Soroker, V., Kamer, J., Schulz, C. M., Francke, W. & Hefetz, A. (2003). Ultrastructural and chemical characterization of egg surface of honeybee worker and queen-laid eggs. Chemoecology 13: 129-134. doi: 10.1007/s00049-003-0238-0

Koedam, D., Velthausz, P.H., Krift, T.V.D., Dohmen, M.R. & Sommeijer, M.J. (1996). Morphology of reproductive and trophic eggs and their controlled release by workers in *Trigona* (*Tetragonisca*) *angustula* Illiger (Apidae, Meliponinae). Physiol. Entomol. 21: 289-296. doi: 10.1111/j.1365-3032.1996.tb00867.x

Koedam, D., Velthuis, H.H.W., Dohmen, M.R. & Imperatriz-Fonseca, V.L. (2001). The behaviour of laying workers and the morphology and viability of their eggs in *Melipona bicolor bicolor*. Physiol. Entomol. 26: 254-259. doi: 10.1046/j.0307-6962.2001.00241.x

Lacerda, L.M. & Simões, Z.L.P. (2006a). Ovos produzidos por rainhas e operárias de *Scaptotrigona depilis* (Hymenoptera, Apidae, Meliponina): morfometria e aspectos relacionados. Iheringia, Sér. Zool. 96: 89-93. doi: 10.1590/S0073-47212006000100016

Lacerda, L.M. & Simões, Z.L.P. (2006b). Effects of internal conditions on the size of eggs laid by queens of *Scaptotrigona depilis* (Moure, 1942) (Apidae, Meliponinae). Sociobiology 47: 85-97. doi: 10.13102.

Pereira, R.A., Morais, MM., Gioli, L.D., Nascimento, F.S., Rossi, M.A. & Bego, L.R. (2006). Comparative morphology of reproductive and trophic eggs in some *Melipona* bees (Apidae, Meliponini). Braz. J. Morphological Sci. 23: 349-354.

Ribeiro, M.F. (2002). Does the queen of *Plebeia remota* (Hymenoptera, Apidae, Meliponini) stimulate her workers to start cell construction after winter? Insect. Soc. 49: 38-40. doi: 10.1007/s00040-002-8276-0

Ribeiro, M.F. & Alves, D.A. (2001). Size variation in *Schwarziana quadripunctata* queens (Hymenoptera, Apidae, Meliponinae). Rev. Etologia 3(1): 59-65.

Ribeiro, M.F., Santos-Filho, P.S. de & Imperatriz-Fonseca, V.L. (2003). Exceptional high queen production in the Brazilian stingless bee *Plebeia remota*. Stud. Neotrop. Fauna Envir. 38: 111-114. doi: 10.1076/snfe.38.2.111.15925.

Ribeiro, M.F., Santos-Filho, P.S. de & Imperatriz-Fonseca, V.L. (2006a). Size variation and egg laying performance in *Plebeia remota* queens (Hymenoptera, Apidae, Meliponinae). Apidologie 37: 653-664. doi: 10.1051/apido:2006046

Ribeiro, M.F., Wenseleers, T., Santos-Filho, P.S. de & Alves, D.A. (2006b). Miniature queens in stingless bees: basic facts and evolutionary hypotheses. Apidologie 37: 191-206. doi: 10.1051/apido:2006023

Sommeijer, M.J., van Zeijl, M. & Dohmen, M.R. (1984). Morphological differences between worker-laid eggs from a queenright colony and a queenless colony of *Melipona rufiventris paraensis* (Hymenoptera: Apidae). Entomol. Berichten, 44: 91-95.

Velthuis, H.H.W. & Sommeijer, M.J. (1991). Morphogenetic hormones in caste polymorphism in stingless bees. In A.P. Gupta (Ed.), Mophogenetic Hormones of Arthropods (pp. 346-382). News Brunswick: Rutgers Univ. Press.

Wenseleers T., Ratnieks, F.L.W., Ribeiro, M.F., Alves, D.A. & Imperatriz-Fonseca, V.L. (2005). Working-class royalty: bees beat the caste system. Biol. Lett. 1: 125-128. doi: 10.1098/rsbl.2004.0281

Wheeler, D. (1996). The role of nourishment in oogenesis. Annu. Rev. Entomol., 41: 407-431. doi: 10.1146/annurev.en.41.010196.002203

Woyke, J. (1994). Comparison of the size of the eggs from *Apis mellifera* L. queens and laying workers. Apidologie, 25: 179-187. doi: 10.1051/apido:19940206.

Woyke, J. (1998). Size change of *Apis mellifera* eggs during the incubation period. J. Apicult. Res. 37: 239-246.

Woyke, J. & Wongsiri, S. (1992). Occurrence and size of laying worker eggs in *Apis florea* colonies. J. Apicult. Res. 31: 124-127.

Woyke, J., Chanchao, C., Wongsiri, S., Wilde, J. & Wilde, M. (2003). Size of eggs from queens of three Asian species and laying workers of *Apis cerana*. J. Apicult. Science 47: 57-71.

Zar, J.H. (1999). Biostatistical Analysis. 4th ed. New Jersey: Prentice Hall, 663 p.

21

Chemical composition, antinociceptive and free radical-scavenging activities of geopropolis from *Melipona subnitida* Ducke (Hymenoptera: Apidae: Meliponini)

SA Souza[1], TLMF Dias[2], TMG Silva[1], RA Falcão[1], MS Alexandre-Moreira[2], EMS Silva[3], CA Camara[1], TMS Silva[1]

1- Universidade Federal Rural de Pernambuco, Recife, PE, Brazil

2- Universidade Federal de Alagoas, Maceió, AL, Brazil

3- Universidade Federal do Vale de São Francisco, Petrolina, PE, Brazil

Keywords

Stingless bees, phenolic, antioxidant.

Corresponding author

Tania Maria Sarmento Silva
Biofito - Depto. de Ciências Moleculares
Universidade Federal Rural de Pernambuco
Rua Dom Manoel de Medeiros, s/n, Dois
Irmãos, Recife, 52171-900, PE, Brazil
E-Mail: sarmentosilva@gmail.com

Abstract

Like many stingless bee species, *Melipona subnitida* Ducke uses geopropolis (a mixture of wax, plant resins, pollen grains and mud) for sealing small crevices in their nest cavities, in order to avoid the entry of air, and for defense against pathogenic microorganisms. The aim of this study was to evaluate the antinociceptive and free radical-scavenging activities of ethanolic extracts of six geopropolis samples from *M. subnitida* and the phenolic fractions obtained by C18-SPE extraction. The *in vivo* antinociceptive activity was analyzed on abdominal constriction induced by acetic acid in mice and *in vitro* free radical-scavenging activities by DPPH and ABTS assays. Additionally we analyzed the chemical composition of the phenolic fractions by HPLC-DAD. The six samples of geopropolis showed variations in the total phenolic content over the period, but not in the chemical profile observed by HPLC-DAD. Geopropolis is a rich source of bioactive compounds as phenolics 6-*O*-*p*-coumaroyl-*D*-galactopyranose, 6-*O*-cinnamoyl-1-*O*-*p*-coumaroyl-β-*D*-glucopyranose, 7-*O*-methyl naringenin, 7-*O*-methyl aromadendrin, 7,4'-di-*O*-methyl aromadendrin, 4'-*O*-methyl kaempferol, 3-*O*-methyl quercetin, 5-*O*-methyl aromadendrin and 5-*O*-methyl kaempferol with potential antioxidant and antinociceptive activities. The antioxidant activity is related to the total phenolic content.

Introduction

Many stingless bee species (Meliponini) store in their nests a large amount of geopropolis, a mixture of wax, plant resins, pollen grains and mud (Nogueira-Neto, 1997). The bees use this material for sealing small crevices in their nest cavities, in order to avoid the entry of air, and for defense against pathogenic microorganisms (Simone-Finstrom & Spivak, 2010). However, despite its popular use in folk medicine, very little is known about its chemical composition and biological activity.

Recently, studies investigating geopropolis from native bees have indicated a potential for bioactive compounds and biological activities. Velikova et al. (2000) analyzed 21 samples of Brazilian geopropolis from 12 different species of stingless bees and observed the presence of compounds such as di- and triterpenes and gallic acid. The same samples showed activity against *Staphylococcus aureus* Rosenbach and cytotoxic activity.

Samples of *Melipona fasciculata* Smith geopropolis showed activity against *Streptococcus mutans* Clarke (Liberio et al., 2011) and antioxidant capacity (Dutra et al., 2014) and eleven compounds were tentatively identified as belonging to the classes of phenolic acids and hydrolysable tannins (gallotannins and ellagitannins). These compounds were responsible for the antioxidant activity and high phenolic content of the geopropolis produced by *M. fasciculata* (Dutra et al., 2014). Geopropolis produced by *Melipona scutellaris* Latreille has been shown to exhibit antimicrobial and antioxidant activities and has anti-inflammatory, antinociceptive and antiproliferative properties (Franchin et al., 2012; Cunha et al., 2013), and benzophenones have been identified as the major compounds (Cunha et al., 2013).

Previous investigations in our laboratory have found that the geopropolis from *Melipona subnitida* Ducke has antioxidant activity. This study led to the isolation and characterization of two phenylpropanoids, one of which was a new compound, and seven flavonoids (Souza et al., 2013). These findings suggested that *M. subnitida* geopropolis is highly bioactive and deserved further study to identify other potential biological activities. Thus, the aim of this study was to evaluate the antinociceptive and free radical-scavenging activities of ethanolic extracts of six geopropolis samples from *M. subnitida* and its phenolic fractions. Additionally, we analyzed the chemical composition of the phenolic fractions obtained by C18-SPE extraction by HPLC-DAD.

Materials and methods

Geopropolis samples and fractionation

For this study, six samples of geopropolis from four *M. subnitida* nests were collected in March 2010 (**1**), July 2011 (**2**), January 2012 (**3**), April 2012 (**4**), June 2012 (**5**) and July 2012 (**6**) at Sítio Riacho Vieirópolis (a semi-arid region), Paraíba State, Brazil. Each sample (200 g) was extracted with 100 mL of ethanol (EtOH) in an ultrasonic water bath. The combined ethanolic extracts were completely evaporated under reduced pressure to a brown residue (2.7 g to 18.4 g). The EtOH extract (100 mg) was dissolved in 2 mL of distilled water, and the solution was adjusted to pH 2.0 by adding concentrated HCl while stirring with a magnetic stirrer at room temperature for 10 min. A C18 cartridge (SPE Strata 1 g, Phenomenex) was sequentially conditioned with 3 mL of MeOH and 6 mL of distilled deionized water without allowing the cartridge to dry. The samples of geopropolis were passed through the cartridge and rinsed with 6 mL of water and the phenolic compounds were eluted with 8 mL of HPLC-grade methanol. The eluate was dried under reduced pressure in a rotatory evaporator at 40 °C to yield 32 to 57 mg of phenolic fraction. These fractions were dissolved in methanol, filtered through a 0.45-μm nylon syringe filter (Whatman) and injected into the HPLC system. The phenolic samples were reconstituted with Tween® 80 and carboxycellulose and also to evaluation for their antinociceptive and antioxidant activities.

Reagents and standards

All reagents used were of analytical grade. Folin-Ciocalteu's phenol reagent, DPPH (1,1-diphenyl-2-picryl hydrazyl), potassium persulfate and Trolox (6-hydroxy-2,5,7,8-tetramethylchroman-2-carboxylic acid) were supplied by Acros Organics (Belgium). ABTS (2,2 azinobis 3-ethylbenzothiazoline-6-sulfonic acid) was purchased from Fluka Chemie GmbH (Switzerland). Ascorbic acid was from Vetec (Brazil). Formic acid (Merck) and methanol (Tedia) were of analytical grade. Dipirone, 3-(4,5-dimethylthiazol-2-yl)-2,5-diphenyltetrazolium (MTT), gallic acid, carboxymethylcellulose-CMC (Sigma); tween® 80 (Sigma-Aldrich, USA) and dimethyl sulfoxide (DMSO) were purchased from Sigma-Aldrich (USA). The compounds 6-*O*-*p*-coumaroyl-*D*-galactopyranose (**1**), 6-*O*-cinnamoyl-1-*O*-*p*-coumaroyl-β-*D*-glucopyranose (**2**), 7-*O*-methyl naringenin (**3**), 7-*O*-methyl aromadendrin (**4**), 7,4'-di-*O*-methyl aromadendrin

(**5**), 4'-*O*-methyl kaempferol (**6**), 3-*O*-methyl quercetin (**7**), 5-*O*-methyl aromadendrin (**8**) and 5-*O*-methyl kaempferol (**9**) had been previously isolated and identified from *M. subnitida* geopropolis (Souza et al., 2013).

HPLC analysis of the phenolic

All chromatographic analyses were performed using a Shimadzu Prominence LC-20AT equipped with a SPD-M20A diode array detector (Shimadzu Corp. Kyoto, Japan). The samples (20 μL) were injected into a Rheodyne 7125i injector with a 20 μL loop. The column heater was set at 40 °C. The chromatographic separation was performed with a Luna Phenomenex C-18 column (250 mm x 4.6 mm x 5 μm). The compounds were separated using a mobile phase consisting of 1% aqueous formic acid (A) and methanol (B) at a flow rate of 1 mL/min. The mobile phase was delivered using the following solvent gradient: 0-10 min, 20-25% B; 10-20 min, 25-60% B; 20-30 min, 60-70% B; 30-35 min, 70-100% B. The injection volume was 20 μL. Chromatograms were recorded at 290 nm and 340 nm. The identification of the compounds was based on their retention times and UV spectra with authentic markers.

Animals

Male and female Swiss mice weighing 20-25 g were used and given access to water and food *ad libitum*. We used six mice per experimental group. The animals were housed at a temperature of 25-28°C with a 12 h light/12 h dark cycle. The procedures described were reviewed and approved by the local Animal Ethics Committee (CEUA UFAL process number 23065.004873/2011-01).

Determination of the total phenolic content

The total phenolic content of the samples was determined with the Folin Ciocalteu reagent, according to the method of Slinkard and Singleton (1977), modified by using gallic acid as a standard phenolic compound. EtOH extracts (100 μL) and phenolic fractions (1 mg/ml) were transferred to an Eppendorf tube with 1 ml. Folin Ciocalteu reagent (20 μL), 820 μL of distilled water were added and the contents of the flask were mixed thoroughly. After 1 min, 60 μL of sodium carbonate (15%) was added and then the mixture was allowed to stand for 2 h. The absorbance was measured at 760 nm with an automatic Biochrom Asys UVM 340 microplate reader (Cambridge, UK). The amount of total phenolic compounds was determined in micrograms of gallic acid equivalents using the equation obtained from the standard gallic acid graph.

DPPH• radical scavenging assay

The free radical-scavenging activity was determined using the DPPH assay, as described previously (Silva et al., 2006) with modifications. The antiradical activity was evaluated using a dilution series to obtain five concentrations (1.0 to 80.0 µg/mL). This process involved mixing the DPPH solution (23.6 µg/mL in EtOH) with the appropriate EtOH extracts and phenolic fractions followed by homogenization. After 30 min, the remaining DPPH radicals were quantified by measuring the absorption at 517 nm with an automatic Biochrom Asys UVM 340 microplate reader (Cambridge, UK).The percentage of inhibition was given by the formula: percent inhibition (%) = [(A0 - A1)/A0] x 100, where A0 was the absorbance of the control solution and A1 was the absorbance in the presence of the sample and standards.

ABTS•+ radical cation decolorization assay

The radical cation decolorization assay was based on the method described by Re et al. (1999) with modifications. ABTS was dissolved in water to yield a final concentration of 7 mM. The ABTS radical cation (ABTS•+) was produced by reacting the ABTS stock solution with 2.45 mM of potassium persulfate (final concentration) and allowing the mixture to stand in the dark at room temperature for 16 h before use. The ABTS•+ solution was diluted to give an absorbance of 0.70 ± 0.05 at 734 nm with ethanol before use. Then, appropriate amounts of the ABTS•+ solution were added into 0.5 mL of the sample solutions in ethanol at five concentrations (1-40 µg/mL). After 10 min, the percentage inhibition of the absorbance at 734 nm was calculated for each concentration with an automatic Biochrom Asys UVM 340 microplate reader (Cambridge, UK), relative to the blank absorbance (EtOH). The capability to scavenge the ABTS•+ radical was calculated using the following equation: ABTS•+ scavenging effect (%) = [(A0-A1/A0) x100], where A0 was the initial concentration of the ABTS•+ and A1 was absorbance of the remaining concentration of ABTS•+ in the presence of sample.

Evaluation of activity ethanol extracts and fractions of geopropolis on abdominal constriction responses caused by acetic acid

Abdominal constrictions (writhes) were induced by the i.p. injection of acetic acid (1.2%) and carried out according to the procedure described previously (Koster et al., 1959; Collier et al., 1968; Fontenele et al., 1996). Mice were treated with EtOH extracts and phenolic fractions (100 mg/kg, i.p.) or Dypirone (10 mg/kg, i.p.) 40 minutes before initiating nociceptive stimulus. Dypirone was used as a positive control and the vehicle (CMC/Tween® 80) (10 mL/kg, i.p.) was used as the negative control (the animals without treatment). The total numbers of writhes, which consisted of constriction of the flank muscles associated with inward movements of the hind limb or with whole body stretching, were counted cumulatively over a 20 min period. The antinociceptive activity was determined as the difference in number of writhes between the control group and the treated group.

Statistical analysis

All analyses were performed in triplicate. The results were expressed as the standard error of the mean (mean ± S.E.M.) and were analyzed using GraphPad Prism 5.0 program (DEMO). Comparisons between groups were made using analyses of variance (ANOVA) followed by Tukey's test. Significance was indicated by a p value ≤0.05. Pearson's correlation test was used to evaluate the correlations.

Results and Discussion

The aim of this study was to evaluate the antinociceptive activity of six samples of *M. subnitida* geopropolis collected over three years. EtOH extracts and the phenolic fractions were evaluated in a model of nociception, and the free radical-scavenging activity was evaluated using the DPPH and ABTS assays. The total phenolic content was determined by the Folin Ciocalteu reagent. In addition, chromatographic profiles were analyzed by HPLC-DAD, and the principal phenolics present in the geopropolis samples were identified.

This study was conducted by an extraction of phenolics using a C18-SPE cartridge as a simpler, less expensive and faster technique compared with the use of liquid-liquid solvent extraction. This technique has been used to determine flavonoid markers in honey (Hadjmohammadi et al., 2009). Interestingly, there is a correlation (r=0.85, p<0.05) between the total phenolic content present in the ethanolic extract and the amount of phenolics extracted by C18-SPE. These samples showed a total phenolic content two times higher when compared with the EtOAc fraction (which is rich in phenolic compounds) obtained by the liquid-liquid extraction of a sample of *M. scutellaris* geopropolis collected in January 2010 (Souza et al., 2013). The phenolic profiles of samples **1-6** also were analyzed by HPLC-DAD. The characterization of these compounds is important because they are associated with a variety of health benefits. The comparative analysis of the chromatograms (Fig 1) shows a similar profile between the six samples obtained by the SPE and the EtOAc fraction (Souza et al., 2013) of geopropolis, again demonstrating that SPE extraction is effective for extraction of phenolics. All phenols (phenylpropanoids and flavonoids) previously identified from EtOAc fraction (Souza et al., 2013) were verified in the samples of this study; the 6-*O*-*p*-coumaroyl-*D*-galactopyranose compounds (**1**), 6-*O*-cinnamoyl-1-*O*-*p*-coumaroyl-β-*D*-glucopyranose (**2**), 7-*O*-methyl naringenin (**3**), 7-*O*-methyl aromadendrin (**4**), 7,4'-di-*O*-methyl aromadendrin (**5**), 4'-*O*-methyl kaempferol (**6**), 3-*O*-methyl quercetin (**7**), 5-*O*-methyl aromadendrin (**8**) and 5-*O*-methyl kaempferol (**9**) were identified (Fig 1).

Further studies are necessary to quantify the compounds identified. The following plant species occur in the region and are resin-producing sources possibly collected by the bees for propolis production: *Myracrodruon urundeuva* Allemão (Anacardiaceae), *Handroanthus impetiginosus* (Mart. & DC.) Mattos (Bignoniaceae), *Jatropha mollissima* (Pohl) Baill.

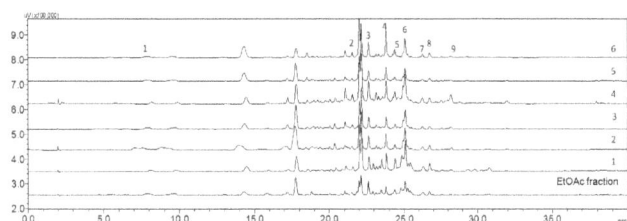

Fig 1. Chromatograms (HPLC-DAD 320 nm) of the phenolic fractions of *Melipona subnitida* geopropolis (1-6) and of EtOAc fraction geopropolis collected in January 2010. Compounds identified were: 6-*O*-*p*-coumaroyl-*D*-galactopyranose (**1**), 6-*O*-cinnamoyl-1-*O*-*p*-coumaroyl-β-*D*-glucopyranose (**2**), 7-*O*-methyl naringenin (**3**), 7-*O*-methyl aromadendrin (**4**), 7,4'-di-*O*-methyl aromadendrin (**5**), 4'-*O*-methyl kaempferol (**6**), 3-*O*-methyl quercetin (**7**), 5-*O*-methyl aromadendrin (**8**) and 5-*O*-methyl kaempferol (**9**)

Table 1. Effects of injections of ethanolic extracts and phenolic fractions of geopropolis on abdominal constrictions induced by acetic acid in mice.

| Samples | Numbers of writhers | | | |
| | EtOH extracts | | Phenolic fractions | |
	Media ± S.E.M.[a]	% inhibition[b]	Media ± S.E.M.[a]	% de inhibition
Control	38.4 ± 2.7	-		
Dipirone	18.8 ± 2.7	29.9 [*]		
1	0.0 ± 0.0	100.0 [***]	4.5 ± 1.0	85.4 [***]
2	0.2 ± 0.2	99.4 [***]	8.2 ± 1.6	73.5 [***]
3	0.2 ± 0.2	99.4 [***]	8.8 ± 2.5	71.4 [***]
4	0.7 ± 0.3	97.5 [***]	3.0 ± 1.9	90.3 [***]
5	0.8 ± 0.6	96.9 [***]	2.0 ± 0.7	93.5 [***]
6	0.2 ± 0.2	99.4 [***]	5.2 ± 3.1	83.2 [***]

[a] Data are expressed as the mean ± SEM, n=6. [b] Symbols indicate significant difference (([*]$P<0.05$ and [***]$P<0.001$, One Way ANOVA followed by Dunnett's test) compared to control group. Control was treated with vehicle (CMC/Tween® 80) (10 ml/kg, i.p.), dypirone 100 mg/kg, i.p. 40 minutes before initiating nociceptive stimulus.

(Euphorbiaceae) and *Anadenanthera colubrina* (Vell.) Brenan (Fabaceae) (Maia-Silva et al., 2012). Other studies to verify the presence of pollen in *M. subnitida* geopropolis are required, because pollen analysis in addition to chemical analysis is a method used to characterize regionally different propolis samples. Pollen types that occur in low frequency in propolis samples can be regarded as an indicator of the botanical species supplying the resin (Matos et al., 2014). It is a good tool for defining the phytogeographical origin of resins and quality of the propolis (Barth et al., 2003). Barth et al. (1999) and Barth and Luz (2003) showed that there is a fairly equal number of pollen grains between the samples of propolis from *Apis* and geopropolis produced by Meliponini, but a wider richness of pollen types is characteristic of geopropolis. In this regard, the Meliponini visits more plant species than the *Apis* bees. Nevertheless, the occurrence of dominant and accessory pollen grains is more frequent in propolis samples, which reflects a higher generalization of honeybees.

Evaluating abdominal constrictions induced by acetic acid was initially used to evaluate the antinociceptive activity of the EtOH extracts (100 mg/kg) of geopropolis and their phenolics fractions (100 mg/kg). The results showed in Fig. 2A and Table 1 demonstrate that the EtOH extract (100 mg/kg), produced inhibition of abdominal constrictions induced by acetic acid in mice (p<0.05), with inhibitions of 96.9% (sample 5) to 100% (sample 1). Phenolic fractions at the same concentration also inhibited the number of writhes (p<0.05) from 71.4% (sample 3) to 93.5% (sample 5), Fig 2B and Table 1. The inhibitory properties of the EtOH extracts and the phenolic fractions versus the abdominal constrictions induced by acetic acid in mice is first suggestion of the antinociceptive potential of these materials. The acetic acid induced constrictions test is a typical model for inflammatory pain that has long been used as a screening tool for the assessment of analgesic properties. The fact that the EtOH extracts showed slightly greater antinociceptive activities than the phenolic fractions suggests that geopropolis contains other compounds responsible for this activity and should be chemically investigated. The phenolic fraction is probably principally responsible for this activity. No reports on antinociceptive activity have been found in the literature for the identified constituents of *M. subnitida* geopropolis.

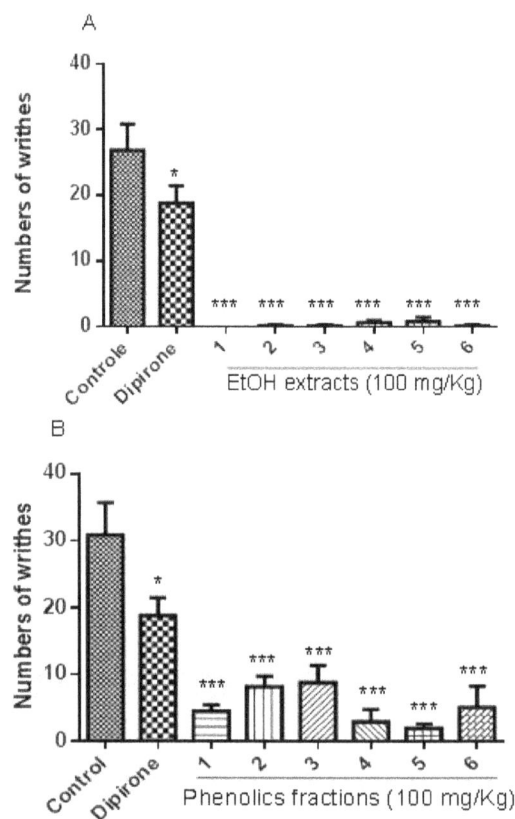

Fig 2. Effects of injections of ethanolic extract of geopropolis and phenolic fractions on abdominal constriction induced by acetic acid in mice. Control groups included the mice treated with only vehicle (negative control) or dypirone (positive control) 40 min before initiating nociceptive stimulus. Data are expressed as the mean ± SEM, n=6. Symbols indicate significant differences (*P<0.05 and ***P<0.001, One Way ANOVA followed by Dunnett's test) compared to the control group.

The free radical-scavenging activities of the EtOH extracts and phenolic fractions from geopropolis are shown in Table 2. The CE_{50} ranged from to 6.99-15.2 μg/mL (ABTS) and 13.3-39.2 μg/mL (DPPH) for the EtOH extracts and 3.2-8.9 μg/mL (ABTS) and 7.5-17.1 μg/mL (DPPH) for the phenolic fractions. The lower EC_{50} value indicates a higher antioxidant activity. The EtOH extracts and phenolic fractions showed a correlation between free radical-scavenging activity and the total phenolic content. The phenolic content ranged from 92.6-201.6 to EtOH extract and 205.5 to 305.3 to phenolic fractions. A correlation between DPPH-ABTS results for the EtOH extracts (r=0.91) and the DPPH-ABTS results for the phenolic fraction (r=0.97) was observed (Table 3).

These results suggest that total phenols, particularly the phenylpropanoids and flavonoids identified in *M. subnitida* geopropolis were responsible for the free radical-scavenging activity. Geopropolis obtained from the other stingless bees showed important antioxidant activities (Silva et al, 2013; Dutra et al, 2014). In early studies other *M. subnitida* products such as the pollen (Silva et al, 2006) and honey showed (Silva et al, 2013) free radical-scavenging activity. The pollen collected by the stingless bees *Melipona rufiventris* Lepeletier (Silva et al, 2009) and honey produced by *Melipona seminigra merrillae* Cockerell (Almeida da Silva et al., 2013) also were reported as having important antioxidant activities.

Table 2. Total phenolic and free radical-scavenging activity of *M. subnitida* geopropolis samples.

Geo-propolis sample	Total phenolic content (mg GAE/g ± SD)		ABTS[a] CE_{50} (μg/mL)		DPPH[a] CE_{50} (μg/mL)	
	EtOH extract	Phenolic fraction	EtOH extract	Phenolic fraction	EtOH extract	Phenolic fraction
1	97.6 ± 5.7	273.9 ± 6.8	15.2± 0.8	4.3 ± 0.1	39.2 ± 0.9	8.4 ± 0.1
2	92.6 ± 8.1	204.5 ± 7.4	13.4 ± 0.7	8.9 ± 0.7	31.7 ± 0.5	17.7 ± 0.2
3	172.6 ± 4.5	305.3 ± 5.0	7.7 ± 0.1	3.4 ± 0.1	15.9 ± 0.4	7.6 ± 0.1
4	150.7 ± 5.1	282.4 ± 1.5	10.3 ± 0.2	4.5 ± 0.2	16.1 ± 0.4	9.8 ± 0.1
5	201.6 ± 4.2	322.4 ± 6.4	6.9 ± 0.3	3.1 ± 0.1	13.3 ± 0.4	7.5 ± 0.1
6	139.3 ± 6.9	261.3 ± 5.8	15.2 ± 0.5	6.0± 0.1	28.9 ± 1.2	10.5 ± 0.1
Ascorbic acid			-		2.8 ± 0.0	2.8 ± 0.4
Trolox			3.21 ± 0.0	3.21 ± 0.0	-	-

[a] Mean value ± standard deviation: n=3, Concentration of antioxidant required to reduce the original amount of the radicals by 50%.

Table 3. Pearson correlation coefficients between the total phenolic content and the antiradical activity DPPH and ABTS.

	DPPH		ABTS	
	EtOH extracts	Phenolic fractions	EtOH extracts	Phenolic fractions
Total Phenolic ContentEtOH extracts	-0.90		-0.85	
DPPHEtOH extracts		0.91		
ABTSEtOH extracts	0.91			
Total Phenolic Content phenolic fractions	-0.94		-0.97	
DPPH phenolic fractions		0.97		
ABTS phenolic fractions	0.97			

Conclusion

The present results from six samples of *M. subnitida* geopropolis collected over three years showed that there is a variation in the total phenolic content over the years but not in the chemical profile. Geopropolis is a rich source of bioactive compounds with potential antioxidant and antinociceptive activities. The antioxidant activity is related to the total phenolic content. The SPE extraction was effective for the extraction of phenolic from *M. subnitida* geopropolis.

Acknowledgments

This work was financially supported by grants from CNPq (CNPq-PPBio 503285/2009-9), FACEPE (Grant no. PRONEM APQ-1232.1.06/10) and CAPES.

References

Almeida-Da-Silva, I.A., Silva, T.M. S., Camara, C.A., Queiroz, N., Magnani, M., Novais, J.S., Soledade, L.E.B., Lima, E.O., De Souza, A.L. & De Souza, A.G. (2013). Phenolic profile, antioxidant activity and palynological analysis of stingless bee honey from Amazonas, Northern Brazil. Food Chem. 141: 3552-3558. doi: 10.1016/j.foodchem.2013.06.072

Barth, O.M. (1998). Pollen analysis of Brazilian propolis, Grana 37: 97-101. DOI: 10.1080/00173130310012512.

Barth, O.M. & Luz, C.F.P. (2003). Palynological analysis of Brazilian geopropolis samples, Grana 42: 121-127. doi: 10.1080/00173130310012512

Collier, H.O.J., Dinneen, J.C., Johnson, C.A. & Schneider, C. (1968). The abdominal constriction response and its suppression by analgesic drugs in the mouse. Brit. J. Pharmacol. 32: 295-

310. Retrieved from: http://www.ncbi.nlm.nih.gov/pmc/articles/PMC1570212/

Cunha, M.G., Franchin, M., Galvão, L.C.C., Ruiz, A.L.T.G., Carvalho, J.E., Ikegaki, M., Alencar, S.M., Koo, H. & Rosalen, P.L. (2013). Antimicrobial and antiproliferative activities of stingless bee *Melipona scutellaris* geopropolis. BMC Complement. Altern. Med. 13: 23. doi: 10.1186/1472-6882-13-23

Dutra, R.P., Abreu, B.V.B., Cunha, M.S., Batista, M.C.A., Torres, L.M.B., Nascimento, F.R.F., Ribeiro, M.N.S. & Guerra, R.N.M. (2014). Phenolic acids, hydrolyzable tannins, and antioxidant activity of geopropolis from the stingless bee *Melipona fasciculata* Smith. J. Agri. Food Chem. 62: 2549-2557. doi: 10.1021/jf404875v

Fontenele, J.B., Viana, G.S.B., Xavier-Filho, J. & Alencar, J.W. (1996). Anti-inflammatory and analgesic activity of a water-soluble fraction from shark cartilage. Braz. J. Med. Biol. Res. 29: 643-646.

Franchin, M., Cunha, M.G., Denny, C., Napimoga M.H., Cunha, T.M., Koo, H., Alencar, S. M., Ikegaki, M. & Rosalen, P.L. (2012). Geopropolis from *Melipona scutellaris* decreases the mechanical inflammatory hypernociception by inhibiting the production of IL-1β and TNF-α. J. Ethnopharmacol. 143: 709-715. doi: 10.1016/j.jep.2012.07.040

Hadjmohammadi, M. R., Nazari, S. & Kamel, K. (2009). Determination of flavonoid markers in honey with SPE and LC using experimental design. Chromatographia 69: 1291-1297. doi: 10.1365/s10337-009-1073-4

Koster, R., Anderson, M. & De Beer, E.J. (1959). Acetic acid for analgesic screening. Fed. Proc. 18: 412-416.

Liberio, S.A., Pereira, A.L.A., Dutra, R.P., Reis, A.S., Araújo, M.J.A.M., Mattar, N.S., Silva, L.A., Ribeiro, M.N.S., Nascimento, F.R.F., Guerra, R.N.M. & Monteiro-Neto, V. (2011). Antimicrobial activity against oral pathogens and immunomodulatory effects and toxicity of geopropolis produced by the stingless bee *Melipona fasciculata* Smith. BMC Complement. Altern. Med. 11:108. doi: 10.1186/1472-6882-11-108

Matos, V.R., Alencar, S.M. & Santos, F.A.R. (2014). Pollen types and levels of total phenolic compounds in propolis produced by *Apis mellifera* L. (Apidae) in an area of the semiarid region of Bahia, Brazil. An. Acad. Bras. Cienc. 86:407-418. doi: 10.1590/0001-376520142013-0109

Maia-Silva, C., Silva, C.I., Hrncir, M., Queiroz, R.T. & Imperatriz-Fonseca, V.L.(2012). Guia de plantas visitadas por abelhas na Caatinga. Fortaleza: Fundação Brasil Cidadão, 191 p.

Nogueira-Neto, P. (1997). Vida e criação de abelhas indígenas sem ferrão. São Paulo: Nogueirapis, 445 p.

Re, R., Pelegrini, N., Proteggente, A., Pannala, A., Yang, M. & Rice-Evans, C. (1999). Antioxidant activity applyying an improved ABTS radical cátion decolorization assay. Free Radical Bio. Med. 26: 1231-1237. doi: 10.1016/S0891-5849(98)00315-3

Silva, E.C.C., Muniz, M.P., Nunomura, R.C.S., Nunomura, S.M. & Zilse, G.A.C. (2013). Phenolic constituents and antioxidant activity of geopropolis from two species of amazonian stingless bees. Quim. Nova. 36:628-633. doi: 10.1590/S0100-40422013000500003

Silva, T.M.S., Camara, C.A., Lins, A.C.S., Agra, F.M., Silva, E.M.S., Reis, I.T. & Freitas B.M. (2009). Chemical composition, botanical evaluation and screening of radical scavenging activity of collected pollen by the stingless bees *Melipona rufiventris* (uruçu amarela). An. Acad. Bras. Cienc. 81: 173-178. doi: 10.1590/S0001-37652009000200003

Silva, T.M.S., Camara, C.A., Lins, A.C.S., Barbosa-Filho, J.M., Silva, E.M.S., Freitas, B.M. & Santos, F.A.R. (2006). Chemical composition and free radical scavenging activity of pollen loads from stingless bee *Melipona subnitida* Ducke. J. Food Composit. Anal. 19:507-511. doi: 10.1016/j.jfca.2005.12.011

Silva, T.M.S., Santos, F.P., Rodrigues, A.E., Silva, E.M.S., Silva, G.S., Novais, J.S., Santos, Francisco, A.R. & Camara, C.A. (2013). Phenolic compounds, melissopalynological, physicochemical analysis and antioxidant activity of jandaíra (*Melipona subnitida*) honey. J. Food Compos. Anal. 29: 10-18. doi: 10.1016/j.jfca.2012.08.010

Slinkard, K. & Singleton, V.L. (1977). Total phenol analyses: automation and comparison with manual methods. Am. J. Enol. Viticult, 28: 49–55.

Simone-Finstrom, M. & Spivak, M. (2010). Propolis and bee health: the natural history and significance of resin use by honey bees. Apidologie 41: 295-311. doi: 10.1051/apido/2010016.

Souza, S.A., Camara, C.A., Silva, E.M.S. & Silva, T.M.S. (2013). Composition and antioxidant activity of geopropolis collected by *Melipona subnitida* (Jandaíra) bees. Evid. Based Complement. Alternat. Med. 2013: 1-5. doi: 10.1155/2013/801383

Velikova, M., Bankovaa, V., Tsvetkovab, I., Kujumgievb, A. & Marcuccic, M.C. (2000). Antibacterial *ent*-kaurene from Brazilian propolis of native stingless bees. Fitoterapia 71: 693-696. doi: 10.1016/S0367-326X(00)00213-6

22

Notes on Ants of the genus *Strumigenys* F. Smith, 1860 (Hymenoptera: Formicidae) in the Arabian Peninsula, with a key to species

Mostafa R. Sharaf [1], Brian L. Fisher[2] & Abdulrahman S. Aldawood[1]

1 - *King Saud University, Riyadh, Saudi Arabia.*

2 - *California Academy of Sciences, San Francisco, California, U.S.A.*

Keywords
Palearctic region, taxonomy, new record, Saudi Arabia, invasive species

Corresponding author
Mostafa R. Sharaf
Economic Entomology Research Unit
EERU - Plant Protection Department
College of Food and Agriculture Science
King Saud University, Riyadh 11451
P. O. Box 2460
E-Mail: *antsharaf@gmail.com*

Abstract

The ant genus *Strumigenys* in the Arabian Peninsula is treated. Three species are recognized, *S. arnoldi* Forel, *S. emmae* (Emery) and *S. membranifera* Emery. The invasive species *S. membranifera* and the Afrotropical species *S. arnoldi* are recorded for the first time from Saudi Arabia and the Arabian Peninsula. A key to the Arabian species based on the worker caste is presented. Biological, ecological and distribution notes for each species are given, as well as a regional distribution map for the three species.

Introduction

The ant genus *Strumigenys* F. Smith, 1860 is one of the largest and most conspicuous genera in the subfamily Myrmicinae (Bolton 2000). The genus is classified in the tribe Dacetini, with 836 valid extant species currently recognized (Bolton 2013), and are distributed worldwide in the tropics, subtropics and warm temperate regions (Bolton 2000, Brown 2000). The majority of species are cryptic soil inhabitants, nesting and foraging in leaf litter, topsoil layers, or wood pieces or stumps embedded in litter and topsoil layers (Brown 1953, Bolton 1983, Dejean 1991), with only a very few exceptional arboreal species (Bolton 1983, 2000). The majority of species with known biology are specialized predators on a broad range of smaller arthropods including: Diplura, Symplyla, Entomobryomorpha, Chilopoda, Pseudoscorpiones, Acarina, Araneae, Isopoda and larvae of many other orders os small Insecta (Wilson 1953, Masuko 1985, Brown 1971, Dejean 1987a, b).

The taxonomic history of the genus is long with numerous contributions including Arnold (1917), Wheeler (1922), Brown (1948, 1949 a, b, c, d, 1953, 1954 a, b), Terayama & Kubota (1989), Baroni Urbani & De Andrade (1994, 2007), Deyrup (1997), Bolton (1983, 1999) and the milestone work (Bolton 2000). More recently, the Indian fauna of the genus was treated by Bharti & Akbar (2013), recognizing 24 species, two of which were newly described and five species recorded for the first time from the country.

Several *Strumigenys* species have spread throughout the world by human commerce (Deyrup & Cover 2009, Wetterer 2011), a phenomenon well documented for many other ant species including *Tapinoma melanocephalum* (Fabricius, 1793) (Williams 1994); *Hypoponera punctatissima* (Roger, 1859), *Linepithema humile* Mayr, 1868, *Paratrechina longicornis* (Latreille, 1802), *Nylanderia vividula* (Nylander, 1846), *N. jaegerskioeldi* (Mayr, 1904), *Monomorium exiguum* Forel, 1894, *Pheidole teneriffana* Forel, 1893, and *Tetramorium caldarium* (Roger, 1857) (Gómez & Espadaler 2006).

Strumigenys membranifera Emery, 1869 and *S. emmae* (Emery 1890) are successful invasive species, with broad distributions (Wetterer 2011). These two species with a third one, *S. rogeri* Emery, 1890, are the most successful invasive *Strumigenys* ants known worldwide. It is thought that all three are of Old World origin but only *membranifera* reaches more temperate areas. *Strumigenys membranifera* is thought to be of African origin and has been reported in a wide range of habitats including cultivated areas, gardens, forest and urban parks (Brown & Wilson 1959, Bolton 1983, 2000). *Strumigenys emmae* is thought to be of Australian origin (Bolton 2000).

For a full diagnosis of the genus *Strumigenys*, see Bolton (1983, 2000) and Baroni Urbani & De Andrade (2007). A brief diagnosis of the genus includes mandibles extended into elongate narrow linear blades that terminate in two preapical teeth, the proximal tooth longer than the distal; in some species mandibles are short, triangular, serially dentate and lack an apical fork of spiniform teeth; lack of apicoscrobal hair; eyes ventrolateral, below the antennal scrobes; petiole node not bidentate dorsally; postpetiole with spongiform appendages present; absence of a basal spongiform pad on the first gastral sternite; specialized body pilosity frequently present.

The first two extensive faunal works on the ant fauna of Kingdom of Saudi Arabia (KSA) (Collingwood 1985) and the Arabian Peninsula (Collingwood & Agosti 1996) recorded no *Strumigenys*. Despite the absence of any record of this cryptic group, these authors predicted *Strumigenys* would be found in the Arabian Peninsula. The first record of a *Strumigenys* from the Arabian Peninsula was by Collingwood & Van Harten (2005) from Al Mukalla, Republic of Yemen, who reported three workers of the tramp species *S. emmae*, collected in Malaise traps. More recently, two workers of *S. arnoldi* Forel, 1913 were collected by one of the authors (BLF) from a forested mountainous area in Al Sarawat Mountains (southwestern region of KSA). These specimens were found nesting in moist soil under a stone. This record was included in El-Hawagryi et al. (2013) treatment of the insect fauna of Al Baha Region, KSA.

In the present study, recent materials of the genus *Strumigenys* from Saudi Arabia and Yemen are studied. Three species, *S. arnoldi*, *S. emmae* and *S. membranifera*, are discussed. An identification key to species is presented based on worker caste to facilitate species recognition. Some notes on ecology, biology and distribution are provided as well as a regional distribution map for the three species.

Materials and Methods

All materials, except a single specimen of *S. arnoldi* which is in the California Academy of Sciences (CASC), are deposited in King Saud University Museum of Arthropods (KSMA), Plant Protection Department, College of Food and Agriculture Sciences, Riyadh, Kingdom of Saudi Arabia. Most specimens were collected by sifting soil and leaf litter

using sifting trays. The two dealated gynes of *S. emmae* were collected using Malaise traps. The distribution range in the Arabian Peninsula was documented based on the new materials, along with previously published data (Yemen specimens) (Collingwood & Van Harten 2005) whose coordinates were not mentioned in the publication, but were obtained from the Google Earth website (www.earth.google.com). The map was created using the ArcGIS 9.2 program, with the help of Prof. Mahmoud S. Abdel-Dayem (King Saud University). Digital colour images were created using a Leica DFC450 digital camera and Leica Application Suite software (ver. 3.8). All images presented herein are available online and can be seen on AntWeb (http://www.antweb.org).

Results

Key to Arabian *Strumigenys*

1 Antenna with 4 segments, the first funicular segment followed directly by the 2-segmented club (pantropical tramp species)..*emmae* **(Emery)**
- Antenna with 6 segments, the first funicular segment separated from the 2-segmented club by two much smaller segments...2
2 Mandibles in full-face view broadest near base and gradually tapering towards the apex; apical fork of mandibles with 2 spiniform teeth; cephalic dorsum pilosity dense, hairs anteriorly curved and scale-like in full-face view (Kenya, Tanzania, Zimbabwe, Saudi Arabia and Yemen)..............*arnoldi* **Forel**
- Mandibles in full-face view with 12 teeth, arranged in a series of 7 larger teeth basally followed by a series of 4 denticles and a small apical teeth; cephalic dorsum pilosity restricted to a single pair of hairs at the highest point of vertex, minute and sparse appressed pubescence present (Cosmopolitan tramp species)..........................*membranifera* **Emery**

Taxonomic Treatment

Strumigenys arnoldi **Forel, 1913**
(Figures 1-3)

Strumigenys arnoldi Forel, 1913: 114 (w.) ZIMBABWE. Afrotropic. See also: Brown, 1954a: 26; Bolton, 1983: 365; Bolton, 2000: 591.

Diagnosis:

Apical fork of mandibles with 2 spiniform teeth; eyes small, maximum diameter less than maximum width of scape; metanotal groove absent; propodeal teeth broadly triangular; spongiform appendages of petiole and postpetiole well developed; cephalic dorsum, in full-face view, with scale-like, dense anteriorly curved hairs; pronotal humeri without flagellate hairs; mesonotum with a single pair of stout standing hairs; ground-pilosity of dorsal mesosoma similar to that of cephalic dorsum but hairs smaller and sparser; petiole, postpetiole and first gastral tergite with stout standing hairs which are swollen and clavate apically; cephalic dorsum finely and densely reticulate-punctate; entire dorsum of mesosoma finely reticulate-punctate, on pronotum this sculpture overlaid by some fine longitudinal rugulation. Color dull yellow to light brown.

Material Examined:

Saudi Arabia, (Al Sarawat Mountains), Al Bahah, Dhi Ain Archaeological Village, 19.9296 °N, 41.44285 °E, 750 m, 20.ix.2011, (B. L. Fisher, leg.), (BLF27515) (2 workers, CASENT0260163, CASENT0260165).

Ecological and Biological notes:

This species was collected from Dhi Ain Archaeological Village (Fig. 4), a cultivated area surrounded by mountains. The field locality includes a mix of native vegetation and cultivated plants including banana (*Musa paradisiaca* L., Family Musaceae), date palm (*Phoenix dactylifera* L., Family, Arecaceae), *Pandanus tectorius* Parkinson (Family Pandanaceae), *Ricinus communis* L. (Family Euphorbiaceae), *Ficus vasta* Forssk. (Family Moraceae), *Acacia* spp. (Family Fabaceae), *Prunus dulcis* (Mill.) D.A.Webb (Family Rosaceae), *Ziziphus spina-christi* (L.) Desf. (Family Rhamnaceae) and lemon orchards (*Citrus limon* (L.) Burm.f., Family Rutaceae). Two specimens were collected by the second author (BLF), found nesting in humid soil under a stone next to a small water stream which was used for irrigation of cultivated plants. The specimens were collected by sifting the soil. Five trips to the same territory were made in an attempt to find additional specimens but no more were collected, indicating the scarcity of the species and the low population in the region.

Distribution:

This is the first record of this species for Saudi Arabia and the Arabian Peninsula. This species is known only from the Afrotropical region, originally described from Zimbabwe (Forel 1913) and recorded from Kenya (Bolton 1983) and Tanzania (Bolton 2000).

Strumigenys emmae (Emery, 1890)
(Figures 5-7)

Epitritus emmae Emery, 1890: 70, pl. 8, fig. 6 (w.) ANTILLES. Neotropic.
Wheeler, 1908: 149 (gyne). Combination in *Quadristruma*: Brown, 1949b: 48; in *Strumigenys*: Bolton, 1999: 1674. Senior synonym of *Strumigenys clypeatus*, *Strumigenys malesiana*, *Strumigenys wheeleri*: Brown, 1949b: 48. See also: Bolton, 1983: 400; Bolton, 2000: 950.

Diagnosis:

Mandibles are a pair of narrow linear outcurved blades, armed with a fork of 2 spiniform teeth; anterior clypeal margin broad, with a feeble median impression; antennae four-segmented; eyes very small with a single ommatidium, situated just above ventral scrobe margin; pronotum slightly flat dorsally; metanotal groove absent; petiole and postpetiole in profile each with moderately developed spongiform appendages; lateral margins of clypeus short and with 2–3 anteriorly curved small spoon-shaped hairs; pronotal humeri each with a straight clavate hair and mesonotum with a similar but shorter pair of hairs; ground-pilosity of mesosomal dorsum in form of numerous scale-like to broadly spoon-shaped hairs; petiole, postpetiole, and gaster with short straight narrowly clavate hairs; dorsum of head behind clypeus reticulate-punctate; dorsal surface of petiole finely punctate to reticulate; postpetiole superficially reticulate to smooth. Color dull yellow to pale brown.

Material Examined:

Yemen, Al Mukalla, 14.533°N, 49.133°E, 11m, 1.ii.2003, (A. V. Harten, leg.), light trap (n=2 dealated gynes, CASENT0906377).

Previous Records:

(Collingwood & Van Harten 2005): Yemen: Ghail Ba Wazir, 14.776111°N, 49.366111°E, xii.2002 (A. V. Harten & M. Hubaishan leg.), Malaise trap (3 workers)

Biology and Habitats:

The Yemeni specimens were collected from Al Mukalla, a coastal city of the Arabian Sea. *Strumigenys emmae* forms small colonies with less than 50 workers and is considered to prefer disturbed habitats such as beach margins (Deyrup & Deyrup 1999).

Distribution:

Strumigenys emmae is a very successful tramp species (Bolton 2000) with a broad distribution including the tropics and subtropics in scattered localities (Brown 1949c), and has spread to at least 28 countries including non-tropical regions (Deyrup & Deyrup 1999). The origin of this species remains uncertain but is thought to be Australian (Bolton 2000). The first reported record from the Peninsula was from Yemen (Collingwood & Van Harten 2005).

Strumigenys membranifera Emery, 1869
(Figures 8-10)

Strumigenys (Trichoscapa) membranifera Emery, 1869: 24, fig. 11 (w.) ITALY. Palearctic.

Emery, 1916: 205 (gyne); Wheeler & Wheeler, 1991: 93. Combination in *Strumigenys (Cephaloxys)*: Emery, 1916: 205; in *Trichoscapa*: Brown, 1948: 113; in *Pyramica*: Bolton, 1999: 1673; in *Strumigenys*: Baroni Urbani & De Andrade, 2007: 123. Senior synonym of *S. foochowensis, S. marioni, S. santschii, S. silvestriana, S. simillima, S. vitiensis, S. williamsi*: Brown, 1948: 114. See also Brown, 1949 a: 6; Wilson, 1953: 483; Bolton, 1983: 319; Bolton, 2000: 322.

Diagnosis:

Mandibles with 12 teeth, arranged in a series of 7 larger teeth basally followed by a series of 4 denticles and a small apical teeth; anterior clypeal margin transverse to broadly feebly convex; eyes small, with only a few ommatidia, situated at ventral scrobe margin; metanotal groove absent; spongiform appendages of pedicel segments massively developed in profile; lateral spongiform appendages of petiolar node large and strongly prominent; lateral and ventral spongiform lobes of postpetiole massive, much larger than exposed area of disc; cephalic dorsum having only minute, sparse appressed pubescence; clypeal margins both anteriorly and laterally lacking projecting hairs; clypeus and lateral margins of head hairless; humeral angles of pronotum without hairs; dorsum of mesosoma bare, with only scattered sparse minute appressed pubescence present; dorsal surfaces of petiole, postpetiole and gaster without hairs but with minute appressed very sparse pubescence; cephalic dorsum reticulate-punctate and dull; sides of mesosoma smooth; propodeal dorsum and declivity smooth; in dorsal view both petiole and postpetiole smooth. Color dull yellow to yellowish brown.

Material Examined:

Saudi Arabia, Qassim, Buraydah, 26.36802°N, 44.03905°E, 653m, 19.x.2013, (Salman S., leg.), (2 workers CASENT0914337, CASENT0914338).

Biology and Habitats:

This species was collected by sifting soil in a date palm orchard at Buraydah (Qassim, central Region, KSA) (Fig.11). Specimens of *S. membranifera* were found inhabiting litter under a date palm tree among fallen decayed and dry dates. The soil was sandy and moist. In 1999, one of the authors (MS) collected a single worker from Abu Swelam Village, 3 km north of El Menia City (Egypt) (unpublished data). The specimen was found nesting under a small stone on a farm, in very compact and humid clay soil. Several Collembola were found in the same habitat and it is possible that *S. membranifera* preys on them as documented by Carlin (1981).

Distribution:

This is the first record of the species from KSA and the Arabian Peninsula. *Strumigenys membranifera* is a successful and widely distributed pantropical tramp species that has spread via human commerce (Wetterer, 2011) through all zoogeographical regions (Bolton 1983, Bolton 2000 & Wetterer, 2011).

Discussion

An interesting aspect of the genus *Strumigenys* in the Arabian Peninsula is its apparent scarcity, which makes the available information on the species limited, and consequently only few specimens are available in museums and collections. This is likely due to the small size and cryptic nature of the species (Bolton 1983, Bolton 1999, Bolton 2000).

The *S. membranifera* records from Buraydah (Qassim, Central Region, KSA) and El Menia (Egypt) were collected in human modified landscapes. Wilson & Hunt (1967) also noted this species in open cultivated fields in Futuna and Wallis Islands. The Buraydah record is an area of date palm trees cultivation, supporting the notion that this species was introduced probably associated with date palm shoots imported from adjacent regions. It seems likely this species could be found in other localities where extensive date palm production occurs, especially the central region and eastern region of the Arabian Peninsula.

Strumigenys arnoldi is an Afrotropical species (Bolton, 2000). The record from Al Sarawat Mountains (Al Bahah. KSA) is further evidence of the Afrotropical faunal affinities of the southwestern Mountains of KSA, supporting results of the faunal studies of the region (Eig 1938; Zohary 1973; Lehrer & Abou-Zied 2008; Doha 2009; Aldawood et al. 2011; Sharaf & Aldawood 2011, 2012; Sharaf et al. 2012a, 2012b; El-Hawagryi et al. 2013; Sharaf & Aldawood 2013).

The lack of previous studies recording *Strumigenys* (Collingwood 1985; Collingwood & Agosti 1996) may reflect their methods of collecting, which are not effective in collecting cryptic groups like dacetine ants. Methods such as sifting soil and leaf litter and using Malaise traps and Winkler bags, are much more effective in collecting these ants (Bestelmeyer et al. 2000).

Acknowledgments

This project was supported by the NSTIP Strategic technologies program, grant number (11-BIO1974-02), in the Kingdom of Saudi Arabia. The authors are indebted to Barry Bolton and Boris Kondratieff for valuable suggestions. The first author is grateful to Tony Hunter for critical reading of earlier draft of the manuscript. Special thanks to Antonius Van Harten for providing specimens of *S. arnoldi* from Yemen, to Lutfi Al Juhany for plants identification, to Michele Esposito,

Erin Prado and Estella Ortega (CAS, San Francisco) for help in photographing species and to Mahmoud S. Abdel-Dayem for providing the map.

References

Aldawood, A.S., Sharaf, M.R. & Taylor, B. (2011). First record of the myrmicine ant genus *Carebara* Westwood, 1840 (Hymenoptera, Formicidae) from Saudi Arabia with description of a new species *C. abuhurayri* sp. n. ZooKeys, 92: 61–69. doi: 10.3897/zookeys.92.770

Arnold, G. (1917). A monograph of the Formicidae of South Africa. Part 3. (Myrmicinae.) Annals of the South African Museum, 14: 271–402.

Baroni Urbani, C. & De Andrade, M.L. (1994). First description of fossil Dacetini ants with a critical analysis of the current classification of the tribe (Amber Collection Stuttgart: Hymenoptera, Formicidae. VI: Dacetini). – Stuttgarter Beiträge zur Naturkunde Serie B (Geologie und Paläontologie), 198: 1–65.

Baroni Urbani, C. & De Andrade, M. L. (2007). The ant tribe Dacetini: limits and constituent genera, with descriptions of new species (Hymenoptera, Formicidae). Annali del Museo Civico di Storia Naturale "Giacomo Doria", 99: 1–191.

Bestelmeyer, B.T.; Agosti, D.; Alonso, L. E.; Brandão, C.R.F.; Brown, W. L.; Delabie, J. H. C.& Silvestre, R. (2000). Field techniques for the study of ground-dwelling ants. In: D. Agosti et al. (eds). Ants, standard methods for measuring and monitoring biodiversity. Biological diversity hand book series. Smithsonian Institution Press, Washington D. C. 280 pp.

Bharti, H.; Akbar, S. A. (2013). Taxonomic studies on the ant genus *Strumigenys* Smith, 1860 (Hymenoptera, Formicidae) with report of two new species and five new records including a tramp species from India. Sociobiology, 60: 387–396.

Bolton, B. (1983). The Afrotropical dacetine ants (Formicidae). Bulletin of the British Museum (Natural History). Entomology, 46:267–416.

Bolton, B. (1999). Ant genera of the tribe Dacetonini (Hymenoptera: Formicidae). Journal of Natural History, 33: 1639–1689.

Bolton, B. (2000). The ant tribe Dacetini. Memoirs of the American Entomological Institute, 65: 1–1028.

Bolton, B. (2013). An online catalog of the ants of the World. Version 1 January 2013 Available from: http://www.antcat.org/catalog/ (Accessed 2 March 2014).

Brown, W.L., Jr. (1948). A preliminary generic revision of the higher Dacetini (Hymenoptera: Formicidae). Transactions of the American Entomological Society, 74: 101–129.

Brown, W.L., Jr. (1949a). Revision of the ant tribe Dacetini. I.

Fauna of Japan, China and Taiwan. – Mushi, 20: 1–25.

Brown, W.L., Jr. (1949b). A few ants from the Mackenzie River Delta. Entomological News, 60: 99.

Brown, W.L., Jr. (1949c). Revision of the ant tribe Dacetini: III. *Epitritus* Emery and *Quadristruma* new genus (Hymenoptera: Formicidae). Transactions of the American Entomological Society, 75: 43–51.

Brown, W.L., Jr. (1949d). A correction. Psyche, 56: 69.

Brown, W.L., Jr. (1953). Revisionary studies in the ant tribe Dacetini. American Midland Naturalist, 50: 1–137.

Brown, W.L., Jr. (1954a). The ant genus *Strumigenys* Fred. Smith in the Ethiopian and Malagasy regions. Bulletin of the Museum of Comparative Zoology, 112: 1–34.

Brown, W.L., Jr. (1954b). The Indo-Australian species of the ant genus *Strumigenys* Fr. Smith: group of *doriae* Emery. Psyche, 60(1953): 160–166.

Brown, W.L., Jr. (1971). The Indo-Australian species of the ant genus *Strumigenys*: group of *szalayi* (Hymenoptera: Formicidae). Pp. 73–86 In: Asahina, S., et al. (eds.) 1971. Entomological essays to commemorate the retirement of Professor K. Yasumatsu. Tokyo: Hokuryukan Publishing Co., Tokyo, Japan. vi + 389 pp.

Brown, W.L., Jr. (2000). Diversity of ants. In: D.Agosti et al. (Eds) Ants. standard methods for measuring and monitoring biodiversity. Biological diversity hand book series. Smithsonian Institution Press, Washington and London, 280 pp.

Brown, W.L., Jr. & Wilson, E.O. (1959). The evolution of the dacetine ants. Quarterly Review of Biology, 34: 278–294.

Carlin, N. F. (1981). Polymorphism and division of labor in the dacetine ant *Orectognathus versicolor* (Hymenoptera: Formicidae). Psyche, 88: 231–244.

Collingwood, C.A. (1985). Hymenoptera: Fam. Formicidae of Saudi Arabia. Fauna of Saudi Arabia, 7: 230–301.

Collingwood, C.A. & Agosti, D. (1996). Formicidae of Saudi Arabia (part 2). Fauna of Saudi Arabia, 15: 300–385.

Collingwood, C.A. &Van Harten, A. (2005). Further additions to the ant fauna (Hymenoptera: Formicidae) of Yemen. Zoology in the Middle East, 35: 73–78.

Dejean, A. (1987a). Étude du comportement de prédation dans le genre *Strumigenys* (Formicidae: Myrmicinae). Insectes Sociaux, 33: 388–405.

Dejean, A. (1987b). Behavioral plasticity of hunting workers of *Serrastruma serrula* presented with different arthropods. Sociobiology, 13: 191–208.

Dejean, A. (1991). Gathering of nectar and exploitation of Aphididae by *Smithistruma emarginata* (Formicidae: Myrmicinae). Biotropica, 23: 207–208.

Deyrup, M. (1997). Dacetine ants of the Bahamas (Hymenoptera: Formicidae). Bahamas Journal of Science, 5: 2–6.

Deyrup, M. & Deyrup, S. (1999). Notes on the introduced ant *Quadristruma emmae* (Hymenoptera: Formicidae) in Florida. Entomological News, 110: 13–21.

Deyrup, M., & S. Cover. (2009). Dacetine ants in southeastern North America (Hymenoptera: Formicidae). Southeastern Naturalist, 8: 191-212.

Doha, S.A. (2009). Phlebotomine sand flies (Diptera, Psychodidae) in different localities of Al-Baha province, Saudi Arabia. Egyptian Academic Journal of Biological Sciences, 1: 31–37.

Eig, A. (1938). Taxonomic studies on the Oriental species of the genus *Anthemis*. Palestine Journal of Botany, 1: 161–224.

El-Hawagryi, M.S., Khalil, M.W., Sharaf, M.R., Fadl, H.H. & Aldawood, A.S. (2013). A preliminary study on the insect fauna of Al-Baha Province, Saudi Arabia, with descriptions of two new species. Zookeys, 274: 1–88. doi: 10.3897/zookeys.274.4529

Emery, C. (1869). Enumerazione dei formicidi che rinvengonsi nei contorni di Napoli con descrizioni di specie nuove o meno conosciute. Annali dell'Accademia degli Aspiranti Naturalisti. Secunda Era, 2: 1–26.

Emery, C. (1890). Studii sulle formiche della fauna Neotropica. Bullettino della Società Entomologica Italiana, 22: 38–80.

Emery, C. (1916). Fauna Entomologica Italiana. I. Hymenoptera-Formicidae. Bullettino della Società Entomologica Italiana, 47: 79–275.

Forel, A. (1913). Fourmis de Rhodesia, etc. récoltées par M. G. Arnold, le Dr. H. Brauns et K. Fikendey. Annales de la Société Entomologique de Belgique, 57: 108–147.

Gómez, K. & Espadaler, X. (2006). Exotic ants in the Balearic Islands. Myrmecologische Nachrichten, 8: 225–233.

Lehrer, A.Z. & Abou-Zied, E.M. (2008). Une espèce nouvelle du genre Engelisca Rohdendorf de la faune d'Arabie Saoudite (Diptera, Sarcophagidae). Fragmenta Dipterologica, 14: 1–4.

Masuko, K. (1985). Studies on the predatory biology of oriental dacetine ants (Hymenoptera: Formicidae). I. Some Japanese species of *Strumigenys, Pentastruma,* and *Epitritus,* and a Malaysian *Labidogenys*, with special reference to hunting tactics in short-mandibulate forms. Insectes Sociaux, 31: 429–451.

Sharaf, M.R. & Aldawood, A.S. (2011). *Monomorium dryhimi* sp. n., a new ant species (Hymenoptera, Formicidae) of the *M. monomorium* group from Saudi Arabia, with a revised key to the Arabian species of the group. ZooKeys, 106: 47–54. doi: 10.3897/zookeys.106.1390

Sharaf, M.R. & Aldawood, A.S. (2012). A new ant species of the genus *Tetramorium* Mayr, 1855 (Hymenoptera, Formicidae) from Saudi Arabia, including a revised key to the Arabian species. PLoS ONE, 7 (2), e30811. doi: 10.1371/journal.pone.0030811

Sharaf, M.R. & Aldawood, A.S. (2013). First occurrence of the *Monomorium hildebrandti*-group (Hymenoptera: Formicidae), in the Arabian Peninsula, with description of a new species *M. kondratieffi* n. sp. Proceedings of the Entomological Society of Washington, 115 (1): 75–84.

Sharaf, M.R., Aldawood, A.S. & El-Hawagry, M.S. (2012a). A new ant species of the genus *Tapinoma* (Hymenoptera, Formicidae) from Saudi Arabia with a key to the Arabian species. ZooKeys, 212: 35–43. doi: 10.3897/zookeys.212.3325

Sharaf, M.R., Aldawood, A.S. & El-Hawagry, M.S. (2012b). First record of the ant subfamily Aenictinae (Hymenoptera, Formicidae) from Saudi Arabia, with the description of a new species. ZooKeys, 228: 39–49. doi: 10.3897/zookeys.228.3559

Terayama, M. & Kubota, S. (1989). The ant tribe Dacetini (Hymenoptera, Formicidae) of Taiwan, with descriptions of three new species. Japanese Journal of Entomology, 57: 778–792.

Wetterer, J. K. (2011). Worldwide spread of the membraniferous dacetine ant, *Strumigenys membranifera* (Hymenoptera: Formicidae). Myrmecological News, 14: 129–135.

Wheeler, G.C. & Wheeler, J. (1991). Instars of three ant species. Psyche, 98: 89–99.

Wheeler, W.M. (1908). The ants of Porto Rico and the Virgin Islands. Bulletin of the American Museum of Natural History, 24: 117–158.

Wheeler, W.M. (1922). The ants of the Belgian Congo. Bulletin of the American Museum of Natural History, 45: 1–1139.

Williams, D.F. (1994). (ed.) Exotic ants. Biology, impact, and control of introduced species. Westview Press. Boulder, Colorado. 332 pp.

Wilson, E.O. (1953). The ecology of some North American dacetine ants. Annals of the Entomological Society of America, 46: 479–495.

Wilson, E.O. & Hunt, G.L. (1967). Ant fauna of Futuna and Wallis Islands, stepping stones to Polynesia. – Pacific Insects, 9: 563–584.

Zohary, M. (1973). Geobotanical foundations of the Middle East. Vols. 1–2. G. Fischer, Stuttgart, Swets & Zeitlinger, Amsterdam, 738 pp.

Figures 1-4; 1-3 *Strumigenys arnoldi* worker, 1 Body in profile, 2 Body in dorsal view, 3 Head in full-face view (casent0260163, Saudi Arabia), Photographer: Estella Ortega, copyright, www.antweb. org, 4 Habitat of *S. arnoldi*, Dhi Ain archaeological village, Al Bahah.

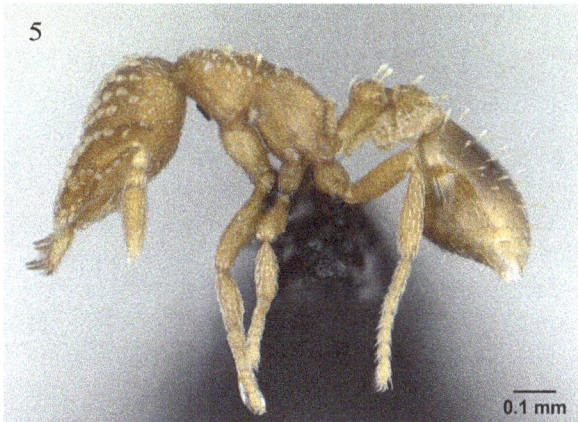

Figures 8-11; 8-10 *Strumigenys membranifera* worker (casent0914338, Saudi Arabia), 8 Body in profile, 9 Body in dorsal view, 10 Head in full-face view, Photographer: Michele Esposito, copyright www.antweb. org, 11 Habitat of *S. membranifera*, Buraydah (Qassim).

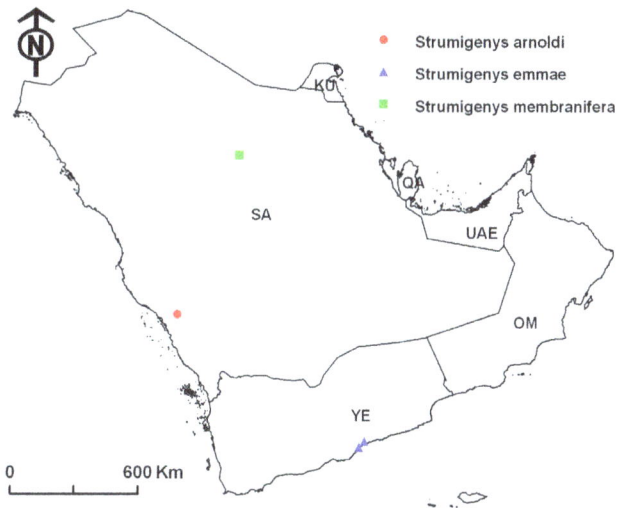

Fig. 12; Distribution map of the genus *Strumigenys* in the Arabian Peninsula. (M.S. Abdel-Dayem map). Abbreviations: (KU) Kuwait, (QA) Qatar, (OM) Oman, (SA) Saudi Arabia, (YE) Yemen, (UAE) United Arab Emirates.

Figures 5-7; *Strumigenys emmae* worker (casent0133445, Mayotte), 5 Body in profile, 6 Body in dorsal view, 7 Head in full-face view. Photographer: Erin Prado, copyright. www.antweb.org

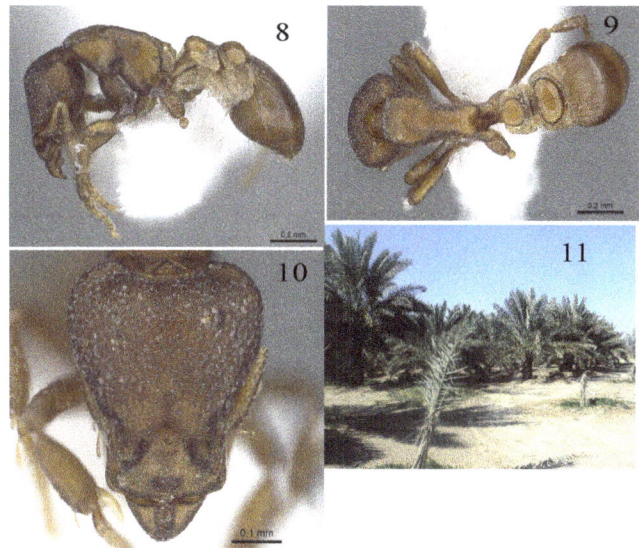

23

Genetic Variability of Stingless Bees *Melipona mondury* Smith and *Melipona quadrifasciata* Lepeletier (Hymenoptera: Apidae) from a Meliponary

JR Koser[1], FO Francisco[2], G Moretto[1]

1 - Universidade Regional de Blumenau, Blumenau, Santa Catarina, Brazil.
2 - Instituto de Biociências, Universidade de São Paulo, São Paulo, São Paulo, Brazil.

Keywords
conservation, heterospecific primers, Meliponini, microsatellites

Corresponding author
Jaqueline Reginato Koser
Departamento de Ciências Naturais
Universidade Regional de Blumenau
Blumenau, Santa Catarina, Brazil
E-Mail: jaquelinekoser@gmail.com

Abstract

The species of stingless bees *Melipona mondury* Smith and *Melipona quadrifasciata* Lepeletier are native to the Atlantic Forest. These species are sensitive to environmental changes and due to habitat loss they are endangered in several Brazilian states. This study aimed to evaluate the genetic variability of populations of these two species at the meliponary of the Regional University of Blumenau through the use of heterospecific microsatellite primers. We collected one worker from 19 colonies of *M. mondury* and from 25 colonies of *M. quadrifasciata*. We found low levels of genetic variability for both species, which may be explained by queen philopatry, intraspecific reproductive parasitism, and/or artificial maintenance of hives. If natural populations of these species are also presenting low genetic variability they might be endangered.

Introduction

The strictly Neotropical stingless bee genus *Melipona* (Hymenoptera: Apidae) comprises more than sixty species in Brazil (Camargo & Pedro, 2013) and is one of the most important insects for pollination in natural and in cultivated areas (Heard, 1999; Slaa et al., 2006). The species *Melipona* (*Michmelia*) *mondury* Smith, 1863 inhabits the Atlantic Forest biome, from Bahia to Rio Grande do Sul states (Camargo & Pedro, 2013). The species *Melipona* (*Melipona*) *quadrifasciata* Lepeletier, 1836 has a wide distribution in southern and southeastern Brazil (Camargo & Pedro, 2013) where it is widely cultivated and especially valued in tomatoes production (Santos et al., 2009; Sarto et al., 2005). In addition, its propolis has medicinal properties (Mercês et al., 2013). Both species are endangered in several Brazilian states (Machado et al., 1998; Marques et al., 2002; Mikich & Bérnils, 2004; Santa Catarina, 2011) due to habitat loss caused by deforestation (Brosi et al., 2007; Brown & Oliveira, 2014).

Nowadays, the Atlantic Forest is extremely fragmented, and bee species suffer from the negative impacts of the interruption of gene flow and decrease in genetic diversity (Freiria et al., 2012). One way to prevent local extinction of these bees is maintaining hives in free-foraging wooden boxes in meliponaries (apiaries for stingless bees). Meliponiculture for crop pollination and for honey extraction has been an encouraging economic practice aligned with sustainable development and educational purposes in several countries (Cortopassi-Laurino et al., 2006).

However, little is known about the genetic variability of hives of native bees in meliponaries (Carvalho-Zilse et al., 2009). Such knowledge is essential for the development of conservation strategies and rational exploitation of native species (Cortopassi-Laurino et al., 2006; Alves et al., 2011). Our aim was to characterize the genetic variability of *M. mondury* and *M. quadrifasciata* maintained in the meliponary of the Regional University of Blumenau.

Material and Methods

We collected one worker from each of 19 colonies of *M. mondury* and from 25 colonies of *M. quadrifasciata* acquired from beekeepers from Vale do Itajaí region, in the Santa Catarina state, Brazil and maintained currently at the meliponary of the Regional University of Blumenau (mRUB) (Fig. 1) localized at 26° 54' 21.81'' S 49° 04' 48.53'' W (Fig.2). Colonies of *M. quadrifasciata* have been maintained at mRUB since 1998 and colonies of *M. mondury* since 2003. These hives were extracted directly from nature in geographically close cities and have not been artificially divided.

Figure 1. Meliponary of the Regional University of Blumenau (mRUB).

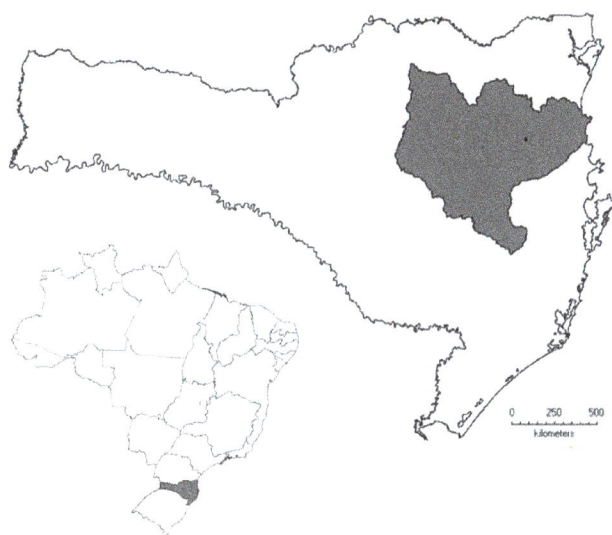

Figure 2. Localization of Santa Catarina state showing the Vale do Itajaí region. Meliponary localization (●).

DNA extraction was performed according to Anderson and Fuchs (1998). For both species, we used 10 microsatellite primer pairs developed for the species *M. bicolor* (Mbi11, Mbi28, Mbi32, Mbi33, Mbi88, Mbi218, Mbi233, Mbi254, Mbi259 and Mbi522) (Peters et al., 1998) and 11 primer pairs developed for *M. mondury* (Mmo03, Mmo06, Mmo08, Mmo10, Mmo11, Mmo15, Mmo19, Mmo20, Mmo21, Mmo22 and Mmo24) (Lopes et al., 2010b). PCR reactions and the annealing temperatures were performed according to Peters et al. (1998) for the Mbi primers and ac-

cording to Lopes et al. (2010b) for the Mmo primers. PCR products were separated by electrophoresis in 12% polyacrylamide gels and stained with silver nitrate for visualization.

The program Arlequin v.3.5.1.3 (Excoffier & Lischer, 2010) was used to calculate allelic richness ($\hat{A}$), observed heterozygosity (H_O), expected heterozygosity (H_E), percentage of polymorphic loci (PPL) and F_{IS} with 10,000 permutations. Hardy-Weinberg Equilibrium (HWE) and Linkage Disequilibrium (LD) were computed using Genepop v.4.1.4 (Rousset, 2008). *P*-values were adjusted with Bonferroni correction (Rice, 1989). The frequencies of null alleles were computed using Cervus 3.0.6 (Kalinowski et al., 2007).

Results

Genetic variability was low for *M. mondury*, but no inbreeding was detected (Table 1). Loci Mbi254 and Mmo19 deviated from HWE even after Bonferroni correction (*P*< 0.0167). LD was detected between Mbi254 and Mmo19 after Bonferroni correction (*P*= 0.0062). The presence of null alleles with a frequency of 29% was found in the locus Mbi254, a heterospecific primer (Table S1). The loci Mmo19 and Mmo21, both specific primers for *M. mondury*, were not affected by null alleles (Table S1).

Melipona quadrifasciata also showed low genetic variability although higher than *M. mondury* (Table 1). Significant inbreeding was detected. Primers developed for *M. bicolor* were more polymorphic in *M. quadrifasciata* than those developed for *M. mondury* (Table S2). Deviation from HWE was detected for the loci Mbi218, Mbi233 and Mmo21 after Bonferroni correction (*P* = 0.0000). LD was not detected (all *P* > 0.05). For *M. quadrifasciata*, the frequency of null alleles was higher than 25% in the following heterospecific primer pairs: Mbi218, Mbi233, Mmo03, Mmo11 and Mmo21 (Table S2). The loci Mbi11, Mbi88, Mbi254 and Mbi259 were not affected by null alleles (Table S2).

Table 1. Genetic variability for *Melipona mondury* and *Melipona quadrifasciata* based on microsatellite data. $\hat{A}$: allelic richness; F_{IS}: inbreeding coefficient; H_E: expected heterozygosity; H_O: observed heterozygosity; N: sample size; PPL: percentage of polymorphic loci.

Species	N	$\hat{A}$	H_O	H_E	PPL	F_{IS}	*P-value*
M. mondury	19	1.60	0.105	0.102	15%	-0.00629	0.569307
M. quadrifasciata	25	2.22	0.129	0.189	50%	0.33493	0.000000

Discussion

Our results indicate low genetic variability in the populations of *M. mondury* and *M. quadrifasciata* maintained at the mRUB. Low genetic variability in native stingless bees is also documented in studies using molecular markers such as RAPD (Tavares et al., 2001), mitochondrial polymorphism (Brito et al., 2013), ISSR (Inter Single Sequence Repeats) (Nascimento et al., 2010; Miranda et al., 2012), and microsatellites (Francisco et al., 2006; Tavares et al., 2007; Carvalho-Zilse et al., 2009; Francini et al., 2009; Alves et al., 2011; Duarte et al., 2011).

A first explanation of this low variability could be the presence of null alleles. In this study, we found high values of null alleles only in one out of three polymorphic loci for *M. mondury* (Mbi254) and in five out of nine polymorphic loci for *M. quadrifasciata* (Mbi218, Mbi233, Mmo03, Mmo11 and Mmo21). In addition, null alleles might be responsible for the high number of monomorphic loci found in both species.

The use of heterospecific primers may also be related to low genetic diversity. In the stingless bee *Plebeia remota*, the same samples analyzed with heterospecific (Francisco et al., 2006) and specific (Francisco et al., 2013) primers showed high divergent values. The same result was found in *Melipona* bees (Lopes et al., 2010a). Cross-species amplification is one of the advantages of microsatellites, and low variability should always be interpreted with caution when using heterospecific primers. However, data obtained from other populations/species with the same primer pairs we used show higher levels of genetic diversity (Tables S3 and S4) suggesting that the low diversity we observed in this work are not due to heterospecific primers only.

Low variability may also be explained by the species natural biology. Queens of most stingless bee species mate with a single male (monandric) (Peters et al., 1999; Palmer et al., 2002) and are known to nidify near maternal nests (Nogueira-Neto, 1954). Their low dispersion increases genetic drift and inbreeding within sub-populations (Hartl & Clark, 2007). A special concern is the maintenance of alleles of the complementary sex determination system in small populations, because inbreeding can lead to the production of diploid males (Cook & Crozier, 1995; Zayed, 2009; Alves et al., 2011). For other stingless bee species, males are the dispersing sex (Cameron et al., 2004; Carvalho-Zilse & Kerr, 2004; Francisco et al., 2013), but no data is available for *M. mondury* and *M. quadrifasciata*. Nevertheless, the location of mRUB near forest fragments might allow the mating of queens from the meliponaries with males from native forests in the surroundings, decreasing their inbreeding probability.

Managed populations of local bees can be considered a reservoir of genetic diversity if they can interbreed with wild populations (Alves et al., 2011). However, to introduce bees from distant populations or another species that could hybridize with local populations might cause outbreeding depression (Lynch, 1991; Waser et al., 2000).

Recent data from Wenseleers et al. (2011) showed intraspecific reproductive parasitism in *M. scutellaris*. If we speculate that this behavior occurs in other *Melipona* species, colonies that were previously unrelated could become related if sibling queens take over these colonies. In these cases, genetic variability would decrease.

Another explanation to account for the low variability may be related to the artificial maintenance of hives through founder events and bottlenecks as already reported for *Apis mellifera* (Sheppard, 1988; Schiff & Sheppard, 1996; Moritz et al., 2007; Delaney et al., 2009; Jaffé et al., 2010; Meixner et al., 2010) and *M. scutellaris* (Carvalho-Zilse et al., 2009; Alves et al., 2011). However, despite low variability, populations can be successfully maintained if a strong care is dispensed over the nests (Alves et al., 2011).

If natural populations of these species are also presenting low genetic variability they might be endangered. Adaptation to environmental changes is dependent on the genetic variations that exist among members of a population. Without variation, populations are more prone to extinction. Therefore, the analysis of natural populations of these species in south Brazil is crucial. Detecting the male flight range is also important since they can prevent genetic isolation in fragmented populations. If meliponaries were able to assist in maintaining the genetic variability of natural populations, they could be used for research and reintroduction programs.

Acknowledgments

We thank the Universidade Regional de Blumenau-FURB for technical assistance. JR Koser was supported by a scholarship from Pipe/Art 170.

References

Alves, D.A., Imperatriz-Fonseca, V.L., Francoy, T.M., Santos-Filho, P.S., Billen, J., & Wenseleers, T. (2011). Successful maintenance of a stingless bee population despite a severe genetic bottleneck. Conservation Genetics, 12: 647-658. doi: 10.1007/s10592-010-0171-z

Anderson, D.L. & Fuchs, S. (1998). Two genetically distincts populations of *Varroa jacobsoni* with contrasting reproductive abilities on *Apis mellifera*. Journal of Apicultural Research, 37: 69-78.

Brito, R.M., Francisco, F.O., Françoso, E., Santiago, L.R. & Arias, M.C. (2013). Very low mitochondrial variability in a stingless bee endemic to Cerrado. Genetics and Molecular Biology, 36: 124-128. doi: 10.1590/S1415-47572013000100018

Brosi, B.J., Daily, G.C. & Ehrlich, P.R. (2007). Bee community shifts with landscape context in a tropical countryside. Ecological Applications, 17: 418-430. doi: 10.1890/06-0029

Brown, J.C. & Oliveira, M. (2014).The impact of agricultural colonization and deforestation on stingless bee (Apidae: Meliponini) composition and richness in Rondônia, Brazil. Apidologie, 45: 172-188. doi:10.1007/s13592-013-0236-3

Camargo, J.M.F. & Pedro, S.R.M. (2013). Meliponini Lepeletier, 1836. In Moure, J.S., Urban, D. & Melo, G.A.R. (Orgs). Catalogue of Bees (Hymenoptera, Apoidea) in the Neotropical Region - online version. Available at http://www.moure. cria.org.br/catalogue. Accessed Jan/18/2014.

Cameron, E.C., Franck, P. & Oldroyd, B. P. (2004). Genetic structure of nest aggregations and drone congregations of the south-east asian stingless bee *Trigona collina*. Molecular Ecology, 13: 2357-2364. doi: 10.1111/j.1365-294X.2004.02194.x

Carvalho-Zilse, G.A. & Kerr, W.E. (2004). Substituição natural de rainhas fisogástricas e distância de vôo dos machos em Tiuba (*Melipona compressipes fasciculata Smith, 1854) e Uruçu (Melipona scutellaris Latreille, 1811) (Apidae, Meliponini*). Acta Amazonica, 34: 649-652. doi: 10.1590/S0044-59672004000400016

Carvalho-Zilse, G.A., Costa-Pinto, M.F.F., Nunes-Silva, C.G. & Kerr, W.E. (2009). Does beekeeping reduce genetic variability in *Melipona scutellaris* (Apidae, Meliponini)? Genetics and Molecular Resources, 8: 758-765.

Cook, J.M. & Crozier, R.H. (1995). Sex determination and population biology in the Hymenoptera. Trends in Ecology and Evolution, 10: 281-286. doi: 10.1016/0169-5347(95)90011-X

Cortopassi-Laurino, M., Imperatriz-Fonseca, V.L., Roubik, D.W., Dollin, A., Heard, T., Aguilar, I., Venturieri, G.C., Eardley, C. & Nogueira-Neto, P. (2006). Global meliponiculture: challenges and opportunities. Apidologie, 37: 275-292. doi: 10.1051/apido:2006027

Delaney, D.A., Meixner, M.D., Schiff, N.M. & Sheppard, W.S. (2009). Genetic characterization of commercial honeybee (Hymenoptera: Apidae) populations in the United States by using mitochondrial and microsatellite markers. Annals of the Entomological Society of America, 102: 666-673. doi: 10.1603/008.102.0411

Duarte, O.M.P., Gaiotto, F.A., Souza, A.P., Mori, G.M. & Costa, M.A. (2011). Isolation and characterization of microsatellites from *Scaptotrigona xanthotricha* (Apidae, Meliponini): A stingless bee in the Brazilian Atlantic Rainforest. Apidologie, 43: 432-435. doi: 10.1007/s13592-011-0109-6

Excoffier, L. & Lischer, H.E.L. (2010). Arlequin suite ver 3.5: A new series of programs to perform population genetics analyses under Linux and Windows. Molecular Ecology Resources, 10: 564-567. doi: 10.1111/j.1755-0998.2010.02847.x

Francini, I.B., Sforça, D.A., Sousa, A.C.B. & Campos, T. (2009). Microsatellite loci for an endemic stingless bee *Melipona seminigra merrillae* (Apidae, Meliponini) from Amazon. Conservation Genetics Resources, 1: 487-490. doi: 10.1007/s12686-009-9113-9

Francisco, F.O., Brito, R.M. & Arias, M.C. (2006). Allele number and heterozigosity for microsatellite loci in different stingless bee species (Hymenoptera: Apidae, Meliponini). Neotropical Entomology, 35: 638-643. doi: 10.1590/S1519-566X2006000500011

Francisco, F.O., Santiago, L.R. & Arias, M.C. (2013). Molecular genetic diversity in populations of the stingless bee *Plebeia remota*: A case study. Genetics and Molecular Biology, 36: 118-123. doi: 10.1590/S1415-47572013000100017

Freiria, G.A., Ruim, J.B., Souza, R.F.D. & Sofia, S.H. (2012). Population structure and genetic diversity of the orchid bee *Eufriesea violacea* (Hymenoptera, Apidae, Euglossini) from Atlantic Forest remnants in southern and southeastern Brazil. Apidologie, 43: 392-402. doi: 10.1007/s13592-011-0104-y

Hartl, D.L. & Clark, A.G. (2007). Principles of Population Genetics. Sunderland: Sinauer Associates, 545 p.

Heard, T.A. (1999). The role of stingless bees in crop pollination. Annual Review of Entomology, 44: 183-206. doi: 10.1146/annurev.ento.44.1.183

Jaffé, R., Dietemann, V., Allsopp, M.H., Costa, C., Crewe, R.M., Dall'olio, R., De La Rúa, P., El-Niweiri, M.A.A., Fries, I., Kezic, N., Meusel, M.S., Paxton, R.J., Shaibi, T., Stolle, E. & Moritz, R.F.A. (2010). Estimating the density of honeybee colonies across their natural range to fill the gap in pollinator decline censuses. Conservation Biology, 24: 583–593. doi: 10.1111/j.1523-1739.2009.01331.x

Kalinowski, S.T., Taper, M.L. & Marshall, T.C. (2007). Revising how the computer program CERVUS accommodates genotyping error increases success in paternity assignment. Molecular Ecology, 16: 1099-1106. doi: 10.1111/j.1365-294X.2007.03089.x

Lopes, D.M., Campos, L.A.O., Salomão, T.M.F. & Tavares, M.G. (2010a). Comparative study on the use of specific and heterologous microsatellite primers in the stingless bees *Melipona rufiventris* and *M. mondury* (Hymenoptera, Apidae). Genetics and Molecular Biology, 33: 390-393. doi: 10.1590/S1415-47572010005000017

Lopes, D.M., Silva, F.O., Fernandes-Salomão, T.M., Campos, L.A.O. & Tavares, M.G. (2010b). A scientific note on the characterization of microsatellite loci for *Melipona mondury* (Hymenoptera: Apidae). Apidologie,41: 138-140. doi: 10.1051/apido/2009067

Lynch, M. (1991). The genetic interpretation of inbreeding depression and outbreeding depression. Evolution, 45: 622-629.

Machado, A.B.M., Fonseca, G.A.B., Machado, R.B., Aguiar, L.M.S. & Lins, L.V. (1998). Livro vermelho das espécies ameaçadas de extinção da fauna de Minas Gerais. Belo Horizonte: Biodiversitas, 605 p.

Marques, A.A.B., Fontana, C.S., Vélez, E., Bencke, G.A.,

Schneider, M. & Reis, R.E. (2002). Lista das espécies da fauna ameaçadas de extinção no Rio Grande do Sul. Porto Alegre: FZB/MCT–PUCRS/PANGEA, 52 p.

Meixner, M.D., Costa, C., Kryger, P., Hatjina, F., Bouga, M., Ivanova, E. & Buchler, R. (2010). Conserving diversity and vitality for honey bee breeding. Journal of Apicultural Research, 49: 85-92. doi: 10.3896/IBRA.1.49.1.12

Mercês, M.D., Peralta, E.D., Uetanabaro, A.P.T. & Lucchese, A.M. (2013). Atividade antimicrobiana de méis de cinco espécies de abelhas brasileiras sem ferrão. Ciência Rural, 43: 672-675.

Mikich, S.B. & Bérnils, R.S. (2004). Livro vermelho da fauna ameaçada no estado do Paraná, http://www.pr.gov.br/iap (accessed date: 4October, 2013).

Miranda, E.A., Batalha-Filho, H., Oliveira, P.S., Alves, R.M.O., Campos, L.A.O. & Waldschmidt, A.M. (2012). Genetic diversity of Melipona mandacaia Smith 1863 (Hymenoptera, Apidae), an endemic bee species from Brazilian Caatinga, using ISSR. Psyche, 2012: 1-6. doi: 10.1155/2012/372138

Moritz, R.F.A., Kraus, F.B., Kryger, P. & Crewe, R.M. (2007). The size of wild honeybee populations (Apis mellifera) and its implications for the conservation of honeybees. Journal of Insect Conservation, 11: 391–397. doi: 10.1007/s10841-006-9054-5

Nascimento, M.A., Batalha-Filho, H., Waldschmidt, A.M., Tavares, M.G., Campos, A.O. & Salomão, T.M.F. (2010). Variation and genetic structure of Melipona quadrifasciata Lepeletier (Hymenoptera, Apidae) populations based on ISSR pattern. Genetics and Molecular Biology, 33: 394–397. doi: 10.1590/S1415-47572010005000052

Nogueira-Neto, P. (1954). Notas bionômicas sobre meliponíneos: III – Sobre a enxameagem. Arquivos do Museu Nacional, 42: 419-451.

Palmer, K.A., Oldroyd, B.P., Quezada-Euán, J.J.G., Paxton, R.J., & May-Itza, W.D.J. (2002). Paternity frequency and maternity of males in some stingless bee species. Molecular Ecology, 11: 2107-2113. doi: 10.1046/j.1365-294X.2002.01589.x

Peters, J.M., Queller, D.C., Imperatriz-Fonseca, V.L. & Strassmann, J.E. (1998). Microsatellite loci from the stingless bees. Molecular Ecology, 7: 783-792.

Peters, J.M., Queller, D.C., Imperatriz–Fonseca, V.L., Roubik, D.W. & Strassmann, J.E. (1999). Mate number, kin selection and social conflicts in stingless bees and honeybees. Proceedings of the Royal Society B, 266: 379-384.

Rice, W.R. (1989). Analyzing tables of statistical tests. Evolution, 43: 223-225.

Rousset, F. (2008). GENEPOP'007: a complete re-implementation of the GENEPOP software for Windows and Linux. Molecular Ecology Resources, 8: 103–106. doi: 10.1111/j.1471-8286.2007.01931.x

Santa Catarina (State), (2011). Conselho Estadual do Meio Ambiente - CONSEMA. Resolução CONSEMA N° 002, de 06 de dezembro de 2011. Reconhece a lista oficial de espécies da fauna ameaçadas de extinção no estado de Santa Catarina e dá outras providências. http://www.fatma.sc.gov.br/images/stories/biodiversidade/resolucao_fauna__002_11_fauna.pdf. (accessed date: 4October, 2013).

Santos, S.A., Roselino, A.C., Hrncir, M. & Bego, L.R. (2009). Pollination of tomatoes by the stingless bee Melipona quadrifasciata and the honey bee Apis mellifera (Hymenoptera, Apidae). Genetics and Molecular Research, 8: 751-757.

Sarto, M.C.L. del, Peruquetti, R.C. & Campos, L.A.O. (2005). Evaluation of the neotropical stingless bee Melipona quadrifasciata (Hymenoptera: Apidae) as pollinator of greenhouse tomatoes. Journal of Economic Entomology, 98: 260-266. doi: 10.1603/0022-0493-98.2.260

Schiff, N.M. & Sheppard, W.S. (1996). Genetic differentiation in the queen breeding population of the western United States. Apidologie, 27: 77–86. doi: 10.1051/apido:19960202

Sheppard, W.S. (1988). Comparative study of enzyme polymorphism in United States and European honey bee (Hymenoptera: Apidae) populations. Annals of the Entomological Society of America, 81: 886–889.

Slaa, E.J., Chaves, L.A.S., Malagodi-Braga, K.S. & Hofstede, F.E. (2006). Stingless bees in applied pollination: practice and perspectives. Apidologie, 37: 293-315. doi: 10.1051/apido:2006022

Tavares, M.G., Dias, L.A.S., Borges, A.A., Lopes, D.M., Busse, A.H.P., Costa, R.G., Salomão, T.M.F. & Campos, L.A.O. (2007). Genetic divergence between populations of the stingless bee uruçu amarela (Melipona rufiventris group, Hymenoptera, Meliponini): Is there a new Melipona species in the Brazilian state of Minas Gerais? Genetics and Molecular Biology, 30: 667-675. doi: 10.1590/S1415-47572007000400027

Tavares, M.G., Ribeiro, E.H., Campos, L.A.O., Barros, E.G. & Oliveira, M.T.V.A. (2001). Inheritance pattern of RAPD markers in Melipona quadrifasciata (Hymenoptera: Apidae, Meliponinae). Journal of Heredity, 92: 279-282. doi: 10.1093/jhered/92.3.279

Waser, N.M., Price, M.V. & Shaw, R.G. (2000). Outbreeding depression varies among cohorts of Ipomopsis aggregata planted in nature. Evolution, 54: 485-491.

Wenseleers, T., Alves, D.A., Francoy, T.M., Billen, J. & Imperatriz-Fonseca, V.L. (2011). Intraspecific queen parasitism in a highly eusocial bee. Biology Letters, 7: 173-176. doi: 10.1098/rsbl.2010.0819

Zayed, A. (2009). Bee genetics and conservation. Apidologie, 40: 237-262.

24

Assessing Sperm Quality in Stingless Bees (Hymenoptera: Apidae)

HM Meneses[1], S Koffler[2], BM Freitas[1], VL Imperatriz-Fonseca[2], R Jaffé[2]

1 - Universidade Federal de Ceará,UFC, Fortaleza, CE, Brazil

2 - Universidade de São Paulo, USP, São Paulo, SP, Brazil

Keywords

sperm viability, sperm morphology, sperm counts, semen, Meliponini.

Corresponding author

Hiara Marques Meneses
Av. Mister Hull 2977, Campus do Picí,
Bloco 808, 60021-970, Fortaleza-CE, Brazil
E-Mail: hiarameneses@gmail.com

Abstract

Although stingless bees have a great potential as commercial pollinators, their exploitation depends on the successful reproduction of colonies on a large scale. To do so, it is essential to develop accurate diagnostic tools that enable a better understanding of the reproductive biology of stingless bees. Sperm counts, sperm morphology and sperm viability (the relative proportion of live to dead sperm), are key parameters assessing semen quality and potential fertilization success. Here we present standardized protocols to assess these three parameters. We used *Scaptotrigona* aff. *depilis* (Moure) as a study model. Semen extractions from the seminal vesicles were found to yield better results when performed in mature rather than in younger males. For morphology and viability analyses, the best semen dilution on Hayes solution was adding 120 µl to the contents of the two seminal vesicles. For sperm counts, however, we recommend a higher dilution (1,000 µl). Sperm viability values were higher when Hayes solution was adjusted to pH 8.7, and when samples were analyzed before 24 hours from collection. Based on these results we present standard protocols, hoping they will be useful to future researchers assessing sperm quality in other stingless bee species.

Introduction

In order to exploit the great potential of stingless bees as commercial pollinators, it is of great importance to improve management and optimize practices to produce colonies on a large scale. Therefore, it is essential to develop accurate diagnostic tools that enable a better understanding of the reproductive biology of stingless bees.

Semen quality is known to determine male reproductive success across taxa (Simmons, 2001). In turn, semen quality can be affected by factors such as the number, morphology and viability of sperm (Garófalo, 1980; Simmons, 2001; den Boer et al., 2009). Few studies have looked at sperm characteristics of bees in general, and even less in stingless bees. Cortopassi-Laurino (1979) counted the number of sperm in males of *Plebeia droryana* (Friese) and found 313,000 ± 102,050 sperm per male. Garófalo (1980) evaluated the amount of semen produced by the males and stored in the female's spermathecae in different species of bees, including eight stingless bees, and found that there is no relationship between social levels and the amount of sperm found in

females and produced by males. Conte et al. (2005) found that the small spermatids are lost during the early stages of spermiogenesis of *Melipona quadrifasciata anthidioides* Lepeletier. Also addressing the spermatogenesis process, Lino-Neto et al. (2008) found that half of the spermatids formed during the spermatogenesis are not transformed into viable sperm cells in *Scaptotrigona xanthotricha* Moure. Pech-May et al. (2012) found a strong positive relationship between the size of male and semen production in *Melipona beecheii* Bennett. Camillo (1971) noted an increase of 2.35 times in the number of sperm of giant *Friesella schrottkyi* (Friese) males (produced occasionally when an unfertilized egg is raised on a royal cell) compared to what is found in common males. Table 1 summarizes previous contributions to the study of stingless bee sperm.

The viability of spermatozoa, or relative proportion of live to dead sperm, is an important parameter to appreciate reproductive success of males (Simmons, 2005), and it is used to analyze semen quality in honeybees (Collins & Donoghue, 1999; Collins, 2004; den Boer et al., 2009; Gençer et al., 2014). It is also commonly used in other organisms to study various aspects of their reproductive biology (Garner et al., 1994; Ball et al., 2001; Paulenz et al., 2002).

Recent advances in the use of flow cytometry to assess sperm traits in social insects allowed significant improvements, decreasing processing times and increasing accuracy (Cornault & Aron, 2008; Paynter et al., 2014). However, flow cytometry equipment is not common across the tropics (where stingless bees occur), and the costs involved in their use and maintenance are substantially higher than those involving fluorescence-based methods. We thus provide standard protocols for fluorescence-based methods, which are cheap and ready available across the tropics, aiming at facilitating further research on stingless bees. Although less accurate than flow cytometry methods, fluorescence-based methods have been widely used and they are well accepted as standard methods (Thomas & Simmons., 2007; den Boer et al., 2008, 2010; Cobey et al., 2013). Moreover, sperm viability obtained with fluorescence microscopy and flow cytometry has been found indistinguishable in some cases (Paynter et al., 2014).

To date, no standard protocol is available to assess sperm quality in stingless bees. Here we fill this gap by presenting appropriate protocols for assessing sperm counts, sperm morphology and sperm viability in stingless bees.

Table 1. Works addressing sperm biology in stingless bees.

Species	Topic	Reference
Friesella schrottkyi (Friese)	Giant males	Camilo et al. (1971)
Plebeia droryana (Friese)	Sperm number	Cortopassi-Laurino (1979)
Friesella schrottkyi (Friese) *Lestrimelitta limao* (Smith) *Melipona marginata* Lepeletier *M. quadrifasciata* Lepeletier *M. rufiventris* Lepeletier *Scaptotrigona postica* (Latreille) *Tetragonisca angustula* (Latreille) *Trigona hyalinata* (Lepeletier)	Sperm produced by males and stored in queen's spermathecae	Garófalo (1980)
Melipona quadrifasciata Lepeletier	Spermatogenesis	Conte et al. (2005)
Scaptotrigona xanthotricha Moure	Spermatogenesis	Lino-Neto et al. (2008)
Melipona beecheii Bennett	Sperm number	Pech-May et al. (2012)

Material and Methods

We used established protocols for sperm counts and sperm viability in honeybees (*Apis mellifera* Linnaeus) (Collins & Donoghue, 1999; Cobey et al., 2013), as a guideline to develop protocols to analyze stingless bee sperm. All tests were conducted in the Bee Laboratory, Department of Ecology, University of São Paulo. Our study model was *Scaptotrigona* aff. *depilis* (Moure), a common stingless bee of Southeastern Brazil. Specimens of each colony were collected and identified with the aid of a specialist (Dr. Silvia R. M. Pedro). Voucher specimens can be found in the Coleção Entomológica Paulo Nogueira-Neto (CEPANN), located in the University of São Paulo, São Paulo, Brazil.

Adjustments were implemented to improve the collection of semen from the seminal vesicles (see Supplementary

Materials for details). Males were collected from male aggregation and from the interior of colonies. Those found in aggregations (Fig 1) were considered mature and originated from different colonies (Paxton et al., 2000). Males collected inside the colonies were only distinguished as freshly emerged males (usually of a lighter color) or adult (darker) males. However, during dissections, we found marked differences between males in testis size and migration of sperm to the seminal vesicles. We thus used this information as a proxy for male age, where males with bigger testis and incomplete semen migration were characterized as immature males.

Fig 1. Aggregation of *Scaptotrigona* aff. *depilis* males.

Different semen dilutions were tested (40, 100 and 200 µl) to facilitate the counting of sperm cells (5 males per treatment). As pH was found to have a profound effect on sperm viability during our initial trials, we experimentally manipulated pH to identify the pH yielding maximal sperm viability (pH tested was 8.4, 8.7 and 9.0, with 3 males per treatment). Likewise, because sperm cells start to die once the ejaculate has been collected, we also tested the effect of time on sperm viability (fresh, 1, 3, 5 and 24 hours after dissection, with 5 males per treatment).

For the sperm counts and sperm morphology analyses we stained sperm using DAPI (Fig 2, see Supplementary Material for details). In order to assess the morphology of sperm cells, black and white images of the DAPI-stained cells were taken (20x magnification). Images were analyzed with the software Image J, adjusting brightness and contrast for better visualization. For the measurement of the sperm head area, each image was adjusted by changing the threshold parameter, which enhances binary contrast. This modification was done until all head area was selected, excluding the sperm tail and background. After this procedure, head area was selected with the automatic selection tool (wand) and measured. Because threshold adjustments may be subjective, training and standardizing the procedure is suggested before the real measurements are taken. Illustrative results are presented for

10 sperm cells from one male (mean ± SE). Sperm counts were estimated by diluting the total amount of sperm of each male 10,000 times and counting cells in three samples of 1 µl (air dried in microscope slides). The samples were also DAPI-stained. Counting began at the right edge of each sample, and continued until de left side of the sample was reached. The mean value obtained from the three samples was multiplied by 10,000, to estimate the total number of sperm per male. Results for sperm counts are presented for six males (mean ± SE).

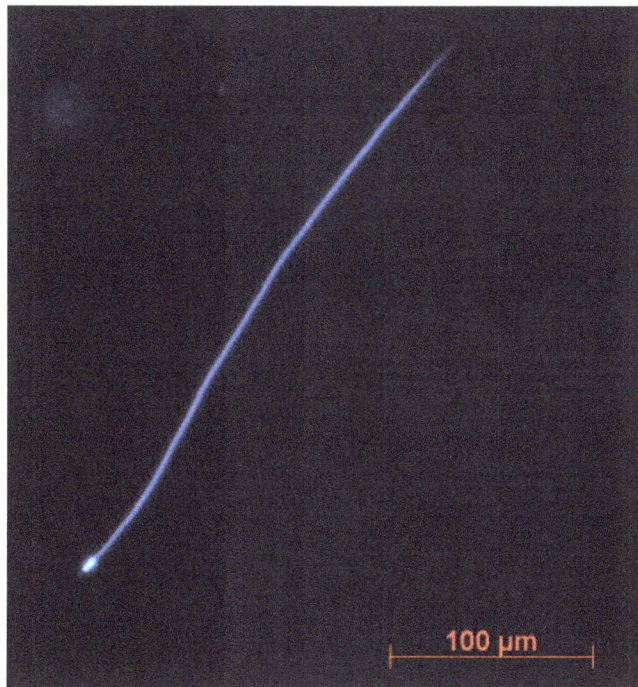

Fig 2. Honeybee (*Apis mellifera*) sperm cell marked with DAPI.

We used the LIVE / DEAD ® Invitrogen Sperm Viability Kit to assess sperm viability. Employing a fluorescence microscope and a cell counter, we proceeded to count the first 400 cells found from the center of the cover slip. Sperm cells were classified as green (live), red (dead) and green / red (dying) (Fig 3). The workflow protocols are summarized in Figs 4 A, B, and C.

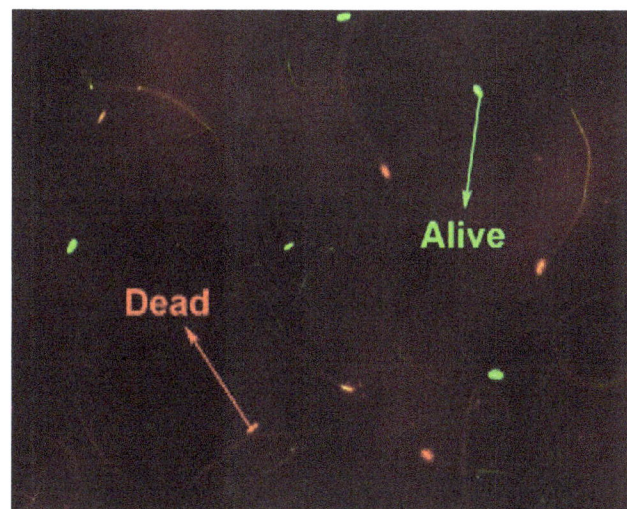

Fig 3. Live (green) and dead (red) honeybee (*Apis mellifera*) sperm.

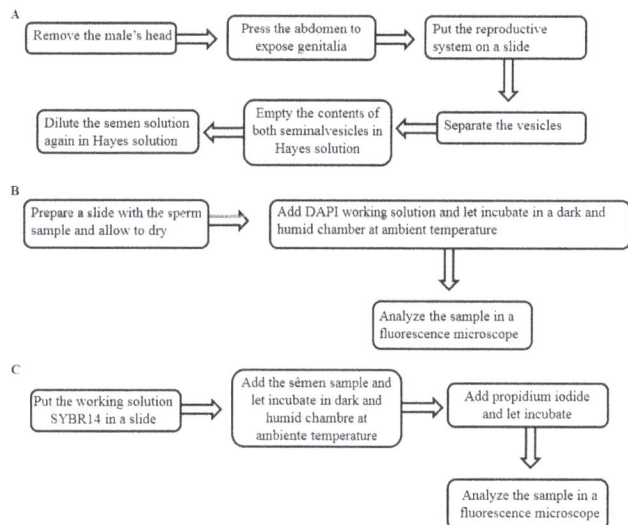

Fig 4 A. Diagram summarizing the workflow of the semen extraction protocol; **B**. Diagram summarizing the workflow of the sperm count and morphology protocol; **C**. Diagram summarizing the workflow of the sperm viability protocol.

Statistical analyses consisted in binomial generalized linear models, as they are the standard method to analyze proportion data (Crawley, 2013), such as sperm viability (dead/alive). Similar analyses are often implemented in studies of sperm viability (den Boer et al., 2009; Stürup et al., 2011). All statistical analyses were implemented in R.

Results

Semen was more easily collected from the seminal vesicles of mature males (Fig 5) rather than younger ones (Fig 6), as semen had already migrated from the testicles to the seminal vesicles in mature males only.

Sperm counts in *S.* aff. *depilis* resulted in an average (± SE) of 1,487,778 ± 36,044 sperm cells per male. When assessing sperm morphology, we found a mean total length of 85.58 ± 1.06 µm, a mean head length of 9.50 ± 0.24 µm, a mean tail length 76.08 ± 0.98 µm and a mean head area of 23.42 ± 0.58 µm². Finally, sperm viability ranged from 52 to 87% (Figs 7 and 8).

We found that the best semen dilution for sperm morphology and viability assays was 100 µl Hayes solution to the contents of the two seminal vesicles, in addition to the initial 20 µl Hayes for emptying the vesicles. This was due to the greater ease during the identification and counting of sperm cells. For sperm counts, however, we recommend a higher dilution (10,000 x) in order to identify single isolated sperm cells.

Sperm viability was significantly affected by pH, being highest when Hayes solution was adjusted to a pH of 8.7 (Fig 6; Table 2). We also found a significant effect of incubation time (Table 3). No difference was found in sperm viability for the initial time intervals tested (fresh, 1, 3 and 5 hours after collection), but a marked decrease in viability was found after 24 hours (Fig 7).

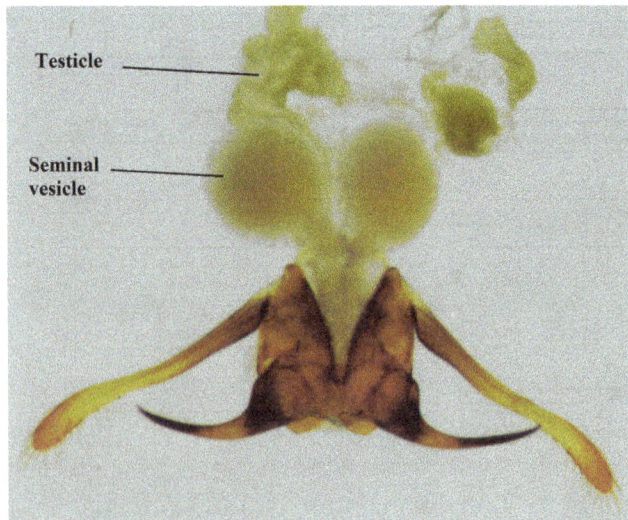

Fig 5. Reproductive system of mature male of *Scaptotrigona* aff. *depilis*.

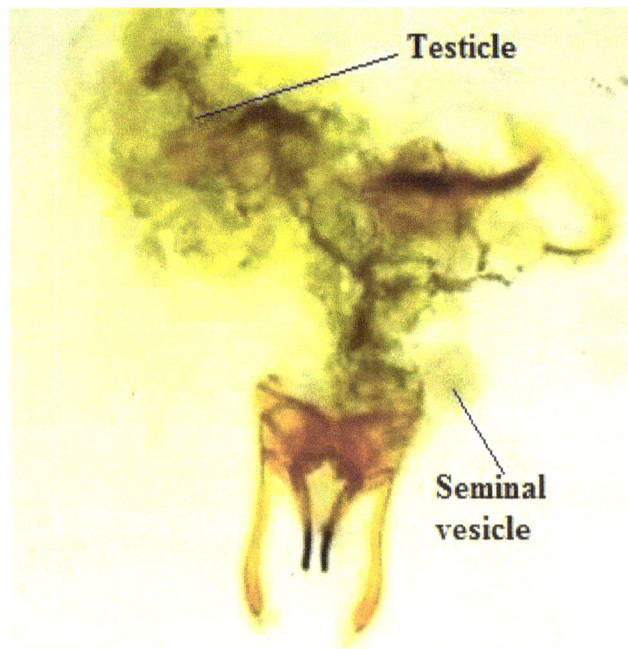

Fig 6. Reproductive system of immature male of *Scaptotrigona* aff. *depilis*.

Table 2. Summary statistics for the effect of pH on sperm viability.

Response	Predictor	Estimate	SE	P-value
Sperm viability	pH 8.4	1.08	0.07	< 0.001
	pH 8.7	0.39	0.10	< 0.001
	pH 9.0	-0.27	0.09	0.003

Discussion

Semen extraction in stingless bees is different than in honeybees. First, stingless bees lack the enormous mucus glands found in honeybees, so the dissection of the reproductive tract is more delicate. Second, the manual collection of the ejaculate from the end phallus is not possible, because the manual eversion of the endophallus fails to stimulate ejaculation. Finally, dissection is more difficult in smaller species, requiring the use of specialized equipment.

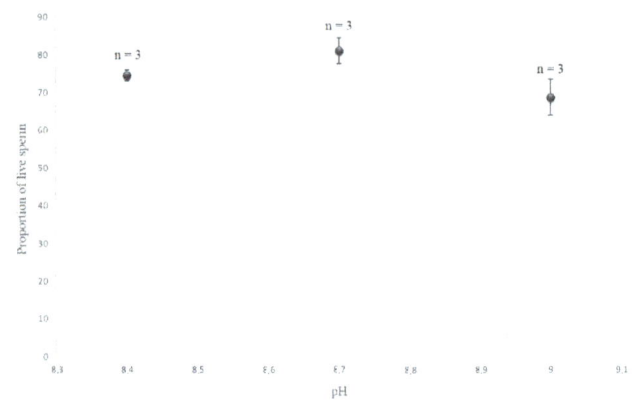

Fig 7. Sperm viability of sperm from *Scaptotrigona* aff. *depilis* diluted in Hayes solution with different pH values.

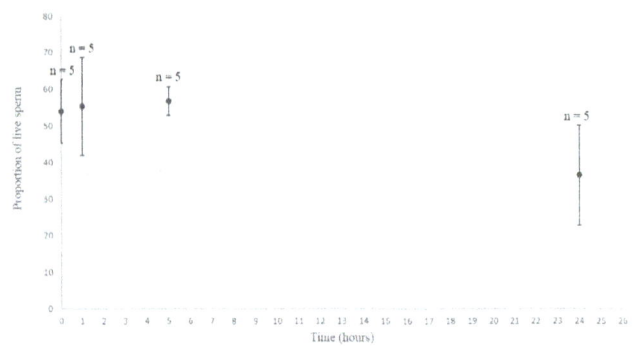

Fig 8. Sperm viability of sperm from *Scaptotrigona* aff. *depilis* assessed on fresh semen, 1, 5 and 24 hours after collection.

Different protocols are available for counting sperm cells. Counting cells using a hemocytometer was described for *A. mellifera* sperm (Cobey et al., 2013; Human et al., 2013), in which a known volume containing unstained cells is analyzed. A drawback of this method is that the counting procedure must be done immediately after dissection. In our protocol, microscope slides are prepared with samples of sperm in a known dilution, and only three samples are counted to estimate the total number of sperm cells. Since samples are dried, the slides can be stored for later analyses, which is an advantage when a high number of males need to be analyzed. Other studies implemented similar techniques successfully (Baer et al., 2006; Stürup et al., 2011).

Our results show that sperm counts, sperm morphology and sperm viability can be effectively assessed

Table 3. Summary statistics for the generalized linear mixed effect model for sperm viability in different intervals between semen collection and viability analysis.

Response	Predictor	Estimate	SE	P-value
Sperm viability	Fresh semen	0.15	0.18	0.41
	1h after collection	0.08	0.26	0.76
	5h after collection	0.12	0.26	0.64
	24h after collection	-0.76	0.26	0.003

in a stingless bee. For sperm morphology and viability, the best semen dilution was found to be 120 μl Hayes solution to the total semen content of a male. For sperm counts, the best dilution factor was 10,000 x. While the pH yielding the highest sperm viability was at 8.7, we found that the best time to assess viability was before 24 h of semen collection. Given the high susceptibility of sperm to manipulation, desiccation, pH and temperature extremes, we recommend that great care must be taken to ensure sperm are analyzed shortly after collection and under the best possible conditions. Based on these results we present standard protocols, hoping they will be useful to future researchers assessing sperm quality in other stingless bee species. Such studies could contribute advance basic knowledge on the reproductive biology of stingless bees, and facilitate their commercial use.

Acknowledgments

We thank Paulo Cesar Fernandes for technical assistance in the lab, Prof. Dr. Ana Castrucci for kindly providing access to her fluorescence microscope and FAPESP for funding (RJ, Process 2012/13200-5).

References

Baer, B., Armitage, S.A.O. & Boomsma. J.J. (2006). Sperm storage induces an immunity cost in ants. Nature 441: 872-875.

Ball, B.A., Medina, V., Gravance, C.G. & Baumber, J. (2001). Effect of antioxidants on preservation of motility, viability and acrosomal integrity of equine spermatozoa during storage at 5°C. Theriogenology 56: 577-589.

Camillo, C. (1971). Estudos adicionais sobre os zangões de *Trigona (Friesella) schrottkyi* (Hym. Apidae). Ciênc. Cult. 29: 279.

Cobey, S.W., Tarpy, D.R. & Woyke, J. (2013). Standard methods for instrumental insemination of *Apis mellifera* queens. In V. Dietemann, J.D. Ellis, P. Neumann (Eds.) The COLOSS BEEBOOK, Volume I: standard methods for *Apis mellifera* research. J. Apicul. Res. 52(4): 1-18.

Collins, A.M. & Donoghue, A.M. (1999). Viability assessment of honey bee, *Apis mellifera* sperm using dual fluorescent staing. Theriogenology 51: 1513-1523.

Collins, A.M. (2004). Sources of variation in the viability of honey bee, *Apis mellifera* L., semen collected for artificial insemination. Invertebr. Reprod. Dev. 45: 231-237.

Conte, M., Lino-Neto, J. & Dolder, H. (2005). Spermatogenesis of *Melipona quadrifasciata anthidioides* (Hymenoptera: Apidae): Fate of the Atypical spermatids. Caryologia 58: 183-188. doi: 10.1080/00087114.2005.10589449.

Cortopassi-Laurino, M. (1979). Observações sobre atividade de machos de *Plebeia droryana* Friese (Apidae, Meliponinae). Rev. Bras. Entomol. 23: 177-191.

Cournault, L. & Aron, S. (2008). Rapid determination of sperm number in ant queens by flow cytometry. Insect. Soc. 55: 283-287.

Crawley, M.J. (2013). The R book. John Wiley & Sons.

den Boer, S.P.A, Boomsma, J.J. & Baer, B. (2008). Seminal fluid enhances sperm viability in the leafcutter ant *Atta colombica*. Behav. Ecol. Sociobiol. 62: 1843-1849. doi: 10.1007/s00265-008-0613-5.

den Boer, S.P.A., Boomsma, J.J. & Baer, B. (2009). Honey bee males and queens use glandular secretions to enhance sperm viability before and after storage. J. Insect. Physiol. 55: 538-543.

den Boer, S.P.A., Baer, B. & Boomsma, J.J. (2010). Seminal fluid mediates ejaculate competition in social insects. Science 327: 1506. doi: 10.1126/science.1184709.

Garner, D.L., Johnson, L.A., Yue, S.T., Roth, B.L. & Haugland, R.P. (1994). Dual DNA staining assessment of bovine sperm viability using SYBR-14 and propidium iodide. J. Androl. 15: 620-629.

Garófalo, C.A. (1980). Reproductive aspect and evolution of social behavior in bees (Hymenoptera, Apoidea). Braz. J. Genet. 3:139-152.

Gençer, H.V., Kahya, Y. & Woyke J. (2014). Why the viability of spermatozoa diminishes in the honeybee (*Apis mellifera*) within short time during natural mating and preparation for instrumental insemination. Apidologie 45: 757-770. doi:10.1007/s13592-014-0295-0.

Human, H., Brodschneider, R., Dietemann, V., Dively, G., Ellis, J., Forsgren, E., Fries, I., Hatjina, F. Hu, F., Jaffé, R., Jensen, A.B., Köhler, A., Magyar, J.P., Özkýrým, A., Pirk, C.W.W., Rose, R., Strauss, U., Tanner, G., Tarpy, D.R., van der Steen, J.J.M., Vaudo, A., Vejsnæs, F., Wilde, J., Williams, G.R. & Zheng, H.Q. (2013). Miscellaneous standard methods for *Apis mellifera* research. J. Apicult. Res. 52.

Lino-Neto, J., Araújo, V.A. & Dolder, H. (2008). Inviability of the spermatids with little cytoplasm in bees (Hymenoptera; Apidae). Sociobiology 51: 163-172.

Paulenz, H., Soderquist, L., Pérez-Pé, R. & Berg, K.A. (2002). Effect of different extenders and storage temperatures on sperm viability of liquid ram semen. Theriognology 57: 823-836.

Paynter, E., Baer-Imhoof, B., Linden, M., Lee-Pullen,T., Heel, K., Rigby, P. & Baer, B. (2014). Flow cytometry as a rapid and reliable method to quantify sperm viability in the honeybee *Apis mellifera*. Cytometry Part A 85: 463-472.

Paxton, R.J. (2000). Genetic structure of colonies and a male aggregation in the stingless bee *Scaptotrigona postica*, as revealed by microsatellite analysis. Insect. Soc. 47: 63-69

Pech-May, F.G., Medina-Medina, L., May-Itzá, W.J., Paxton, R.J. & Quezada-Euán, J.J.G. (2012). Colony pollen reserves affect body size, sperm production and sexual development in males of the stingless bee *Melipona beecheii*. Insect. Soc. 59: 417-424.

Simmons, L.W. (2001). Sperm competition and its evolutionary consequences in the insects. Priceton: Princeton Univ. Press.

Simmons L.W. (2005) Sperm viability matters in insect sperm competition. Curr. Biol. 15: 271-275. doi: 10.1016/j.cub.2005.01.032

Stürup, M., den Boer, S., Nash, D., Boomsma, J. & Baer, B. (2011). Variation in male body size and reproductive allocation in the leafcutter ant *Atta colombica*: estimating variance components and possible trade-offs. Insectes Soc. 58: 47-55.

Thomas, M.L. & Simmons, L.W. (2007). Male crickets adjust the viability of their sperm in response to female mating status. Am. Nat. 170: 190-195. DOI:10.1086/519404

25

Distance and habitat drive fine scale stingless bee (Hymenoptera: Apidae) community turnover across naturally heterogeneous forests in the western Amazon

TM Misiewicz[1,2], E Kraichak[2,3] , C Rasmussen[4]

1 - University of California, Berkeley, Berkeley, California

2 - The Field Museum, Chicago, IL, USA

3 - Kasetsart University, Bangkok, Thailand

4 - Aarhus University, Aarhus, Denmark

Keywords

Meliponini, beta-diversity, distance-based redundancy analysis, Peru, white-sand forest.

Corresponding author

Tracy M. Misiewicz
University of California, Berkeley
Department of Integrative Biology
3060 Valley Life Sciences Building
#3140, Berkeley, California 94720
E-Mail: tracymisiewicz@gmail.com

Abstract

High tree species richness in the western Amazon has been attributed to heterogeneous soils, which harbor edaphic specialist trees. While rapid transitions in tree communities are well documented across these variable soils, few studies have investigated the role of habitat heterogeneity in structuring animal communities. Stingless bees are taxonomically diverse and important natural pollinators in Neotropical forests. However, little is known about their community structuring at local scales in naturally heterogeneous environments. We systematically sampled stingless bee communities found across three paired sites that included adjacent patches of white-sand and non-white-sand forest in the lowland Amazonian region of Loreto, Peru. We sought to understand: (1) How stingless bee species richness and abundance differ among white-sand and non-white-sand habitats and (2) The relative influence of fine scale geographic distance and habitat type in structuring stingless bee communities. We found that species richness did not differ between habitats and that species abundances were highest in white-sand habitats. Community analyses for sampling sites pooled across all months demonstrated that location and soil type played a significant role in structuring bee communities and that community turnover may be more strongly influenced by distance in white-sand habitats than non-white sand habitats. Our results suggest that distance and habitat play an important role in driving stingless bee community turnover at fine scales and that the interaction between habitat and geographic distance may promote higher stingless bee community turnover in white-sand habitats than non-white sand habitats.

Introduction

The lowland amazon basin is notable for its exceptionally high levels of species diversity across a wide range of taxa (Erwin, 1988; Kress et al., 1998; Pitman et al., 2001). High levels of species richness have been attributed to extreme habitat heterogeneity with particular emphasis on the mosaic of soil types found across the Western Amazon in Peru (Terborgh, 1985; Whitney & Alvarez, 1989). These variable soils create a patchwork of forest types differentiated by plant communities that have strong edaphic associations with nutrient poor white-sand soil patches and the more fertile brown-sand and clay soils which surround them (Tuomisto & Ruokolainen,

1994; Fine et al., 2010). White-sand forests differ from surrounding forests in that they harbor significantly less species richness, are shorter in stature, and experience higher temperatures on average below the canopy due to increased light penetration than forests found on surrounding soils (Medina & Cuevas, 1989; Fine et al., 2010). In the western Amazon white-sand forests exist as small habitat islands, usually no larger than a few square hectares (Fine et al., 2005).

While the majority of research examining the role of edaphic variation in species turnover in the Amazon has been centered on tree communities (Tuomisto & Ruokolainen, 1994; Fine et al., 2010; but see Alvarez Alonso et al., 2013) similar habitat specialization may also be present in animals. Because forests constitute the primary habitat and food source for many forest dwelling animals we expect that abrupt changes in floristic composition across habitats may in turn drive turnover in animal communities.

Animals that provide pollination services play a particularly important role in tropical ecosystems where the majority of trees are reliant on animal interactions for pollen transfer (Bawa, 1990). While turnover in bee communities across forest fragments and agriculturally modified landscapes has been well studied in relation to crop production (Tylianakis et al., 2005; Jha & Vandameer, 2010) the role of naturally heterogeneous habitats in driving species turnover in the lowland Amazon has largely been neglected (but see Abrahamczyk et al., 2010).

Stingless bees (Hymenoptera: Apidae: Meliponini) are an important taxon for studies of biodiversity and species turnover in the lowland Amazon because they are highly diverse (*ca* 500 species) with their center of diversity found in the Neotropics (*ca* 400 species) (Michener, 2013). Additionally, they are the most important native providers of pollination services in the Amazon, making them essential for ecosystem functioning (Engel & Dingemans-Bakels, 1980; Roubik, 1995).

Most stingless bee species are considered to be generalist pollinators and they exhibit a wide range of variation in nesting habits across species. Nests are usually arboreal or subterranean and are constructed using diverse construction materials including mud, wood pulp, feces, and plant exudates (Schwarz, 1948; Roubik, 1989). Foraging distance away from nest site is dependent on the size of the bee with distances ranging from less than 500 m to 2 km (Kuhn-Neto et al., 2009). Given the variation in nest site preferences between lineages relatively little attention has been paid to the fine scale distribution and ecology of Neotropical stingless bees and no studies have investigated species turnover across naturally occurring environmental gradients in undisturbed forest sites.

Furthermore, because the movement of animal pollinators directly influences the distance, direction and degree of pollen dispersal, they ultimately determine the spatial pattern of gene movement within and among plant populations (Garcia et al., 2007). If pollinators are restricted in their foraging area due to habitat preference (Dieckmann, 2004) then the question of ecological specialization in bee communities may be of particular interest to plant ecologists as well.

In this study we simultaneously examined the effect of habitat and distance in structuring stingless bee communities at a local scale. We systematically sampled native bee communities found across three paired sites that included adjacent patches of white-sand and non-white-sand forest across more than 100 km in the lowland Amazon in the region of Loreto, Peru in order to answer two questions (1) How does stingless bee species richness and abundance differ among white-sand and non-white-sand habitats? (2) What is the relative influence of fine scale geographic distance and habitat type in structuring stingless bee communities?

If stingless bees are generalist pollinators with relatively large foraging ranges, we expect that geographic distance will play a greater role in structuring bee communities than habitat type providing that trees exhibit similar flowering phenology across habitat types. Alternatively, if bees prefer floral resources provided by soil specialist trees, nesting sites that are more common in one particular habitat (*i.e.* large vs. small stems or clay vs. sandy soil in the case of subterranean nesters) or environmental differences such as temperature or predation risk then we may find that habitat type plays a stronger role than distance in structuring stingless bee communities.

Materials and methods

Study sites

Three primary study areas, each containing adjacent white-sand and non-white-sand forest patches, were established in the region of Loreto, Peru (Fig 1; Table 1). Area one and two are located within the Allpahuayo Mishana National Reserve in the Nanay River watershed and area three is located approximately 100km to the south in the Ucayali River watershed. We consider each forest patch a sampling site.

Sampling design

All trapping was conducted using bee pan traps. These traps are easily standardized and avoid collector bias (Westphal et al., 2008). Traps were created using 12-oz clear plastic soup bowls painted fluorescent blue, fluorescent yellow, or white in order to account for variation in color preference among bee species. Four trapping stations consisting of six bee pan traps were established in white-sand and non-white-sand forest sites at each of the three areas for a total of 24 trapping stations across six collecting sites (Fig 1). Within each collecting site each trapping station was established 200-250 m distant from any other trapping station.

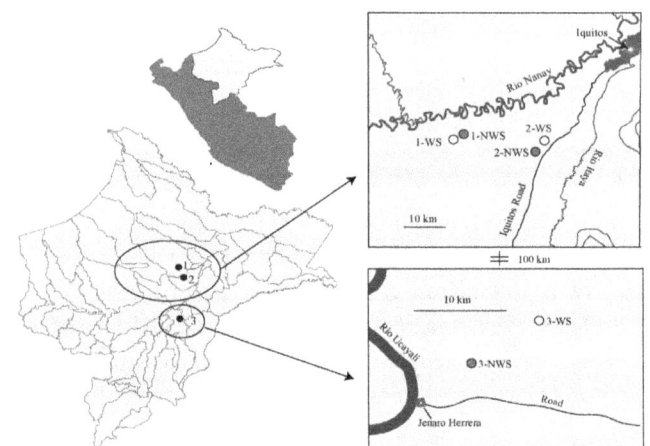

Fig 1. Sample sites and soil types where stingless bees were sampled in the region of Loreto, Peru. Numbered points represent the three areas where adjacent white-sand and non-white-sand forests were found. Each individual sampling site is displayed in the inset. Grey circles represent non-white-sand forests and white circles represent white-sand forests.

Table 1. Geographic coordinates, number of individuals, number of genera, and number of species collected at each trapping site as well as total abundance (A) and richness (S) for each habitat type and collection site.

		White-Sand						Non-White-Sand				Totals per site	
Site	Lat.	Long.	Ind. (N)	Gen. (N)	spp. (N)	Site	Lat.	Long.	Ind. (N)	Gen. (N)	spp. (N)	A	S
1-WS	-3.91	-73.55	215	8	16	1-NWS	-3.90	-73.55	92	9	15	307	20
2-WS	-3.95	-73.40	112	5	6	2-NWS	-3.97	-73.42	236	9	12	348	14
3-WS	-4.86	-73.61	415	9	16	3-NWS	-4.88	-73.64	31	6	9	446	20
A			742			A			359				
S			27			S			24				

Each trapping station contained two sets of yellow, blue and white traps. One set was suspended one meter above the ground with each individual bowl spaced at a distance of five meters to avoid bowl competition (Droege, 2010). The second set of bowls was suspended at a height of 15–20 m in the canopy directly above the ground traps. Bowls were filled with six ounces of soapy water solution (one tsp blue Dawn brand soap per two liters of water).

Pan traps were set out at each site once per month for 24 h between March-July 2010. While Loreto, Peru exhibits very little seasonality our sampling period extended from the high water season, when rivers rise substantially, through the low water season. All trapped specimens were collected in the field and transferred to 96% ethanol. Specimens were separated, pinned and identified and have been deposited at the Essig Museum of Entomology at University of California, Berkeley. Identification of all specimens was done by C.R. by direct comparison with a large synoptic collection of Peruvian stingless bees previously identified by J.M.F. Camargo.

General diversity

We assessed the effectiveness of our sampling method using species accumulation curves and the Chao estimator (Chao, 1987). Rarefaction curves were calculated using the individual-based species matrix and the species accumulation curve and Chao estimates were calculated using the 'specacum' and 'specpool' functions respectively within the package 'vegan' (Oksanen et al., 2013). We assessed the dominance structure within our dataset by ranking the relative abundance of each species using a regular base plot (Magurran, 2004).

Fine scale community structure

We investigated the influence of trap height, trap color, month collected, soil type and trapping site on species richness and abundance using generalized linear models (Dobson, 1990). The full model was constructed using trap height, trap color, month collected, soil type and trap site as a fixed effect. Poisson distribution and logarithmic link function were selected as an error distribution and link function, as it has shown to work well with

count data (Bolker et al., 2009). The significance of each explanatory variable was then tested using Chi-square test to compare the reduction of deviance from the residual deviance (Hastie & Pregibon, 1992). We then identified the best-fitting model using backward step-wise model selection with AIC criterion, using function "stepAIC" in package MASS.

Broad scale community structure

In order to assess the broad scale effects of soil type and location as defined by the three sampling areas (Fig 1), we combined all monthly collection data for each of the four trap stations within each of the six collecting sites for a total of 24 sample units. We transformed the data by first applying square root to the abundance data followed by standardization using the Wisconsin method (Bray & Curtis, 1957), which reduces the effect of overly dominant species in the data set and controls for sampling effort at each trap site (Legendre & Gallagher, 2001). We first quantified relative contributions of soil type and location to the community structure by performing variance partitioning with the function 'varpart'. The analysis partitions the explained variation in community structure into different components based on the studied environmental factors (Borcard et al., 1992). Then, in order to visualize results and specifically test the significance of effects of soil and location, as well as their interaction in driving community structure, we used a distance-based Redundancy Analysis (db-RDA) with soil and location as constraints. This method allowed us to carry out constrained ordinations using non-Euclidean distance measures (Gower, 1966; Gower, 1985; Legendre & Anderson, 1999; Legendre & Legendre, 2012). Our distance matrix was created using Bray-Curtis distance, which only accounts for shared presences between two sites (Anderson et al., 2011), and the redundancy analysis was carried out using the 'capscale' function (Anderson & Willis, 2003). The significance of constrained ordination was assessed using a permutation test for Constrained Correspondence Analysis (Legendre et al., 2011; Legendre & Legendre, 2012) using the function 'anova.cca'. The P-value is calculated by comparing the observed F-value with the values from 999 permutations of community data.

We also used a Permutational Multivariate Analysis of Variance (PERMANOVA) to further test the effects of soil, location, and their interaction using the function 'adonis'. This analysis is analogous to parametric Multivariate Analysis of Variance (MANOVA), but has been shown to be more robust for community data, as the P-value is derived from permutation, as opposed to the comparison against a known distribution (Anderson, 2001).

Finally, using the function 'mantel' we implemented Mantel tests (Legendre & Legendre, 2012) individually on white-sand and non-white-sand populations to determine if geographic distance was more important in structuring bee communities in one habitat or the other. All functions for community analysis are available the R package 'vegan.' All statistical analyses were carried out in R 3.0.3 (R Development Core Team 2013).

Results

General diversity

We trapped a total of 1109 bees representing three families, 17 genera and 39 species. All but three taxa were Apidae (Appendix S1). Thirty-one species (79%) were identified to the species level and the remaining eight were sorted to morphospecies. Eight of the collected specimens representing six species were identified as solitary bees and were therefore discarded from the dataset for further analysis. All other specimens for the remaining 33 species were stingless bees and were included in all analyses (Table 1).

The species accumulation curve approached, but did not reach an asymptote (Fig 2) suggesting that our sampling was adequate but not exhaustive. The estimated species richness across all habitats and sites was 38.6 species (Standard Error: ±3.85) meaning we captured approximately 77 percent to 95 percent of the estimated total number of stingless bee species.

Plebeia minima (Gribodo) and *Plebeia* sp. A were particularly abundant in our data set (N=584, N=235). 29 species were represented by medium to low abundances. Six species were represented by singletons (Table 2; Fig 3). Preliminary analyses showed little to no effect when singletons were omitted from the data. As a result, all further analyses were carried out with the inclusion of singletons. A total of 19 species (N=1,048) were found in both white-sand and non-white-sand habitats, five species (N=13) were found only in non-white-sand habitats and nine species (N=40) were found only in white-sand habitats (Table 2). Total abundance (A) and species richness (S) for each habitat type and sampling location are reported in Table 1.

Fine scale community structure

Soil type and trapping site had a significant effect on species richness and abundance, while trap height, and month

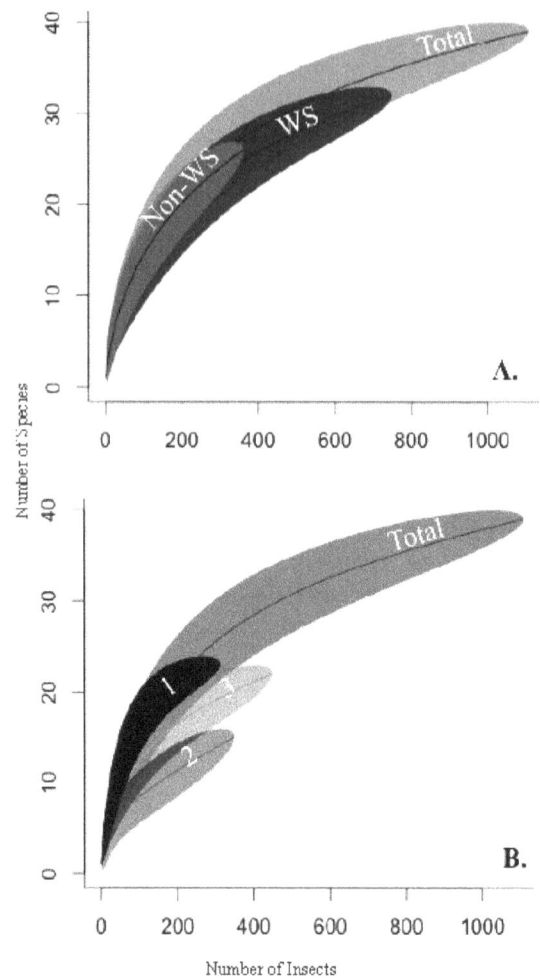

Fig 2. Species accumulation curves for stingless bees. A. Species accumulation curves for white sand habitat, terrace habitat and total across habitats. B. Species accumulation curves by sampling sites 1, 2 and 3 and total across all sampling sites.

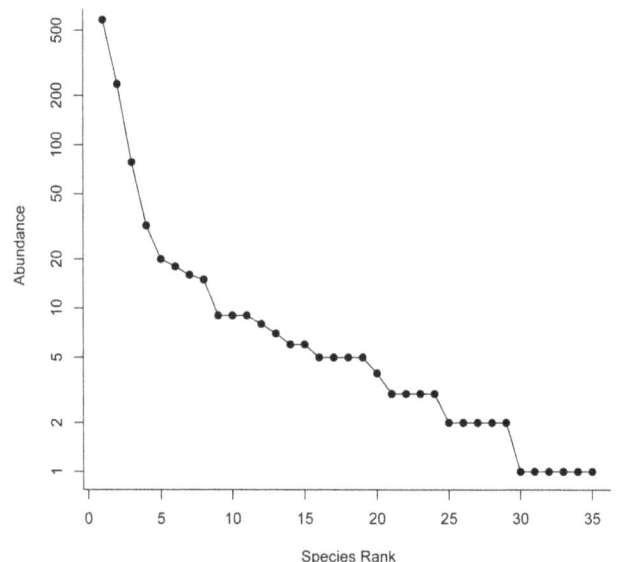

Fig 3. Species abundance distribution for all stingless bee species for all sites sampled.

Table 2. Bee species collected by soil type

Species	N ind. collected on WS	N ind. collected on non-WS
Leurotrigona pusilla	1	1
Leurotrigona muelleri	0	1
Melipona bradleyi	2	0
Melipona crinite	5	0
Melipona gr. *rufiventris*	1	0
Nannotrigona melanocera	73	5
Nannotrigona schultzei	4	1
Nogueirapis butteli	3	5
Partamona epiphytophila	3	2
Partamona testacea	0	1
Plebeia minima	342	242
Plebeia sp. A	200	35
Plebeia sp. B	18	0
Plebeia sp. C	3	1
Plebeia sp. D	13	3
Plebeia sp. E	1	0
Ptilotrigona lurida	7	2
Ptilotrigona pereneae	6	0
Scaura latitarsis	1	0
Scaura tenuis	0	6
Schwarzula coccidophila	4	11
Schwarzula timida	33	1
Tetragona clavipes	6	3
Tetragona dissecta	2	0
Tetragona gr. *dorsalis*	1	8
Tetragona handirschii	0	3
Tetragonisca angustula	3	0
Trigona amalthea	0	1
Trigona cilipes	0	7
Trigona guianae	3	2
Trigona williana	2	1
Trigonisca bidentata	4	15
Trigonisca gr. *ceophloei*	1	1

Table 3. Results of the generalized linear models for the effects of soil type, location, month, trap height and trap color on stingless bee richness and abundance at the smallest scale.

	Bee species richness			Bee abundance		
	df	Deviance	P	df	Deviance	P
Null model	252	108.92		252	1326.7	
Soil type	1	6.37	0.010	1	62.55	<0.001
Location	2	12.45	0.002	2	64.01	<0.001
Month	1	1.59	0.200	1	49.38	<0.001
Trap height	1	0.17	0.670	1	21.75	<0.001
Pan color	2	0.27	0.870	2	2.72	0.25

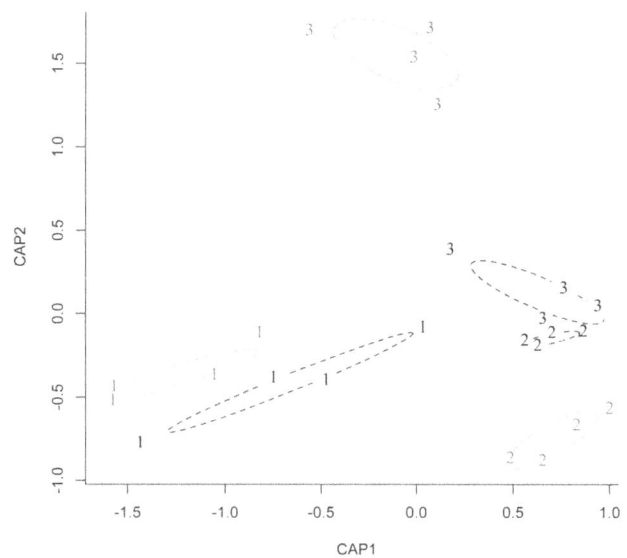

Fig 4. Results from distance-based redundancy analysis. Gray ellipses correspond to white-sand habitat and black ellipses correspond to non-white-sand habitat. 1, 2 and 3 refer to sampling sites. CAP refers to 'Constrained Analysis of Principal Coordinates'.

collected only significantly affected on abundance (Table 3). The multiple linear model that best explained species richness included soil type and trap site (ΔAIC=6.28). The model that best explained species abundance included soil type, month collected, trapping site and height, but it was only marginally better than the full model with the trap color (ΔAIC=0.96). Mantel tests suggests that geographic distance is correlated with species turnover in both habitat types (white-sand, r=0.4, P<0.001; non-white-sand, r=0.3, P=0.02).

Discussion

To our knowledge this is the first study to examine the role of geographic distance and forest type in structuring stingless bee communities at local scales across a naturally heterogeneous landscape in the Western Amazon. We found that both location and habitat were important in structuring stingless bee communities even over extremely small spatial scales. Our results were consistent with other tropical bee studies, which demonstrate changes in bee communities across a variety of spatial scales and environmental gradients (Tylianakis et al., 2005; Abrahamczyk et al., 2011; Batista Matos et al., 2013). While the total amount of variance explained by these two factors seems relatively small these results are in line with other studies of community turnover particularly those using natural gradients.

White-sand forests exist as small patches or habitat 'islands' surrounded by a matrix of non-white-sand forest. Accordingly, factors such as migration, colonization and local extinction may play a stronger role in structuring white-sand bee communities than non-white-sand bee communities. If non-white-sand habitat is less favorable for bee species found in white-sand forests then these communities may experience higher levels of isolation due to the compounded effect of habitat and distance. In this case metacommunity dynamics could play an important role in increasing turnover among white-sand forests effectively amplifying the effect of geographic distance and habitat alone.

While our results demonstrated that stingless bee community structure is influenced by location, habitat type also plays significant role particularly at very fine geographic scales. We found that variation across sampling sites was driven in part by soil type however, species specific to one forest type tended to be rare in our collections, making it difficult to discern between true habitat specificity and insufficient sampling. Species abundances were much higher in white-sand-forests than non-white-sand forests suggesting that while many stingless bee species utilize both forest types habitat preferences may dictate where they are more commonly found. Floral and nesting resources are both important in structuring bee communities across habitats (Tependino & Stanton, 1981; Petanidou & Ellis, 1996). Fierro et al. (2012) found that stingless bee species show preferences for particular tree taxa, which commonly provide ideal nesting sites, as well as species-specific foraging behavior suggesting that turnover in tree diversity likely drives changes in stingless bee distributions. While we did not quantify differences in floral or nesting resource availability between habitat types in this study marked differences in floristic composition, forest structure, microclimate and abiotic resources are likely driving differences in these neighboring bee communities.

Habitat based differences in stingless bee communities may also reinforce tropical tree specialization across habitat boundaries. Many tree species that are endemic to white-sand forest patches in Peru have congeners associated with parapatric non-white-sand forests (Fine et al., 2010) and divergent natural selection across adjacent white-sand and non-white-sand habitats has been shown to play an important role in maintaining boundaries between ecologically divergent tree populations (Misiewicz & Fine, 2014). If pollinators forage less frequently outside of their preferred habitat type they may indirectly limit pollen flow between ecologically divergent plant populations increasing reproductive isolation.

This study indicates that geographic distance, forest type and the interaction between the two are important in structuring stingless bee communities supporting the hypothesis that dispersal processes such as migration and colonization interact with niche specialization in determining local patterns of community composition.

Acknowledgements

The authors gratefully acknowledge the Ministerio del Ambiente of Peru for providing research and export permits, Carlos Rivera of SERNANP-Allpahuayo-Mishana and the Instituto de Investigaciones de la Amazonía Peruana (IIAP) for institutional support. This research was financially supported by National Science Foundation DEB-1311117 (T.M.M). The authors thank Julio Sanchez and Italo Mesones for assistance in the field and Paul V.A. Fine for his insightful comments on the manuscript.

References

Abrahamczyk, S., Steudel, B. & Kessler, M. (2010). Sampling Hymenoptera along a precipitation gradient in tropical forests: The effectiveness of different coloured pan traps. Entomol. Exp. Appl. 137: 262-268. doi:10.1111/j.1570-7458.2010.01063.x

Abrahamczyk, S., Kluge, J., Gareca, Y., Reichle, C. & Kessler, M. (2011). The influence of climatic seasonality on the diversity of different pollinator groups. PLoS ONE. doi: 10.1371/journal.pone.0027115.

Alvarez Alonso, J., Metz, M.R. & Fine, P.V.A. (2013). Habitat specialization by birds in western Amazonian white-sand forests. Biotropica 45: 365-372. doi: 10.1111/btp.12020

Anderson, M.J. (2001). A new method for non-parametric multivariate analysis of variance. Austral Ecol. 26: 32-46. doi: 10.1111/j.1442-9993.2001.01070.pp.x

Anderson, M.J. & Willis, T.J. (2003). Canonical analysis of principal coordinates: a useful method of constrained ordination for ecology. Ecology 84: 511–525. doi: 10.1890/0012-9658

Anderson, M.J., Crist, T.O., Chase, J.M., Vellend, M., Inouye, B.D., Freestone, A.L., Sanders, N.J., Cornell, H.V., Comita, L.S., Davies, K.F., Harrison, S.P., Kraft, N.J.B., Stegen, J.C. & Swenson, N.G. (2011). Navigating the multiple meanings of β diversity: a roadmap for the practicing ecologist. Ecol. Lett. 14: 19-28. doi: 10.1111/j.1461-0248.2010.01552.x.

Batista Matos, M.C., Sousa-Souto, L., Almeida, R.S. & Teodoro, A.V. (2013). Contrasting patterns of species richness and composition of solitary wasps and bees (Insecta: Hymenoptera) according to land-use. Biotropica. 45: 73-79. doi: 10.1111/j.1744-7429.2012.00886.x

Bawa, K.S. (1990). Plant-pollinator interactions in tropical rain forests. Annu. Rev. Ecol. Evol. Syst. 21: 299-422. doi:10.1146/annurev.es.21.110190.002151

Bolker, B.M., Brooks, M.E., Clark, C.J., Geange, S.W., Poulsen, J.R., Stevens, M.H.H. & White, J.S.S. (2009). Generalized linear mixed models: a practical guide for ecology and evolution. Trends Ecol. Evol. 24:127-135. doi:10.1016/j.tree.2008.10.008

Borcard, D., Legendre, P. & Drapeau, P. (1992). Partialling out the spatial component of ecological variation. Ecology 73: 1045-1055. doi: 10.1023/A:1009693501830

Bray, J.R. & Curtis, J.T. (1957). An ordination of the upland forest communities of southern Wisconsin. Ecol. Monogr. 27: 325-349. doi: 10.2307/1942268

Chao, A. (1987). Estimating the population size for capture-recapture data with unequal catchability. Biometrics 43: 783–791.

Corbet, S., Fussell, M., Ake, R., Fraser, A., Gunson, C., Savage, A. & Smith, K. (1993). Temperature and the pollinating activity of social bees. Ecol. Entomol. 18: 17-30. doi: 10.1111/j.1365-2311.1993.tb01075.x

Dick, C.W., Etchelecu, G. & Austerlitz, F. (2003). Pollen dispersal of tropical trees (*Dinizia excelsa*: Fabaceae) by native insects and African honeybees in pristine and fragmented Amazonian rainforest. Mol. Ecol. 12: 753-764. doi: 10.1046/j.1365-294X.2003.01760.x

Dieckmann, U. & Doebeli, M. (2004). Adaptive dynamics of speciation: Sexual populations. In U. Dieckmann, J.A.J. Metz, M. Doebeli & D. Tautz (Eds.), Adaptive Speciation (pp. 76-111). Cambridge: Cambridge University Press.

Dobson, A.J. (1990). An Introduction to Generalized Linear Models. London: Chapman and Hall.

Droege, S., Tepedino, V.J., Lebuhn, G., Link, W., Minckley, R.L., Chen, Q. & Conrad, C. (2010). Spatial patterns of bee captures in North American bowl trapping surveys. Insect Conserv. Diver. 3: 15-23. doi: 10.1111/j.1752-4598.2009.00074.x

Engel, M.S. & Dingemans-Bakels, S.F. (1980). Nectar and pollen resources for stingless bees (Meliponinae, Hymenoptera) in Surinam (South America). Apidologie. 11: 341-350. doi: 10.1051/apido:19800402

Erwin, T.L. (1988). The tropical forest canopy: The heart of biotic diversity. In E.O. Wilson (Ed.), Biodiversity (pp. 123-129). Washington DC: National Academy Press.

Fierro, M.M., Cruz-López, L., Sánchez, D., Villanueva-Gutiérrez, R. & Vandame, R. (2012). Effect of biotic factors on the spatial distribution of stingless bees (Hymenoptera: Apidae, Meliponini) in fragmented neotropical habitats. Neotrop. Entomol. 41: 91-104. doi: 10.1007/s13744-011-0009-5

Fine, P.V.A., Garcia-Villacorta, R., Pitman, N.C.A., Mesones, I. & Kembel, S.W. (2010) A floristic study of the white-sand forests of Peru. Ann. Mo. Bot. Gard. 97: 283-305. doi: 10.3417/2008068

Garcia, C., Jordano, P., & Godoy, J.A. (2007). Contemporary pollen and seed dispersal in a Prunus mahaleb population: patterns in distance and direction. Mol. Ecol. 16: 1947-1955. doi: 10.1111/j.1365-294X.2006.03126.x

Gower, J.C. (1966). Some distance properties of latent root and vector methods used in multivariate analysis. Biometrika 53: 325-338. doi: 10.1093/biomet/53.3-4.325

Gower, J.C. (1985). Properties of Euclidean and non-Euclidean distance matrices. Linear Algebra Appl. 67: 81-97. doi: 10.1016/0024-3795(85)90187-9

Hastie, T.J. & Pregibon, D. (1992). Generalized linear models. In J.M. Chambers & T.J. Hastie (Eds.), Statistical Models in S. Wadsworth & Brooks/Cole.

Jha, S. & Vandermeer, J.H. (2010). Impacts of coffee agroforestry management on tropical bee communities. Biol. Conserv. 143: 1423-1431. doi: 10.1016/j.biocon.2010.03.017

Kress, J., Heyer, W.R., Acevedo, P., Coddington, J.A., Cole, D., Erwin, T.L., Meggers, B.J., Pogue, M.G., Thorington, R.W., Vari, R.P., Weitzman, M.J. & Weitzman, S.H. (1998).

Amazon biodiversity: Assessing conservation priorities with taxonomic data. Biodivers. Conserv. 7: 1577-1587. doi: 10.1023/A:1008889803319

Kuhn-Neto, B., Contrera, F.A.L., Castro, M.S. & Nieh, J.C. (2009). Long distance foraging and recruitment by a stingless bee, *Melipona mandacaia*. Apidologie 40: 472-480. doi: 10.1051/apido/2009007

Legendre, P. & Anderson, M.J. (1999). Distance-based redundancy analysis: testing multispecies responses in multifactorial ecological experiments. Ecol. Monogr. 69: 1-24. doi: 10.1890/0012-9615(1999)069[0001:DBRATM]2.0.CO;2

Legendre, P. & Gallagher, E.D. (2001). Ecologically meaningful transformations for ordination of species data. Oecologia 129: 271-280. doi: 10.1007/s004420100716

Legendre, P. & Legendre, L. (2012). Numerical Ecology. Amsterdam: Elsevier Science.

Legendre, P., Oksanen, J. & Ter Braak, C.J.F. (2011). Testing the significance of canonical axes in redundancy analysis. Methods Ecol. Evol. 2: 269-277. doi: 10.1111/j.2041-210X.2010.00078.x

Misiewicz, T.M & Fine, P.V.A. (2014). Evidence for ecological divergence across a mosaic of soil types in an Amazonian tropical tree: *Protium subserratum* (Burseraceae). Mol. Ecol. 23: 2543-2558. doi: 10.1111/mec.12746.

Magurran, A.E. (2004). Measuring biological diversity. Oxford: Blackwell Publications.

Medina, E. & Cuevas, E. (1989). Patterns of nutrient accumulation and release in Amazonian forests of the upper Rio Negro basin. In J. Proctor (Ed.), Mineral nutrients in tropical forest and savanna ecosystems (pp. 217-240). Oxford: Blackwell Scientific.

Michener, C.D. (2013). The Meliponini. In P. Vit, S.R.M. Pedro & D.W. Roubik (Eds.), Pot-honey: a legacy of stingless bees (pp. 3-18). New York: Springer.

Oksanen, J., Blanchet, F.G., Kindt, R., Legendre, P., Minchin, P.R., O'Hara, R.B., Simpson, G.L., Solymos, P., Henry, M., Stevens, H., & Wagner, H. (2013). Vegan: Community Ecology Package. R package version 2.0.9. Http://CRAN.R-project.org/package=vegan.

Petanidou, T. & Ellis, W.N. (1996) Interdependance of native bee faunas and floras in changing Mediterranean communities. In A. Matheson, S.L. Buchmann, C.O'Toole, P. Westrich & I.H. Williams (Eds), The conservation of bees. Linnean Society Symposium Series 18 (pp. 201-226). London: Academic Press.

Pitman, N.C.A., Terborgh, J., Silman, M.R., Núñez, P., Neill, D.A., Cerón, C.E., Palacios, W. & Aulestia, M. (2001). Dominance and distribution of tree species in upper Amazonian terra firme forests. Ecology 82: 2101-2117. doi: 10.2307/2680219

Roubik, D.W. (1989). Ecology and natural history of tropical bees. Cambridge: Cambridge University Press.

Roubik, D.W. (1995). Pollination of Cultivated Plants in the Tropics, Bulletin 118. Rome: Food and Agriculture Organization of the United Nations.

Ruokolainen, K., Tuomisto, H., Macia, M.J., Higgins, M.A. & Yli-Halla, M. (2007). Are floristic and edaphic patterns in Amazonian rain forests congruent for trees, pteridophytes and Melastomataceae? J. Trop. Ecol. 23: 13-25. doi: 10.1017/S0266467406003889

Schwarz, H.F. (1948). Stingless bees (Meliponinae) of the western hemisphere. Bull. Am. Mus. Nat. Hist. 90: 1-546. doi: 10.1590/S1519-69842004000400003

Tependino, V.J. & Stanton, N.L. (1981). Diversity and competition in bee-plant communities on short-grass prairie. Oikos 36: 35-44.

Terborgh, J. (1985). Habitat selection in Amazonian birds. In M.L. Cody (Ed.), Habitat selection in birds (pp. 311-338). New York: Academic Press.

Tuomisto, H. & Ruokolainen, K. (1994). Distribution of Pteridophyta and Melastomataceae along an edaphic gradient in Amazonian rain forest. J. Veg. Sci. 5: 25-34. doi: 10.2307/3235634

Tylianakis, J.M., Klein, A.M. & Tscharntke, T. (2005) Spatiotemporal variation in the diversity of Hymenoptera across a tropical habitat gradient. Ecology 86: 3296-3302. doi: 10.1890/05-0371

Westphal, C., Bommarco, R., Carre, G., Lamborn, E., Morison, N., Petanidou, T., Potts, S.G., Roberts, S.P.M, Szentgyorgyi, H., Tscheulin, T., Vaissiere, E., Woyciechowski, M., Biesmeijer, J.C., Kunin, W.E., Settele, J. & Steffan-Dewenter, I. (2008) Measuring bee diversity in different European habitats and biogeographical regions. Ecol. Monogr. 78: 653-671. doi: 10.1890/07-1292.1

Whitney, B.M. & Alvarez, A.J. (1998) A new *Herpsilochmus* antwren (Aves: Thamnophilidae) from northern Amazonian Peru and adjacent Ecuador: the role of edaphic heterogeneity of tierra firme forest. Auk. 115: 559–576.

26

Polygyny, Inbreeding and Wingless Males in the Malagasy Ant *Cardiocondyla shuckardi* Forel (Hymenoptera, Formicidae)

J Heinze,[1] A Schrempf,[1] T Wanke[1], H Rakotondrazafy,[2] T Rakotondranaivo[2], B Fisher[2,3]

1 - Universität Regensburg, Biologie I, 93040 Regensburg, Germany.

2 - Madagascar Biodiversity Center, PBZT Tsimbazaza, 101 Antananarivo, Madagascar.

3 - California Academy of Sciences, San Francisco, USA.

Keywords

inbreeding, reproductive tactics, ergatoid males, colony structure, Madagascar.

Corresponding author:
Jürgen Heinze
Universität Regensburg, Biologie I,
93040, Regensburg, Germany
E-Mail: juergen.heinze@ur.de

Abstract

The ant genus *Cardiocondyla* exhibits a fascinating diversity of its reproductive biology, with winged and wingless males, long-winged and short-winged queens, strict monogyny and facultative polygyny with or without queen fighting. Here we report on the previously unstudied Malagasy ant *C. shuckardi*. We describe the nesting habits, male morphology and colony structure of this species. Furthermore, based on the genotypes from three microsatellite loci we document a very high incidence of sib-mating.

Introduction

The myrmicine ant genus *Cardiocondyla* comprises about 100 species of minute to small ants that are widely distributed throughout Africa, Europe, and Asia (Seifert, 2003). Colony composition and the reproductive behavior of queens and males vary tremendously among species, making *Cardiocondyla* an ideal system to investigate the evolution of social structures and alternative reproductive tactics in ants.

Because of its ancestral male diphenism with winged disperser males and wingless, "ergatoid" males (Kugler, 1983), *Cardiocondyla* has become a model system for the investigation of alternative reproductive tactics in male ants (e.g., Oettler et al., 2010; Schwander & Leimar, 2011). Winged males resemble typical males of other ant species in morphology, reproductive physiology, and behavior: they are peaceful, disperse shortly after eclosion, and use their limited sperm supply to inseminate young queens from other colonies. Wingless males, in contrast have strong shear- or sickle-shaped mandibles, are relatively long-lived (several

weeks to one year, Yamauchi et al., 2006) and, uniquely among social Hymenoptera, have lifelong spermatogenesis (Heinze & Hölldobler, 1993). Wingless males rarely if ever disperse, and in many species engage in lethal fighting with rivals and attempt to monopolize mating with all female sexuals emerging from their natal nests (Stuart et al., 1987; Kinomura & Yamauchi, 1987; Heinze et al., 1998). Within *Cardiocondyla*, winged disperser males have been lost convergently in at least two clades, and in several species fighting among wingless males has been replaced by territoriality or mutual tolerance (Frohschammer & Heinze, 2009; Lenoir et al., 2007). Queen behavior is similarly variable. Facultative polygyny, i.e., the peaceful coexistence of several fertile queens per nest, appears to be the ancestral state, from which queen-queen fighting (Yamauchi et al., 2007; Heinze & Weber, 2011) and obligate monogyny (a single queen per nest) have evolved (Schrempf & Heinze, 2007).

Previous investigations have focused on the social organization and life history of species of Southeast Asian or Central European origin (Kugler, 1983; Kinomura & Yamauchi, 1987; Heinze et al., 1998; Lenoir et al., 2007), but little is known about the African "*C. shuckardi* group" (sensu Seifert, 2003). This group is of particular interest as it is phylogenetically situated between the *C. nuda* clade (with facultative polygyny and male fighting at least in *C. mauritanica* and *C. kagutsuchi*) and a clade consisting of several monogynous species with only wingless, mutually tolerant males (Oettler et al., 2010).

One of the six presently recognized species of the *C. shuckardi* group, *C. venustula* Wheeler, 1908 has recently been studied in populations introduced to Kaua'i (Hawai'i; Frohschammer & Heinze, 2009) and Puerto Rico (J. Heinze and Susanne Jacobs, unpubl.), and in a native population in South Africa (Heinze et al., 2013). As yet, only wingless males have been found, with some particularly large males from South Africa combining the typical morphology of wingless males with the presence of ocelli and vestigial wings without otherwise approaching the morphology of typical winged ant males. *C. venustula* males appear to defend small territories and fight against males that intrude into their home range. Colonies were at least temporarily polygynous, in contrast to Wheeler's report on monogyny (Wheeler, 1908).

C. shuckardi Forel, 1891 was described from Imerina, the central highlands of Madagascar. According to B. Seifert (pers. comm.) it appears to be restricted to this island. Records of "*C. shuckardi*" from other parts of Africa (Samways, 1983; van Hamburg et al., 2004; Hita Garcia et al., 2009; Kone et al., 2012), the Arab peninsula (Collingwood & Agosti, 1996) and Iran (Ghahari & Collingwood, 2011) are probably misidentifications of other species of the *C. shuckardi* group owing to the often extremely close morphological similarities among species of *Cardiocondyla* (e.g., Seifert, 2003, 2008). Here we describe the results of a field study on the occurrence and colony composition of *C. shuckardi* on the outskirts of Antananarivo, Madagascar, and describe the wingless males of this species.

Material and Methods

Study area

We identified potentially suitable collecting sites from previous records of *C. shuckardi* captured in Malaise and pit-fall traps as listed on www.antweb.org. In March 2012, we visited four sites: the garden of the Madagascar Biodiversity Centre at Tsimbazaza, Antananarivo (18° 55' 57" S, 47° 31' 31.8" E, 1284 m), a sandy threshing floor and paths in front of the palace of Ilafy (18° 51' 14.5" S, 47° 33' 56.8" E, 1356 m), the edge of a eucalypt plantation at Ambohidrabiby (18° 45' 55" S, 47° 36' 37" E, 1381 m), and a gravel path branching off Route Nationale 2 at Mandraka Park (18° 54' 20" S, 47° 53' 35" E, 1369 m). We followed foragers of *C. shuckardi* to their nest entrances in the soil. Nests were carefully excavated, adults and brood were collected into plastic vials with an aspirator. All specimens were transferred to the Madagascar Biodiver-

sity Centre in Antananarivo, and censused under a binocular microscope. Ovaries of queens were dissected following Buschinger and Alloway (1978).

Population genetics

We investigated the suitability of primers previously developed to amplify seven microsatellite loci in other species of *Cardiocondyla* (CE2-3A, CE2-4A, CE2-4E, CE2-5D, CE2-12D, Lenoir et al., 2005; CARD8 and CARD21, Schrempf et al., 2005) for the determination of the genetic structure of *C. shuckardi* colonies. Though only three loci exhibited some variability, we analyzed the genotypes of 4 to 20 workers from each of 14 colonies (total 143 workers) and 1 to 6 dealate queens from 11 colonies (total 31 queens) from Ilavy, 10 workers and 4 or 5 queens from each of two colonies from Mandraka, and 10 workers and 5 queens each from one colony from Ambohidrabiby and one colony from Tsimbazaza to obtain an estimate of colony and population structure of this ant (233 total individuals). Specimens included both old and young queens, most of which still had wings when collected. We could not determine the genotypes at all loci in some individuals, and three individuals were removed because their CE2-12D genotypes appeared to have alleles not found anywhere else. We used Relatedness 4.02 (Goodnight & Queller, 1999) to estimate inbreeding coefficients and nestmate relatedness in colonies from Ilafy. Confidence intervals were obtained by jackknifing by loci. From the inbreeding coefficient we calculated the frequency of sib-mating following Suzuki and Iwasa (1980).

Morphology of males

Six males of *C. shuckardi* were mounted on points and inspected under a Wild M10 binocular microscope with a Wild 1.6x planapochromatic objective at a magnification of 160-200x. We measured head width, scape length, eye diameter, mesonotum width and length (Weber's length), petiole width, and postpetiole width. In addition, we counted the number of funicular segments.

Results

Occurrence and colony composition

C. shuckardi appears to be a common ant in degraded, open patches of grassland, in sandy areas along the edges of unpaved paths, and in parks and gardens in the central highlands of Madagascar. Nests were particularly dense and easy to locate in the regularly watered gardens of Madagascar Biodiversity Centre and near a parched ditch at Ilafy. Solitary workers were seen foraging over distances of more than two meters, and in a few cases we also observed pairs of workers engaged in tandem running, a typical behavior for *Cardiocondyla* (Wilson, 1959; Heinze et al., 2006). Colony nests consisted of pea-sized chambers in sandy soil, under stones, roots, or between pebbles in the upper 15 cm of the ground. Most nest entrances were surrounded by conspicuous middens of corpses of other ants, predominantly *Pheidole.*

In total, we censused 62 colonies (32 from Ilafy, 16 from Tsimbazaza, 10 from Ambohidrabiby, 4 from Mandraka). Individual nests contained between one and seven dealate queens and up to 85 workers. In the most intensively studied site, Ilafy, six nests were queenless, 11 had one queen, and the remaining 15 nests had two to seven queens (26 queenright colonies: median, quartiles 1, 2, 4). Nests contained 10 to 80 workers (median, quartiles, 30.5, 22, 40). Colony composition was similar in the other localities. Upon collection, 14 colonies contained 1 to 10 winged female sexuals, seven colonies contained a single wingless male each, and one colony contained two wingless males.

We observed numerous solitary, wingless queens moving outside the nest, suggesting that at least a fraction of young queens disperse after mating and shedding their wings in their natal nests. Dissection showed that the ovaries of all queens in two multi-queen colonies (all five queens from a colony from Tsimbazaza, five of 12 queens from a colony from Ambohidrabiby) contained developing and mature eggs and had a filled spermatheca. From ovarian status it appeared that three or four of these 10 queens had only recently begun to produce eggs. The ovaries of the other queens contained clearly visible corpora lutea and/or two or three mature eggs, suggesting that they had been fertile for a longer period. Multi-queen colonies therefore appear to be truly polygynous.

Population genetics

Only three of the seven tested *Cardiocondyla* primers were suitable for the genetic analysis of colony structure. In 183 workers and 50 queens from 18 colonies, we found seven alleles at CE2-3A (82 – 94bp), 5 alleles at CE2-12D (143, 159, 163, 165, 167) and two alleles at CARD8 (128 and 130bp). The number and variability of loci allows for only a crude analysis of colony structure. Nevertheless, our analysis clearly suggests a high frequency of inbreeding (Table 1). In the most intensively studied population, Ilafy, observed heterozygosities of workers and queens were much lower than expected heterozygosities at all three loci, resulting in inbreeding coefficients F_{IS} of 1.000 at Card 8, 0.808 at CE2-3A, and 0.559 at CE2-12D (overall F_{IS} 0.685 ± SE 0.133, corresponding to an average percentage of sib-mating of 89.7%). The presence of null alleles cannot be ruled out for Card 8. Ignoring this locus, the inbreeding coefficients at CE2-3A and CE-12D alone give an average percentage of sib-mating of 94.4% and 83.5%, respectively. A similar excess of homozygotes was observed in colonies from the other collection sites (Table 1).

Eleven of 18 colonies contained worker genotypes that were not compatible with single mating by a single queen.

Several genotype combinations suggest the coexistence of several matrilines rather than multiple queen matings. For example, one worker and one queen from the Mandraka 2 colony had a genotype at all three loci that did not overlap with those of other workers and queens.

Average nestmate relatedness in the 14 colonies from Ilafy was 0.784 ± 0.112, ranging in individual colonies from 0.487 ± 0.079 to 1 ± 0 in colonies in which all workers had the same genotype at all loci. Excluding the almost invariable locus Card8 relatedness did not change the result (0.787 ± 0.098, range 0.475 ± 0.069 to 1± 0). Relatedness among workers was 0.766 ± 0.108, ranging from 0.356 ± 0.0.082 to 1 ± 0 (excluding Card 8 0.771 ± 0.095; range 0.353 ± 0.069 to 1 ± 0). The high inbreeding coefficient and the frequent occurrence of more than two worker genotypes per colony suggests that this value does not reflect monogyny and monandry. Instead, many colonies appear to be composed of several inbred lineages of workers. In contrast, queens showed much less variation than workers: all but two queens from Ilafy had the same homozygous genotype at loci Card8 and CE2-3A and nestmate queens had one or two different genotypes at CE2-12D. Queen relatedness in the five multi-queen colonies was 0.949 ± 0.034 (excluding Card 8 0.848 ± 0.076).

Morphology of males

Wingless males closely resemble those of *C. venustula* in morphology, coloration, and size (Fig 1). Compared to wingless males of other *Cardiocondyla* species, but similar to those of *C. venustula*, wingless males of *C. shuckardi* are relatively large (Weber's length 0.62–0.67 mm; thorax width 0.35–0.37 mm). They have large heads (head width 0.49–0.53mm) with strongly sclerotized mandibles and relatively small eyes (eye diameter 0.10–0.11 mm). The antennae resemble those of workers (scape length 0.39–0.41) and the funiculus consists of 10 to 11 segments (Fig. 2), with the typical fusion of segments previously reported from other wingless *Cardiocondyla* males (Seifert, 2003). The pronotal shoulders are well-developed but appear to be less angular than in *C. venustula* from South Africa (Fig. 1). Petiole and postpetiole width ranged from 0.16–0.19 mm and 0.25–0.28 mm, respectively, with a median ratio between petiole and postpetiole width of 0.65. Several males showed injuries such as missing legs (as in the *C. venustula* male in Fig. 1), and one male was found decapitated. This suggests that wingless males of *C. shuckardi* engage in fights for the monopolization of mating with nestmate female sexuals similar to those engaged in by their counterparts in other species.

Discussion

Cardiocondyla shuckardi appears to be rather common in the Madagascar highlands. Nest density was particularly high in places with regularly high humidity, i.e., in gardens or along dry ditches. This matches findings from other

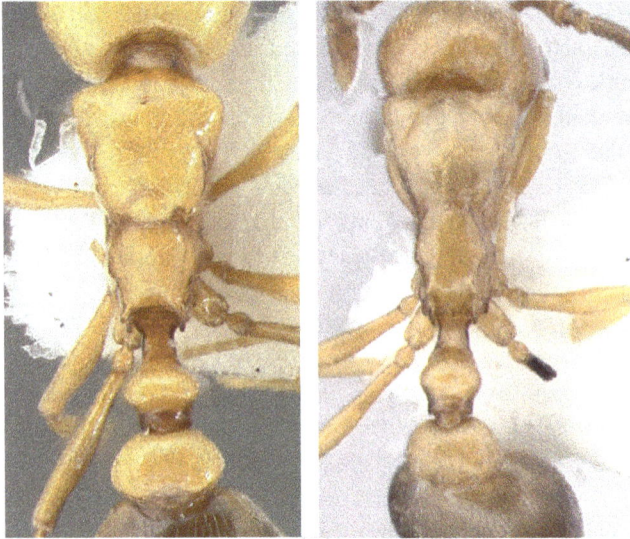

Figure 1 - Dorsal view of the alitrunk of wingless males of the ants *Cardiocondyla shuckardi* (left, from Ilafy) and *C. venustula* (right, from uThukela valley, South Africa; same ant as in Heinze et al., 2013) (photos by Christiana Klingenberg).

Cardiocondyla species, which build nests in humid patches in xeric environments, i.e., on sandy river banks or beaches (Seifert, 2003; Lenoir, 2006; Oettler et al., 2010). We collected the colonies from cavities in the uppermost 10 cm of the soil after heavy rains, but it is likely that the ants move deeper underground during drier periods. For example, *C. elegans* nests reach to a depth of 40 cm (Lenoir, 2006) and nest chambers of the desert species *C. ulianini* were found as deep as 1.5 m (Marikovsky & Yakushkin, 1974). Colonies of *C. shuckardi* were small, with fewer than 100 workers, as is typical for *Cardiocondyla* (Heinze, 1999; Oettler et al., 2010). Nevertheless, because of the high density of nests, foragers of *C. shuckardi* appeared to be among the most abundant ants in our collecting sites. The species presumably plays a considerable role in anthropogenically disturbed, open habitat, such as plantations and gardens. Its ecology therefore appears to resemble that of *C. venustula* in South Africa (e.g., Samways, 1983; van Hamburg et al., 2004).

Our study clarifies two important aspects of the reproductive biology of this species. First, its colonies are at least temporarily facultatively polygynous, and second, males are wingless and resemble those of the related species *C. venustula* in size and morphology (Frohschammer and Heinze 2009). In the studied specimens of *C. shuckardi*, pronotal shoulders appeared to be less angular than in the presently available males of *C. venustula*, but because of the limited sample, the large variation of *C. venustula* males (Heinze et al., 2013) and the lack of males from other species of the *C. shuckardi* group, it would be premature to define universal diagnostic features. Injuries in the examined males of *C. shuckardi* and the simultaneous presence of two males in field colonies suggest that males may attack and damage young rivals but do not always engage in lethal fighting with other adult males. Hence, male behavior appears to be similar to that of *C. ve-*

nustula and *C. mauritanica* males (Frohschammer and Heinze, 2009; Heinze et al., 1993). Whether *C. shuckardi* males defend small "territories" within the nests against other males, as observed in *C. venustula* (Frohschammer and Heinze, 2009), remains to be determined.

As in other species of this genus, analyzing the genetic structure of colonies and populations of *C. shuckardi* was difficult because of the extremely low variability of genetic markers. Only three of seven microsatellite loci were variable to some extent. This obviously prevents us from making conclusions about queen mating frequencies or fine-scaled population and colony structure. Nevertheless, the available data clearly suggest that inbreeding is extremely common in *C. shuckardi*. From the inbreeding coefficient we estimate that more than 80% of all matings involve full sibs. This matches the condition in monogynous *C. batesii, C. elegans,* and *C. nigra*, where 50–80% of all matings involve brothers and sisters (Schrempf et al., 2005; Lenoir et al., 2007, Schrempf, 2014). Dissection data and the co-occurrence of up to four different worker genotypes in colonies of *C. shuckardi* indicate that several mothers may contribute to the offspring of single colonies. Furthermore, the genotypes indicate that individuals may occasionally be exchanged among nests. One queen and one worker from Mandraka had a multilocus-genotype different from those of all other studied nestmates, suggesting the adoption of alien queens as in *C. elegans* (Lenoir et al., 2007).

The combination of frequent inbreeding, polygyny, and queen or worker adoption results in a high nestmate relatedness value, which presumably would be much lower if corrected for inbreeding (e.g., Pamilo, 1985). However, such a correction is not yet possible in *C. shuckardi*, as the mating frequency of queens is unknown and single males may mate with multiple queens. Regardless of the exact value of relatedness, the genotype patterns suggest that queens mate in their natal nests with brothers and then disperse and seek adoption in other nests, in a manner similar to what has been observed in related species (Schrempf et al., 2005; Lenoir et al., 2007). More detailed studies on ants of the *C. venustula* group will help to better understand their peculiar life history.

Acknowledgments

The research was made possible through a collecting permit to B.L. Fisher and funding of Deutsche Forschungsgemeinschaft (He 1623/34). We thank Sandra Theobald and Julia Giehr for their help with genotyping and Christiana Klingenberg, Naturkundemuseum Karlsruhe, for z-stack photographs of the males.

References

Buschinger, A. & Alloway, T.M. (1978). Caste polymorphism in *Harpagoxenus canadensis* M. R. Smith (Hym. Formicidae). Insectes Sociaux, 5: 339-350. doi: 10.1007/BF02224298

Collingwood, C.A. & Agosti, D. (1996). Formicidae (Insecta: Hymenoptera) of Saudi Arabia (part 2). Fauna of Saudi Arabia, 15: 300–385.

Frohschammer, S. & Heinze, J. (2009). Male fighting and "territoriality" within colonies of the ant Cardiocondyla venustula. Naturwissenschaften, 96: 159–163. doi: 10.1007/s00114-008-0460-2

Ghahari, H. & Collingwood, C.A. (2011). A study on the ants (Hymenoptera: Formicidae) of southern Iran. Calodema, 176: 1–5.

Goodnight, K.F. & Queller, D.C. (1999). Relatedness 5.0 [software]. Goodnight Software.

Heinze, J. (1999). Male polymorphism in the ant Cardiocondyla minutior (Hymenoptera: Formicidae) Entomologia Generalis, 23: 251–258.

Heinze, J. & Hölldobler, B. (1993). Fighting for a harem of queens: physiology of reproduction in Cardiocondyla male ants. Proceedings of the National Academy of Science USA, 90: 8412–8414.

Heinze, J. & Weber, M. (2011). Lethal sibling rivalry for nest inheritance among virgin ant queens. Journal of Ethology, 29: 197–201. doi: 10.1007/s10164-010-0239-8

Heinze, J., Kühnholz, S., Schilder, K. & Hölldobler, B. (1993). Behavior of ergatoid males in the ant, Cardiocondyla nuda. Insectes Sociaux, 40: 273–282. doi: 10.1007/BF01242363

Heinze, J., Hölldobler, B. & Yamauchi, K. (1998). Male competition in Cardiocondyla ants. Behavioral Ecology and Sociobiology, 42: 239–246. doi: 10.1007/s002650050435

Heinze, J., Cremer, S., Eckl, N. & Schrempf, A. (2006). Stealthy invaders: the biology of Cardiocondyla tramp ants. Insectes Sociaux, 53: 1–7. doi: 10.1007/s00040-005-0847-4

Heinze, J., Aumeier, V., Bodenstein, B., Crewe, R.M. & Schrempf, A. (2013). Wingless and intermorphic males in the ant Cardiocondyla venustula. Insectes Sociaux, 60: 43–48. doi: 10.1007/s00040-012-0263-5

Hita Garcia, F., Fischer, G., Peters, M.K., Snelling, R.R. & Wägele, J.W. (2009). A preliminary checklist of the ants (Hymenoptera: Formicidae) of Kakamega Forest (Kenya). Journal of East Africa Natural History, 98: 147–165. doi: 10.2982/028.098.0201

Kinomura, K. & Yamauchi, K. (1987). Fighting and mating behaviors of dimorphic males in the ant Cardiocondyla wroughtoni. Journal of Ethology, 5: 75–81. doi: 10.1007/BF02347897

Kone, M., Konate, S., Yeo, K., Kouassi, P.K. & Linsenmair, K.E. (2012). Changes in ant communities along an age gradient of cocoa cultivation in the Oumé region, central Côte d'Ivoire. Entomological Science, 15: 324–339.

Kugler J. (1983). The males of Cardiocondyla Emery (Hymenoptera: Formicidae) with the description of the winged males of Cardiocondyla wroughtoni (Forel). Israel Journal of Entomology, 17: 1–21.

Lenoir, J.-C. (2006). Structure sociale et stratégie de reproduction chez Cardiocondyla elegans. PhD thesis, Université de Tours, France.

Lenoir, J.-C., Schrempf, A., Lenoir, A., Heinze, J. and Mercier, J.-L. (2005). Five polymorphic microsatellite markers for the study of Cardiocondyla elegans (Hymenoptera: Myrmicinae). Molecular and Ecological Resources, 5: 565–566. doi: 10.1111/j.1471-8286.2005.00989.x

Lenoir, J.-C., Schrempf, A., Lenoir, A., Heinze, J. & Mercier, J.-L. (2007). Genetic structure and reproductive strategy of the ant Cardiocondyla elegans: strictly monogynous nests invaded by unrelated sexuals. Molecular Ecology, 16: 345–354. doi: 10.1111/j.1365-294X.2006.03156.x

Marikovsky, P.I. & Yakushkin, V.T. (1974). Muravei Cardiocondyla uljanini Em., 1889 i sistematicheskoye polosheniye "parasiticheskovo muravya Xenometra". Izvestiya Akademii Nauk Kazakhskoi SSR Seriya Biologicheskikh, 3: 57–62.

Oettler, J., Suefuji, M. & Heinze, J. (2010). The evolution of alternative reproductive tactics in male Cardiocondyla ants. Evolution, 64: 3310–3317. doi: 10.1111/j.1558-5646.2010.01090.x

Pamilo, P. (1985). Effect of inbreeding on genetic relatedness. Heredity, 103: 195–200.

Samways, M.J. (1983). Community structure of ants (Hymenoptera: Formicidae) in a series of habitats associated with citrus. Journal of Applied Ecology, 20: 833–847. doi: 10.2307/2403128

Schrempf, A. (2014) Inbreeding, multiple mating and foreign sexuals in the ant Cardiocondyla nigra (Hymenoptera: Formicidae). Myrmecological News, 20: 1-5.

Schrempf, A. & Heinze, J. (2007). Back to one: consequences of derived monogyny in an ant with polygynous ancestors. Journal of Evolutionary Biology, 20: 792–799. doi: 10.1111/j.1420-9101.2006.01235.x

Schrempf, A., Reber, C., Tinaut, A. & Heinze, J. (2005). Inbreeding and local mate competition in the ant Cardiocondyla batesii. Behavioral Ecology and Sociobiology, 57: 502–510. doi: 10.1007/s00265-004-0869-3

Schwander, T. & Leimar, O. (2011). Genes as leaders and followers in evolution. Trends in Ecology and Evolution, 26: 143–151. doi: 10.1016/j.tree.2010.12.010

Seifert, B. (2003). The ant genus Cardiocondyla (Hymenoptera: Formicidae) - a taxonomic revision of the elegans, bulgarica, batesii, nuda, shuckardi, stambuloffii, wroughtonii, emeryi, and minutior species groups. Annalen des Naturhistorischen Museums in Wien B, 104: 203–338.

Seifert, B. (2008). Cardiocondyla atalanta Forel, 1915, a cryptic sister species of *Cardiocondyla nuda* (Mayr, 1866) (Hymenoptera: Formicidae). Myrmecological News 11: 43–48.

Stuart, R.J., Francoeur, A. & Loiselle, R. (1987). Lethal fighting among dimorphic males of the ant, *Cardiocondyla wroughtonii.* Naturwissenschaften 74: 548–549. doi: 10.1007/BF00367076

Suzuki, Y. & Iwasa, Y. (1980). A sex ratio theory or gregarious parasitoids. Research on Population Ecology, 11: 366–382. doi: 10.1007/BF02530857

van Hamburg, H., Andersen, A.N., Meyer, W.J. & Robertson, H.G. (2004). Ant community development on rehabilitated ash dams in the South African Highveld. Restoration Ecology, 12: 552–558. doi: 10.1111/j.1061-2971.2004.00421.x

Wheeler, W.M. (1908). The ants of Puerto Rico and the Virgin Islands. Bulletin of the American Museum of Natural History, 24: 117–158.

Wilson, E.O. (1959). Communication by tandem running in the ant genus *Cardiocondyla*. Psyche, 66: 29–34.

Yamauchi, K., Ishida, Y., Hashim, R. & Heinze, J. (2006). Queen-queen competition by precocious male production in multiqueen ant colonies. Current Biology, 16: 2424–2427. doi: 10.1016/j.cub.2006.10.007

Yamauchi, K., Ishida, Y., Hashim, R. & Heinze, J. (2007). Queen-queen competition and reproductive skew in a *Cardiocondyla* ant. Insectes Sociaux,. 54: 268–274. doi: 10.1007/s00040-007-0941-x

Table 1 - Genotypes of workers and queens of the ant *Cardiocondyla shuckardi* from 18 colonies from four populations on Madagascar (Ilafy, Ambohidrabiby, Tsimbazaza, Mandraka) at three microsatellite loci (1a:Card8, CE2-3A; 1b: CE2-12D). For each locus, the number of workers / queens with a certain genotype is given.

Table 1a

	Card8			CE2-3A												
	128/128	128/130	130/130	82/84	84/84	84/86	84/94	86/86	86/90	86/92	86/94	88/88	90/90	92/92	92/94	94/94
Ilafy 1	19/6							15/6	1/0				2/0	1/0		
Ilafy 2	10/4							9/4								
Ilafy 3	9/3		1/0					13/3			6/0				1/0	
Ilafy 4	19/5					1/0		5/4	2/0	1/0						
Ilafy 5	10/5					1/0		9/5								
Ilafy 6	10/-							10/								
Ilafy 7	9/1							9/1								
Ilafy 8	10/1							10/1								
Ilafy 9	8/1												8/1			
Ilafy 10	4/-				4/-											
Ilafy 11	10/1				4/1		6/0									
Ilafy 12	10/1							8/1			2/0					
Ilafy 13	8/3							8/3								
Ilafy 14	4/-				4/-											
Ambo	8/5	1/0	1/0	1/0		1/3		3/2		3/0	1/0	1/0				
Tsimb 1	10/5							10/5								
Mand 1			10/5					1/0	1/1						1/1	6/3
Mand 2	1/1		9/3					1/1	0/1					0/1	6/1	3/0

Table 1 b

					CE2-12D				
	143/143	143/159	159/159	159/167	163/163	163/165	165/165	165/167	167/167
Ilafy 1					2/2	8/3	1/0		1/0
Ilafy 2					1/0	3/0	3/2		
Ilafy 3				2/0				1/1	16/2
Ilafy 4			1/0	4/0					¾
Ilafy 5							10/5		
Ilafy 6								3/-	7/-
Ilafy 7						3/0	2/0	3/1	
Ilafy 8			10/1						
Ilafy 9							1/0	2/1	5/0
Ilafy 10			1/-	1/-				1/-	1/-
Ilafy 11							10/1		
Ilafy 12								4/1	6/0
Ilafy 13							2/0	2/1	3/2
Ilafy 14									4/-
Ambo							4/5		
Tsimb 1									8/2
Mand 1	7/3	1/1	1/1						
Mand 2	4/0	3/2	2/1						1/1

27

Defensive repertoire of the stingless bee *Melipona flavolineata* Friese (Hymenoptera: Apidae)

TM Nunes[1], LG Von Zuben[1], L Costa[2], GC Venturieri[2]

1 - Universidade de São Paulo, SP, Brazil

2 - Empresa Brasileira de Pesquisa Agropecuária, Embrapa Amazônia Oriental, Belém, PA, Brazil

Keywords

Meliponini, pheromones, mandibular gland, *Lestrimelitta limao*

Corresponding author

Túlio M. Nunes
NPPNS, Departamento de Física e Química, FCFRP, Universidade de São Paulo, SP, Brazil
E-Mail: tulionunes@usp.br

Abstract

Despite the loss of the sting apparatus, Meliponini (stingless) bees have not lost their ability to defend themselves. Several defensive strategies have been described for the group, including biting and resin deposition. Defensive behavior can be mediated by chemical communication, for example through the use of alarm pheromones. The Stingless bee species *Melipona flavolineata* Friese is an important species for meliponiculture in Brazil, especially in the Amazon region. In order to improve the current management methods for the species, this study aimed to describe the range of defensive strategies used by the stingless bee *M. flavolineata* towards inter and intraspecific chemical signals known to trigger defensive responses in related species, namely the head secretions of the robber bee *Lestrimelitta limao* (Smith) and the mandibular gland extract of conspecifics *M. flavolineata* workers. The stimuli provoked different defensive reactions. The head secretions of the robber bee repelled returning foragers, elicited the enclosing of the nest entrance tube with batumen balls and the agglomeration of workers outside the box. In contrast, the mandibular gland extract elicited aggression towards the pheromone deposition site, transport of resin and generalised agitated flights. Our results confirm the role of the mandibular gland as a source of alarm pheromone in this species and the chemical triggering of a specific defensive response to the known cleptoparasite *L. limao*.

Introduction

The Meliponini bees are a diverse group of social insects comprising over 400 species (Michener, 2000; Camargo & Pedro, 2007). These bees are called stingless bees as a result of their loss of the sting apparatus and consequent inability to defend themselves by stinging, as commonly observed in other bees (Wilson, 1971). Despite the lack of sting, the Meliponini bees have developed a series of defensive mechanisms, including biting and resin deposition among others, which are triggered and modulated by inter and intraspecific chemical and visual stimuli (Wilson, 1971; Wittmann et al., 1990; Schorkopf et al., 2009).

Stingless bee colonies store a great amount of resources which have to be defended from a variety of predators and parasites, including many species of insects and also vertebrates (Nogueira-Neto, 1997). The diversity of defence mechanisms described for the group is vast and ranges from narrow nest entrances to chemical weapons (Roubik et al., 1987; Couvillon et al., 2008). The most commonly described defensive behavior is the specific biting of hairs and vulnerable body regions such as the eyes and ears (Wilson, 1971; Wittmann, 1985). Biting behavior is usually linked to distressful sounds (Wilson 1971). Another characteristic defensive mechanism is the deposition of plant resins on the potential predator (Greco et al., 2010). Resin deposits are a common feature on different species of stingless bees' nests and in the presence of a disturbance the bees carry this material with their legs or mandibles and deposit it on the predator (Sakagami, 1982). Still, some species make use of caustic chemical secretions as described in the genus *Oxytrigona* (Roubik et al., 1987).

Among natural enemies of stingless bees are the kleptobiotic stingless bees belonging to the *Lestrimelitta* genus. The robber bees invade other stingless bee species' nests and pillage their reserves, carrying wax, honey, pollen and mostly larval food from recently enclosed reproductive cells (Sakagami et al., 1993). The stingless bees have developed a range of defensive behaviors against the attack of *Lestrimelitta* species (Wittmann et al., 1990; Nunes et al., 2008; Grüter et al., 2012). Some species respond non-aggressively to the attack of *Lestrimelitta* bees by hiding at the periphery of the nest and underneath brood combs (Michener, 1946; Sakagami, et al., 1993). Other species strongly react to the attacks, biting the legs and wings of the invasive species, depositing resin and closing the nest entrance (Wittmann, 1985; Nunes et al., 2008).

Beyond the defensive mechanisms triggered by interspecific stimuli, intraspecific chemical communication is also thought to initiate aggressive responses (Cruz & López et al., 2007; Schorkopf et al., 2009). The mandibular glands of stingless bees are regarded as a source of intraspecific communication substances eliciting alarm responses (Smith & Roubik 1983; Cruz & López et al., 2007; Schorkopf et al., 2009). The species analysed showed highly volatile gland contents (Schorkopf et al., 2009). In the presence of its secretion, workers attacked the pheromone source, avoided the food sources and exhibited agitated behavior, which consists in vertical flights at high speed and intense buzzing sounds (Smith & Roubik 1983; Johnson et al., 1985; Cruz & López et al., 2007; Schorkopf et al., 2009).

The stingless beekeeping is a rapidly growing activity in the tropical areas of the globe (Cortopassi-Laurino et al., 2006; Contrera et al., 2011). These bees provide a wide range of economic opportunities such as the extraction of natural products like honey and resin and also improve the production of local crops through pollination (Venturieri, 2008; Oliveira et al., 2012). In the Brazilian Amazon region, the management of the stingless bee *Melipona flavolineata* Friese is an economic alternative for small land holders (Venturieri et al., 2003; Magalhães & Venturieri, 2010). Even though its economic value has been shown, the correct management of this species requires detailed behavioural analysis. Understanding the means by which the species defends itself from a wide range of predators and parasites is essential to develop this culture in scale. We thus analysed the range of defensive reaction of the stingless bee *M. flavolineata,* evaluating workers' responses towards inter and intraspecific chemical signals known for triggering defensive response in related species, the head secretions of robber bees (*Lestrimelitta limao* (Smith)) and conspecific mandibular glands contents.

Materials and Methods

Bees and study site

The tests were conducted at "Cocal do Tauá", municipality of Santo Antônio do Tauá, Pará, Brazil (1°05'17.54"S 48°15'20.82"O) in August 2014, between 9:00 and 14:00h.

For the tests, we used 14 colonies of *M. flavolineata* conditioned in wooden boxes specially designed for this species (Venturieri, 2008).

Behavioral tests

To evaluate the defensive responses of *M. flavolineata* workers colonies were randomly allocated to three groups: (1) *L. limao* head extract (2) *M. flavolineata* mandibular gland extract and (3) pure solvent (Dichloromethane), the latter to avoid any ambiguous result due to a reaction against the solvent odor. The *L. limao* group consisted of an extract of macerated *L. limao* heads (three bee-equivalent, i.e. the amount of compound used for each colony was equivalent to the content of three workers of *L. limao*) in dichloromethane. In the mandibular gland group, colonies were treated with an extract of *M. flavolineata* dissected mandibular glands (four gland-equivalent). Each treatment was applied directly at the entrance tube.

Following the treatment, workers behavior repertoire was recorded for 5 minutes. In order to evaluate the repellence effect of each treatment group, the number of foragers entering the nest was recorded, if no bees entered the nest during this time, we recorded the time the first forager returned. The defensive response of workers was evaluated by recording the total number of bees exiting the nest and gathering at the nest entrance around the treatment site. Also, the presence or absence of the following behavioral patterns has been recorded: attacks to the extract deposition site, presence of workers carrying resin in their hind legs and overall agitation (*i.e.* workers increasing the flight speed, performing vertical flights and noticeable buzzing sound). After 10 minutes from the treatment, the colony was opened in order to evaluate its general state, presence of resin and batumen balls deposits.

Statistical analyses

In order to analyse the number of workers entering the nest five minutes following the treatment and the number of workers leaving the nest and gathering at its entrance, a generalized linear model (GLM) with Poisson error distribution was used to assess differences among treatments, with the level of statistical significance established at $\alpha < 0.05$. The number of workers entering the nest and total number of workers gathering at the box were regarded as dependent variables and treatment (with three levels: solvent control, mandibular gland and *L. limao* head extract) as categorical independent variable. The multiple pair-wise comparisons among treatments were made using Tukey contrasts.

Results and Discussion

Workers of *M. flavolineata* showed a clearly distinct behavioral pattern in response to the control group and treatments (the cephalic extract of the robber bee and the mandibular glands extract). Workers in the control group did not present any visible change in the behavior after the solvent was deposited at the nest entrance, showing no generalized agitation and workers returning normally to the nest. The comparison between colonies treated with the cephalic extract of *L. limao* and the control group showed a significantly lower number of workers returning to the nest in five minutes following the treatment indicating a repellent effect of the cephalic extract (Poisson GLM: Z = - 7.50, p<0.001, n = 10; Fig 1).

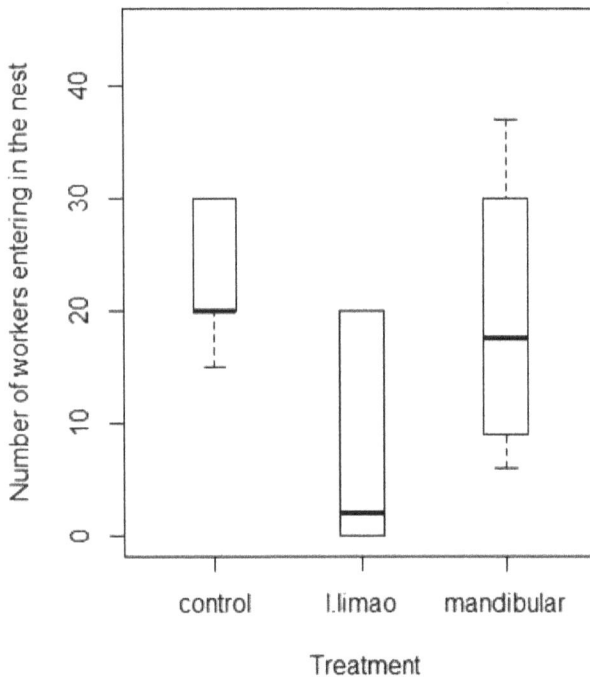

Fig. 1 – Box-and-whisker plots of the number of *Melipona flavolineata* workers entering the nest during five minutes following the treatment. Box plots show the median, 25&75% percentiles. Whiskers show all data excluding outliers (dots).

The duration of the repellent effect varied across the colonies and the maximum latency observed was 8 minutes until the first worker entered the nest. Following the repellent effect, the workers from inside the nest initiated an intense vibration and a conspicuous buzzing sound. Next, the workers started to leave the nest in great numbers and to agglomerate in the box surrounding its entrance (*i.e.* the site of the extract deposition). The maximum number of workers observed crowding at the nest entrance was 223 in one colony treated with the cephalic extract of the robber bee. There was a significantly higher number of workers crowding at the nest

in the colonies treated with the cephalic extract of *L. limao* when compared to the control group (Poisson GLM: Z = 7.78, p<0.001, n = 10; Fig 2). The crowding workers displayed intense vibration of the thorax and wings and frequent rubbing of the last pair of legs to the tip of the abdomen (Fig 3). After a few minutes of this behavioral display, the workers ceased the vibration and returned to the interior of the nest.

Ten minutes after the start of the treatment, observations inside the colonies showed workers dragging small batumen balls (1-5mm) to the entrance tube. Batumen balls deposits were observed in all the nests and usually took place on the side of the internal entrance tube aperture and were associated with ventilation openings (Fig 4). The deposition of batumen balls at the nest entrance as a reaction to *L. limao* attacks have been observed many times during natural raids (Venturieri, pers. observation). The entrance blockage using batumen balls has also been described for the species *Melipona paraensis* Ducke and this artefact was found inside of the nest of the tropical species *Melipona seminigra merrilae* Cockerell and *Melipona crinita* Moure & Kerr (Portugal-Araujo, 1978) showing that this defensive strategy is shared by other species of this genus.

The colonies treated with the mandibular gland extract showed a different response from those in the control group or treated with the cephalic extract of the robber bee. The number of bees returning to the nest during five minutes following the treatment decreased in this group when compared to the control group (Poisson GLM: Z = - 3.34, p<0.001, n = 9; Fig 1). This reduction however was smaller than the one observed in the colonies treated with the cephalic extract of the robber bee (Poisson GLM: Z = 4.40, p<0.001, n = 9; Fig 1). The workers returning to the hive were not repelled by the pheromone source but instead became agitated performing

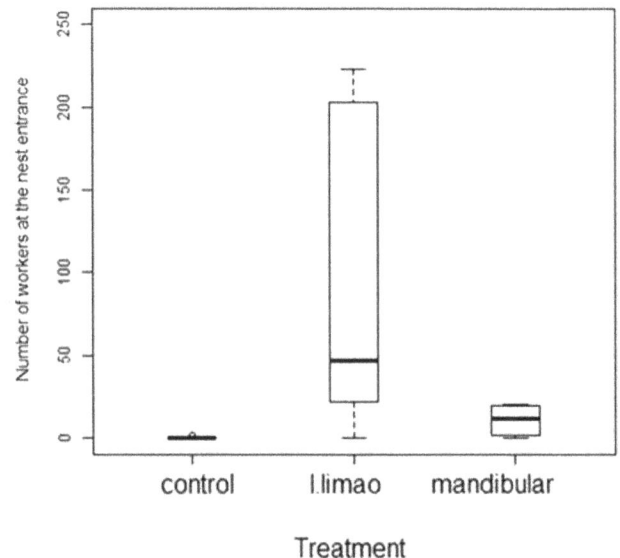

Fig. 2 – Box-and-whisker plots of the number of *Melipona flavolineata* workers gathering outside the nest box during five minutes following the treatment. Box plots show the median, 25&75% percentiles. Whiskers show all data excluding outliers (dots).

quick angular flights and an intense buzzing sound. Following the treatment with the mandibular gland extract, several workers left the nest and stayed outside the box, gathering around the nest entrance. The maximum number of workers agglomerated outside the box near its entrance was higher when compared to the control group (Poisson GLM: $Z = 4.55$, $p<0.001$, $n = 9$; Fig 2) but significantly lower in comparison with the robber bee cephalic extract group (Poisson GLM: $Z = -13.95$, $p<0.001$, $n = 9$; Fig 2).

Fig 3 – Workers of *Melipona flavolineata* gathering outside the nest box surrounding the entrance after the deposition of *L. limao* cephalic extract.

Generalized agitation with workers flying quickly in angular flights and noticeable buzzing sounds occurred after both treatments but not in the control (Table 1). Some behavioural displays could only be observed in colonies treated with the mandibular gland extract, for example workers leaving the hive carrying resin in their corbicula and similar behaviour happening in neighbouring colonies (Table 1, Fig 5). Similarly, attacks to the pheromone site only were observed in the mandibular gland extract group (Table 1).

The results described here demonstrate the repellent effect of *M. flavolineata* workers in the presence of cephalic secretions of *L. limao*. The avoidance of the nest under attack

Fig 4 – Batumen balls deposit inside a *Melipona flavolineata* nest.

Fig 5 – Workers of *Melipona flavolineata* carrying resin in their corbicula after the treatment with mandibular gland extract of

by returning foragers has been reported for different species of stingless bees (Michener 1946; Kerr 1951; Sakagami & Laroca, 1963). The tests demonstrated that the behavioral responses are mediated by interspecific chemical signals which are secreted by the cephalic glands of the robber bee. The cephalic extract contains chemical secretions of the mandibular glands and the labial glands. The composition of the two glands differs substantially and the source of the repellent effect remains to be investigated (von Zuben, 2012).

Table 1 – Percentage of *Melipona flavolineata* colonies responding to the different pheromone treatments. CE, cephalic extracts of the robber bee *L. limao*. GM, mandibular gland extracts of conspecifics *M. flavolineata* workers.

Behaviour	Control	Treatment	
		CE	MG
Generalised agitation	-	80	75
Attack to pheromone site	-	-	75
Workers leaving the nest carrying resin	-	-	100
Workers from neighbouring colonies leaving the nest carrying resin	-	-	100
Resin deposits inside of the nest	100	100	100
Batumen ball deposits inside of the nest	100	100	100

The volatility of the robber bee cephalic secretions together with the gradual shift from avoidance of the secretion source to a defensive response (*i.e.* workers agitated around the entrance) suggests a dose-dependent reaction towards the robber bee compounds. Three bee-equivalent extracts had a repellent effect. The highly volatile aspect of the compounds results in a quick decrease of its quantity at the treatment site. A few minutes following the treatment with the cephalic extract, the reaction observed changed from hiding and avoiding the source of the stimuli to generalized agitation, noticeable buzzing and workers leaving the colony and gathering at the treatment site. This shift in the behavior suggests that workers of *M. flavolineata* can use different defence strategies against

L. limao attacks in response to varying concentration of the secretions. The concentration of the cephalic extract at the moment it was applied mimics the presence of a substantial number of robber bee workers in the entrance tube which is avoided by the hosts. Smaller amount of this secretion reflects the presence of a few individuals and trigger more aggressive reaction from the hosts.

The host species could stop an attack with relatively few casualties in the presence of a small group of the robber bees but not in the presence of a large group. Avoiding a battle in the presence of a large number of invaders results in the loss of storage resources but it prevents a significant worker loss, which are necessary for the hive re-construction after the attack. The mixed defensive strategy to avoid large groups of invaders and to attack small groups might be associated with a better cost-benefit for this specie than the strategy to always fight observed in some other species (e.g. *Tetragonisca angustula* (Latreille)).

The tests with co-specific mandibular gland secretions suggest its use as an alarm pheromone for this species. Attacks of the pheromone source, agitated angular flights and workers leaving the nest with resin on the hind legs shows the use of these glandular compounds in triggering defensive responses. The results for this species corroborate the idea of mandibular gland as the source of alarm pheromones in stingless bees (Smith & Roubik, 1983; Cruz & López et al., 2007; Schorkopf et al., 2009).

There were clear differences between the responses towards inter and intraspecific chemical secretions. While the interspecific signals (i.e. head extract of the robber bee) provoked a repellent effect, the intraspecific pheromone (mandibular gland extract) elicited generalised agitation of foragers arriving in the nest. Moreover, there were specific responses in the intra-specific mandibular gland extract group such as workers carrying resin that could not be observed in the robber bee extract group. The differences described support the idea that workers of *M. flavolineata* do not use alarm pheromones from mandibular gland during the attack of *L. limao,* but instead react directly towards the heterospecific signals.

Acknowledgments

The authors are thankful for the financial support from BIONORTE (CNPQ/FAPESPA) provided to GCV and FAPESP (TMN Proc. 2011/22991-3).

References

Camargo, J.M.F. & Pedro, S.R.M. (2007). Meliponini Lepeletier, 1836. In J.S. Moure, J.S., D. Urban & G.A.R. Melo (Eds), Catalogue of bees (Hymenoptera, Apoidea) in the Neotropical region. Curitiba: Sociedade Brasileira de Entomologia.

Contrera, F.A.L., Menezes, C. & Venturieri, G.C. (2011). New horizons on stingless beekeeping (Apidae, Meliponini). Rev. Bras. Zootecn. 40: 48-51.

Cortopassi-Laurino, M., Imperatriz-Fonseca, V.L., Roubik, D.W., Dollin, A., Heard, T., Aguilar, I., Venturieri, G.C., Eardley, C. & Nogueira-Neto, P. (2006). Global meliponiculture: challenges and opportunities. Apidologie 37: 275-292.

Couvillon, M.J., Wenseleers, T., Imperatriz-Fonseca, V.L., Nogueira-Neto, P. & Ratnieks, F.L.W. (2008). Comparative study in stingless bees (Meliponini) demonstrates that nest entrance size predicts traffic and defensivity. J. Evolution. Biol. 21: 194-201. doi: 10.1111/j.1420-9101.2007.01457.x

Cruz & López, L., Aguilar, S., Malo, E., Rincón, M., Guzman, M. & Rojas, J.C. (2007). Electroantennogram and behavioral responses of workers of the stingless bee *Oxytrigona mediorufa* to mandibular gland volatiles. Entomol. Exp. Appl. 123(1): 43-47. doi: 10.1111/j.1570-7458.2007.00522.x

Greco, M.K., Hoffmann, D., Dollin, A., Duncan, M., Spooner-Hart, R. & Neumann, P. (2010). The alternative Pharaoh approach: stingless bees mummify beetle parasites alive. Naturwissenschaften 97(3): 319-323. doi: 10.1007/s00114-009-0631-9

Grüter, C., Menezes, C., Imperatriz-Fonseca, V.L. & Ratnieks, F.L.W. (2012). A morphologically specialized soldier caste improves colony defense in a neotropical eusocial bee. P. Natl. Acad. Sci. USA 109: 1182-1186. doi: 10.1073/pnas.1113398109

Johnson, L.K., Haynes, L.W., Carlson, M.A., Fortnum, H.A. & Gorgas, D.L. (1985). Alarm substances of the stingless bee, *Trigona silvestriana*. J. Chem. Ecol. 11: 409-416. doi: 10.1007/BF00989552.

Kerr, W.E. (1951). Bases para o estudo da genética de populações dos Hymenoptera em geral e dos Apinae sociais em particular. An. ESALQ 8: 219-354.

Magalhães, T.L. & Venturieri, G.C. (2010). Aspectos econômicos da criação de abelhas indígenas sem ferrão (Apidae: Meliponini) no nordeste paraense. Série documentos EMBRAPA 364: 1-36.

Michener, C.D. (1946). Notes on Panamanian species of stingless bees. J. New York Entomol. Soc. 54: 179-197.

Michener, C.D. (2000). The bees of the world. Baltimore: John Hopkins University Press.

Nogueira-Neto, P. (1997). Vida e criação de abelhas indígenas sem ferrão. São Paulo: Nogueirapis.

Nunes, T.M., Nascimento, F.S., Turatti, I.C., Lopes, N.P. & Zucchi, R. (2008). Nestmate recognition in a stingless bee: does the similarity of chemical cues determine guard acceptance? Anim. Behav. 75:1165-1171. doi: 0.1016/j.anbehav.2007.08.028

Oliveira, P.S., Müller, R.C.S., Dantas, K.D.G.F., Alves, C.N., Vasconcelos, M.A.M.D. & Venturieri, G.C. (2012). Phenolic acids, flavonoids and antioxidant activity in honey of *Melipona fasciculata*, *M. flavolineata* (Apidae, Meliponini) and *Apis mellifera* (Apidae, Apini) from the Amazon. Quim. Nova 35: 1728-1732.

Portugal-Araujo, V.D. (1978). Um artefato de defesa em colônias de Meliponineos. Acta Amaz.

Roubik, D.W., Smith, B.H. & Carlson, R.G. (1987). Formic acid in caustic cephalic secretions of stingless bee, Oxytrigona (Hymenoptera: Apidae). J. Chem. Ecol., 13: 1079-1086. doi: 10.1007/BF01020539.

Sakagami, S.F., Roubik, D.W. & Zucchi, R. (1993). Ethology of the robber stingless bee, *Lestrimelitta limao* (Hymenoptera: Apidae). Sociobiology 21: 237-277.

Sakagami, S.F. (1982) Stingless bees. In H.R. Hermann (Ed.), Social Insects (pp. 361-423), New York: Academic Press.

Sakagami, S.F. & Laroca, S. (1963). Additional Observations on the habits of the cleptobiotic stingless bees, the genus *Lestrimelitta* Friese (Hymenoptera, Apoidea). J. Fac. Sci. Hokkaido Univ. Series Zool. 15: 319-339.

Schorkopf, D.L.P., Hrncir, M., Mateus, S., Zucchi, R., Schmidt, V.M. & Barth, F.G. (2009). Mandibular gland secretions of meliponine worker bees: further evidence for their role in interspecific and intraspecific defence and aggression and against their role in food source signalling. J. Exp. Biol. 212: 1153-1162. doi: 10.1242/jeb.021113.

Smith, B.H. & Roubik, D.W. (1983). Mandibular glands of stingless bees (Hymenoptera: Apidae): Chemical analysis of their contents and biological function in two species of *Melipona*. J. Chem. Ecol. 9: 1465-1472.

Venturieri, G.C., Raiol, V.D.F.O. & Pereira, C.A.B. (2003). Avaliação da introdução da criação racional de *Melipona fasciculata* (Apidae: Meliponina), entre os agricultores familiares de Bragança-PA, Brasil. Biota Neotropica 3: 1-7.

Venturieri, G.C. (2008) Caixa para a criação de uruçu-amarela *Melipona flavolineata* Friese, 1900. Embrapa Amazônia Oriental. Comunicado técnico.

Venturieri, G.C. (2008). Criação de abelhas indígenas sem ferrão. Embrapa Amazônia Oriental. 2nd Ed.

Zuben, L.G. (2012) Determinantes bionômicos e eco-químicos do cleptoparasitismo de *Lestrimelitta limao* Smith (Hymenoptera: Apidae, Meliponini), Monographs: Universidade de São Paulo.

Wilson, E.O. (1971) The insect societies. Cambridge: Harvard University Press.

Wittmann, D. (1985) Aerial defense of the nest by workers of the stingless bee *Trigona (Tetragonisca) angustula* (Latreille) (Hymenoptera: Apidae). Behav. Ecol. Sociobiol. 16: 111-114. doi: 10.1007/BF00295143

Wittmann, D., Radtke, R., Zeil, J., Lübke, G. & Francke, W. (1990). Robber bees (*Lestrimelitta limao*) and their host chemical and visual cues in nest defense by *Trigona (Tetragonisca) angustula* (Apidae: Meliponinae). J. Chem. Ecol. 16: 631-641. doi: 10.1007/BF01021793.

28

Diapause in Stingless Bees (Hymenoptera: Apidae)

CF dos Santos, P Nunes-Silva, R Halinski, B Blochtein

Pontifícia Universidade Católica do Rio Grande do Sul, Porto Alegre, RS, Brazil.

Keywords
Plebeia, provisioning and oviposition process, reproductive diapause, temperature, photoperiodism.

Corresponding author
Charles Fernando dos Santos
Depto. de Biodiversidade e Ecologia,
Faculdade de Biociências,
Pontifícia Universidade Católica do Rio Grande do Sul
Av. Ipiranga, 6681, Partenon
Porto Alegre, RS, Brazil
E-Mail: chasanto@gmail.com

Abstract
Extreme environmental conditions may negatively affect the development of animals. Insects show a wide range of adaptive behaviors that have allowed them to respond successfully to adverse climatic conditions by temporarily interrupting some of their activities or development. One such behavior is diapause. Diapause can be defined as a gradual and progressive interruption in development or ontogeny of an organism in any phase of its lifecycle in order to survive cyclic unfavorable environmental conditions. This review presents an overview of the current knowledge of diapause in stingless bees and describes the various studies on this subject. It focuses on *Plebeia* species, the most studied genus in this regard. In this group of bees, provisioning and oviposition behavior ceases in autumn/winter, a so-called reproductive diapause. Besides the cessation of brood-rearing activity, other behaviors, such as foraging, are also modified. The mechanisms that induce the reproductive diapause are still unclear, but evidence points to temperature and photoperiodism as the main drivers of this behavior.

Diapause - a strategy for surviving unfavourable environmental conditions

Extreme environmental conditions may negatively affect the development of certain animals and many have evolved behavioral strategies in order survive. Insects present a wide range of adaptive behaviors that have allowed them to respond successfully to adverse climatic conditions by temporarily interrupting some of their activities or development in the face of, for example, extreme temperatures (cold or warm) (Tauber & Tauber, 1976; Denlinger, 1986; Kort, 1990; Tatar & Yin, 2001; Koštál, 2006). Such behavioral responses have variously been termed dormancy, quiescence, hibernation, aestivation and diapause (Tauber & Tauber, 1976; Denlinger, 1986). However, conceptual differences exist among these terms relating to the kind of environmental variable and whether the response is merely immediate (non-programmed) or rhythmical (endogenously and genetically programmed) (Tauber & Tauber, 1976; Denlinger, 1986, 2002; Koštál, 2006).

In a broader sense, dormancy is a term embracing several others (e.g., quiescence, hibernation, aestivation and diapause) and refers to any temporary arrest in behavior performed by an insect in response to any unfavorable climatic condition (Tauber & Tauber, 1976; Denlinger, 1986). However, the terms quiescence and diapause have also frequently been used as synonyms of dormancy, although both are actually different kinds of dormancy (Tauber & Tauber, 1976; Denlinger, 1986). Basically, both quiescence and diapause are temporary interruptions in the developmental cycle or activity of an organism in response to environmental adversity. Quiescence can be defined as any sudden, non-cyclic response, whereas diapause is a strictly cyclic or rhythmical response, i.e., genetically pre-programmed (Tauber & Tauber, 1976; Denlinger, 1986).

A concise and clear definition of diapause is a gradual and

progressive interruption in development or ontogeny of any organism in some phase of its life cycle in order to survive unfavorable environmental conditions that occur cyclically (Tauber & Tauber, 1976; Denlinger, 1986; Kort, 1990; Tatar & Yin, 2001; Koštál, 2006). Among the major insect orders, including hemimetabolous and holometabolous ones, diapause behavior has been best observed in Orthoptera, Diptera, Coleoptera, Lepidoptera, Hemiptera and Hymenoptera (Tauber & Tauber, 1976; Herman, 1981; Denlinger, 1986; Lefevere et al., 1989; Kort, 1990; Greenfield & Pener, 1992; van Benthem et al., 1995; Kipyatkov et al., 1997; Pick & Blochtein, 2002b). Diapause behavior may be divided into at least three stages: (1) pre-diapause, (2) diapause, and (3) post-diapause, during which insects exhibit a series of behavioral, biochemical and morphological changes (Denlinger, 2002; Koštál, 2006). Furthermore, diapause behavior may occur at any stage of the insect life cycle: egg, larval, pupal (or nymph) or adult (Tauber & Tauber, 1976; Herman, 1981; Denlinger, 1986; Lefevere et al., 1989; Kort, 1990; Greenfield & Pener, 1992; van Benthem et al., 1995; Kipyatkov et al., 1997; Pick & Blochtein, 2002b). When diapause occurs in sexually active insects, breeding is usually compromised due to the interruption of certain physiological processes related to: (1) female oogenesis, (2) the activity of male accessory glands, or (3) the reproductive behavior of both sexes (Kimura, 1988; Tatar & Yin, 2001). Diapause in sexually active insects has, therefore, been termed reproductive diapause, a period during which insects may temporarily cease egg-laying and searching for sexual partners (Kimura, 1988; Tatar & Yin, 2001).

Predicting the arrival of adverse environmental conditions

Since diapause is a rhythmical biological strategy to survive unfavorable environmental conditions, it requires mechanisms that anticipate the arrival and ending of such conditions. Indeed, many insects have been found to effectively decode certain environmental cues in order to enter and/or terminate diapause behavior (Tauber & Tauber, 1976; Denlinger, 1986; Kort, 1990; Tatar & Yin, 2001; Koštál, 2006). Such cues include the amount of food available in the environment, the relative humidity, the parasite load, and the physiological state of host plants (e.g., solute concentration, age, limb senescence) (Tauber & Tauber, 1976; Derr, 1980; Denlinger, 1986; Greenfield & Pener, 1992; Takagi & Miyashita, 2008; Togashi, 2014).

However, the most important environmental cues used by insects to prepare for the arrival or ending of adverse environmental conditions are photoperiodism and temperature (Tauber & Tauber, 1976; Tauber & Kyriacou, 2001; Saunders, 2012, 2014). Photoperiod, i.e., daylength, is a pivotal cue because of its seasonality and invariability (Tauber & Tauber, 1976; Tauber & Kyriacou, 2001; Saunders, 2012, 2014). Temperature is also a significant environmental cue, although it may fluctuate seasonally from year to year (Tauber & Tauber, 1976; Tauber & Kyriacou, 2001; Saunders, 2012, 2014). Nevertheless, both photoperiod and temperature often exert significant synergistic power influencing an insect's response (Kimura, 1988; Vaz Nunes & Saunders, 1989; Chen et al., 2014; Saunders, 2014).

Why study diapause in bees?

Diapause has been observed amongst insects which play diverse ecological roles, e.g., herbivorous agricultural pests (Adedokun & Denlinger, 1985) and pollinators (e.g., van Benthem et al., 1995; Goulson, 2010). These insects may exert social and economic impacts in many regions of the world. However, beneficial insects acting as pollinators currently add high economic value to agriculture globally (€153 billion; Gallai et al., 2009), in addition to their contribution to the reproduction of wild plants (Gallai et al. 2009; Ollerton et al. 2011).

It is known that many bee species undergo diapause at some point in their life cycle, including solitary bee species, such as *Megachile rotundata* (Fabricius), *Nomia melanderi* (Cockerell), *Osmia rufa* (Linnaeus), and *Osmia lignaria* Say (Hsiao & Hsiao, 1969; Bosch et al., 2010; Fliszkiewicz et al., 2012; Wasielewski et al., 2013). Amongst the social bees, honey bees (*Apis mellifera* Linnaeus) show certain behavioral adaptations for surviving low temperatures, e.g., the storage of honey, the decrease or complete absence of broods during autumn and winter, increased longevity, and reduced metabolism of the workers (Seeley, 1985). Bumblebees (*Bombus* Latreille) have queens that survive the winter in diapause, and are certainly the most commonly investigated in this respect (Plowright & Laverty, 1984; Beekman et al., 1998; Goulson, 2010). While there are many aspects of diapause in bees that could be dealt with here, our focus is to review diapause in stingless bees, a subject that has still been little studied.

Stingless bees and the diapause

Stingless bees are found in tropical and subtropical regions all around the world (Sakagami, 1982). They usually nest in holes in trees, in the soil, and even in cavities in human constructions, but some also construct aerial nests (Nogueira-Neto, 1997). Their nests are perennial and they are active throughout the year (Sakagami, 1982; Nogueira-Neto, 1997). However, some species may show quite a remarkable change in activities at the colony as a diapausal response to adverse climatic conditions.

Most descriptions in the literature of diapause in stingless bees have concerned the genus *Plebeia* Schwarz, e.g., *Plebeia droryana* (Friese), *Plebeia emerina* (Friese), *Plebeia wittmanni* Moure & Camargo (Friese), *Plebeia nigriceps* (Friese), *Plebeia remota* (Holmberg), *Plebeia julianii* Moure, and *Plebeia saiqui* (Friese) (Juliani, 1967; Terada et al., 1975; Imperatriz-Fonseca & Oliveira, 1976; Kleinert-Giovannini, 1982; Wittmann, 1989; van Benthem et al., 1995; Pick & Blochtein, 2002b; Ribeiro et al., 2003; Witter et al. 2007; Alves et al., 2009; Nunes-Silva

Table 1. Biological and ecological features presented by stingless bees (Hymenoptera: Apidae: Meliponini) during the reproductive diapause in autumn and/or winter in southern South America. Locality: PR=Paraná, Brazil.

Species	Overwintering strategy	Nest architecture	Worker size [3]	Number of workers per nest	Months in diapause	Maximal worker age (days)	Temperature (°C)	Locality	References
Plebeia droryana	Obligatory	Involucrum	3.5 - 4.7	2,000 – 3,000	3	–	11.7[7]; 9.8[7]	São Paulo Rio Grande do Sul	Terada et al. (1975); Blochtein (pers. obs.)
Plebeia emerina	Obligatory	Involucrum	4.0 - 4.5	–	3	107	11.7[7]; 12.16[8]	São Paulo Rio Grande do Sul	Kleinert-Giovannini, (1982); Santos et al. (2009)
Plebeia julianii	Obligatory	Pillars	3.0	300[4]	3 – 4	–	11.1[7]; 9.8[7]	Paraná Rio Grande do Sul	Juliani (1967); Witter et al. (2007)
Plebeia wittmanni	Obligatory	Pillars	4.5	–	3	274	9.8[7]	Rio Grande do Sul	Wittmann (1989)
Plebeia saiqui	Obligatory	Involucrum	4.0 - 5.0	7000[5]	2.5 – 6	174	11.0[8]	Rio Grande do Sul	Pick and Blochtein (2002 a) Pick and Blochtein (2002 b)
Plebeia nigriceps	Obligatory	Pillars	3.5	200[4]	3 – 4	–	9.8[7]	Rio Grande do Sul	Witter et al. (2007)
Plebeia remota	Obligatory	Pillars	2.75 - 4.0	2,000 – 5,000[6]	2 – 5	–	11.3[8]	São Paulo	van Benthem et al. (1995); Ribeiro et al. (2003); Nunes-Silva et al. (2010)
Melipona obscurior	Facultative	Involucrum	7.1	–	2 – 6	120	9.8[7]	Rio Grande do Sul	Borges & Blochtein (2006); Blochtein et al. (2008)
Melipona quadrifasciata	Facultative	Involucrum	8.6 - 9.6	300 – 400	2 – 3 (?)	–	11.7[7]	São Paulo	Nogueira-Neto (1997)
Melipona bicolor shencki	–[2]	Involucrum	8.9 - 9.1	–	3	–	9.8[7]	Rio Grande do Sul	Blochtein et al. (2008)
Trigona ventralis hoozana [1]	Facultative (?)	Involucrum	–	10,000	–	–	5.0	Chhiayi County, Taiwan	Sung et al. (2008)

(1) East Asian (may reach mountainous regions up to 2,500 m above sea level);

(2) Reproductive diapause not detected, but marked decrease in egg-laying in winter;

(3) Measures (in mm) made in a reference bee from the collection of the Science and Technology Museum of Pontifícia Universidade Católica do Rio Grande do Sul (PUCRS);

(4) Drumond et al. (1998); (5) Witter 2007;(6)van Benthem et al., 1995;

(7) Average minimum temperature (Instituto Nacional de Metereologia [INMET], 2014);

(8) Minimum temperature registered during diapause according to mentioned reference.

et al., 2010). However, diapause has also occasionally been observed or suspected in *Melipona quadrifasciata* Lepeletier, *Melipona obscurior* Moure and *Trigona ventralis hoozana* Strand (Kleinert-Giovannini & Imperatriz-Fonseca, 1986; Nogueira-Neto, 1997; Borges & Blochtein, 2006; Sung et al., 2008) (Table 1). Due to the more frequent occurrence of diapause in *Plebeia* species compared to the other genera of stingless bees, this paper mainly describes the diapause of *Plebeia* species in southern Brazil, with reference to diapause in other stingless bees whenever possible.

In this sense, we know little about whether other stingless bee species in subtropical regions of South America also show any facultative or obligatory reproductive diapause. Many stingless bee species occur in this region, such as *Melipona quinquefasciata* Lepeletier, *Melipona quadrifasciata* Lepeletier, *Mourella caerulea* Friese, *Paratrigona subnuda* Moure, *Scaptotrigona bipunctata* (Lepeletier), *Scaptotrigona depilis* (Moure), *Schwarziana quadripunctata* (Lepeletier), *Tetragonisca fiebrigi* (Schwarz) and *Trigona spinipes* (Fabricius) (Camargo & Pedro, 2013). Not all of these species are managed in these areas due to their aggressiveness, rarity or fragility when kept in hives, making observations of the provisioning and oviposition process (POP) difficult.

The colony reproductive phase

Plebeia spp. occurs from the northernmost parts of Mexico to the southern parts of South America (Camargo & Pedro, 2013). In Brazil, this genus is represented by at least 19 species (Camargo & Pedro, 2013), although many species remain to be identified and described (Silveira et al., 2002; Camargo & Pedro, 2013). Nine of these 19 *Plebeia* species occur in the southern parts of Brazil (Silveira et al., 2002; Camargo & Pedro, 2013), and seven of them are known to undergo diapause (see above). Within the tribe Meliponini, *Plebeia* is considered a phylogenetically basal group among the Neotropical stingless bees (Camargo & Pedro, 1992). A remarkable feature of this group is the aggressive and physically intense queen-worker interactions (Sakagami, 1982; Zucchi, 1993). These interactions occur predominantly during the POP, which is highly ritualized and complex in *Plebeia* spp. (Sakagami, 1982; Zucchi, 1993).

Briefly, the POP can be defined as the construction of brood cells and provisioning of larval food by nurse workers, followed by queen oviposition and the subsequent sealing of the brood cells by nurse workers (Sakagami, 1982; Zucchi, 1993). In *Plebeia*, POP is characterized by a large number of brood cells being built simultaneously until they reach the collar stage, a phase that may last from 3.5 to 7 hours (occasionally less than 30 minutes) (van Benthem et al., 1995; Drumond et al., 1996, 1997, 1998, 2000). During this stage, some workers may lay trophic and/or reproductive eggs on the comb or in the periphery of a brood cell, and such eggs are often eaten by the queen (van Benthem et al., 1995; Drumond et al., 1996, 1997,

1998, 2000). While some brood cells are still in the collar stage, workers on the comb display intense agitation, apparently stimulated by the presence of the queen (van Benthem et al., 1995; Drumond et al., 1996, 1997, 1998, 2000). Subsequently, many workers start to load larval food into the new brood cells, and after a certain volume has been reached, the queen quickly lays her eggs onto the food and the cells are sealed by workers (van Benthem et al., 1995; Drumond et al., 1996, 1997, 2000, 1998).

Hence, in the colony reproductive phase (outside of diapause) there is a clear sequence of brood cell building, provisioning and egg-laying, which differs according to species. Furthermore, there is usually very little, if any, involucrum cover over the brood cell area (van Benthem et al., 1995; Drumond et al., 1996, 1997, 1998, 2000).

It is usually thought that stingless bees have a rather low ability to effectively thermoregulate their nests. In general, these bees build lamellae of wax and propolis, often called cerumen involucra, which partially or entirely cover the brood combs during the cold season (Nogueira-Neto, 1997). Such nest architecture provides thermal insulation that passively protects brood combs and the adult bee population during unfavorable climatic conditions, such as unduly low temperatures (Jones & Oldroyd, 2006). However, such passive thermoregulation in stingless bees is not nearly as effective as the active thermoregulation provided by honey bees, for example, which use contractions of their thoracic muscles to produce heat inside their nests (Jones & Oldroyd, 2006).

The addition of wax and propolis lamellae by stingless bees may be directly related to their need for facultative or obligatory reproductive diapause. It has also been suggested that their reduced ability to effectively thermoregulate their nests and their reproductive diapause may be related to body size (Blochtein et al., 2008). In stingless bee inhabiting southern Brazil, there is a progressive increase in worker size from *Plebeia* spp. (with obligatory diapause), through *M. obscurior* (with facultative diapause) to *Melipona bicolor schencki* Gribodo (with no dia-

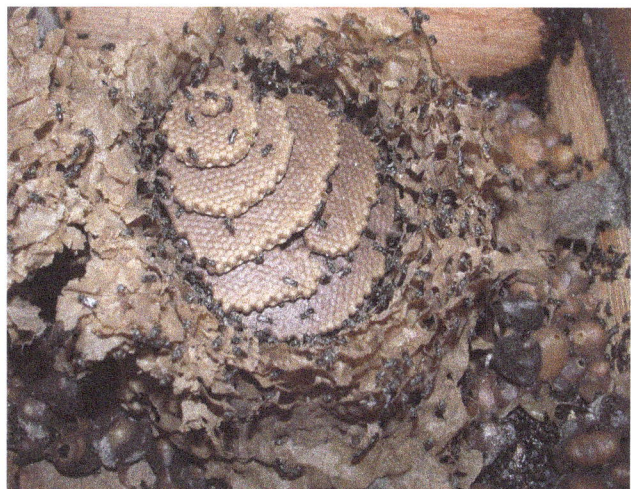

Fig. 1. Colony reproductive phase: Plebeia emerina (Hymenoptera: Apidae: Meliponini) nest showing many workers over brood combs, some brood cells being built in the periphery. Involucrum layers can already be seen around combs.

Fig. 2. Diapause phase: Plebeia emerina (Hymenoptera: Apidae: Meliponini) nest showing brood combs entirely covered by involucrum layers. Below right: some honey pots.

pause) (Blochtein at al., 2008) (Table 1). It is thought that in cold regions, such as southern Brazil, there could be a relationship between bee body size and their ability to withstand low temperatures (Blochtein et al., 2008).

The diapause phase

Many populations of *Plebeia* spp. (*P. droryana, P. emerina, P. julianii, P. nigriceps, P. remota, P. saiqui* and *P. wittmanni*) in southern South America (Table 1) may be exposed to quite rigorous winters that may reach temperatures at or below zero in June to July (Instituto Nacional de Metereologia [INMET], 2014). In preparation for these cold winters, the construction rate of brood cells falls gradually until it stops entirely. Consequently, the queen's egg-laying is interrupted in many *Plebeia* colonies during the milder autumn months of March to May (Juliani, 1967; Terada et al., 1975; Imperatriz-Fonseca & Oliveira, 1976; Kleinert-Giovannini, 1982; Wittmann, 1989; van Benthem et al., 1995; Pick & Blochtein,

2002b; Ribeiro et al., 2003; Alves et al., 2009; Nunes-Silva et al., 2010). This kind of diapause in stingless bees has been called reproductive diapause (Juliani, 1967; Terada et al., 1975; van Benthem et al., 1995; Pick & Blochtein, 2002b; Ribeiro et al., 2003; Alves et al., 2009; Nunes-Silva et al., 2010).

One of the more conspicuous signs that a given colony of *Plebeia* spp. is preparing for reproductive diapause is the progressive construction of a multi-layered involucrum covering the brood comb area, as observed in *P. droryana* and *P. emerina* (Drumond et al., 1996) and in *P. remota* (Ribeiro et al., 2003) (Figs 1 and 2). In other cases, a substantial increase in the number of cerumen pillars may occur, also over the brood combs, as in *P. remota* (Ribeiro et al., 2003) (Fig 3a, b; Table 1). At the peak of reproductive diapause, when all of the brood has already emerged ("winter workers"), brood cells or combs are no longer constructed and no combs are found in the nest (Ribeiro et al., 2003). Many workers and the queen can be seen clustered in specific regions of the nest, under or over food pots or among the pillars (or lamellae), where they exhibit lethargic movements or remain completely immobile (Ribeiro et al., 2003). These "winter workers" are longer lived than the workers emerging at other times (Table 1) (Blochtein et al., 2008; Nunes-Silva et al., 2010).

During reproductive diapause, queens of *P. remota* may lose weight, as evidenced by their reduced physogastry (Ribeiro et al., 2003). This lower queen physogastry observed during diapause has been ascribed to a regression in ovarian development due to the cessation of egg-laying (Ribeiro et al., 2003). However, little is known about this phenomenon in stingless bees, although it is possible that glycogen and body fat reserves also significantly decrease during this period. Reproductive diapause in stingless bees has been recorded as lasting from 2 up to 6 months (Juliani, 1967; van Benthem et al., 1995; Pick & Blochtein, 2002b; Ribeiro et al., 2003; Alves et al., 2009; Nunes-Silva et al., 2010) (Table 1). However, colonies inhabiting the same locality may not always commence their reproductive diapause in synchrony (Juliani, 1967; van Ben-

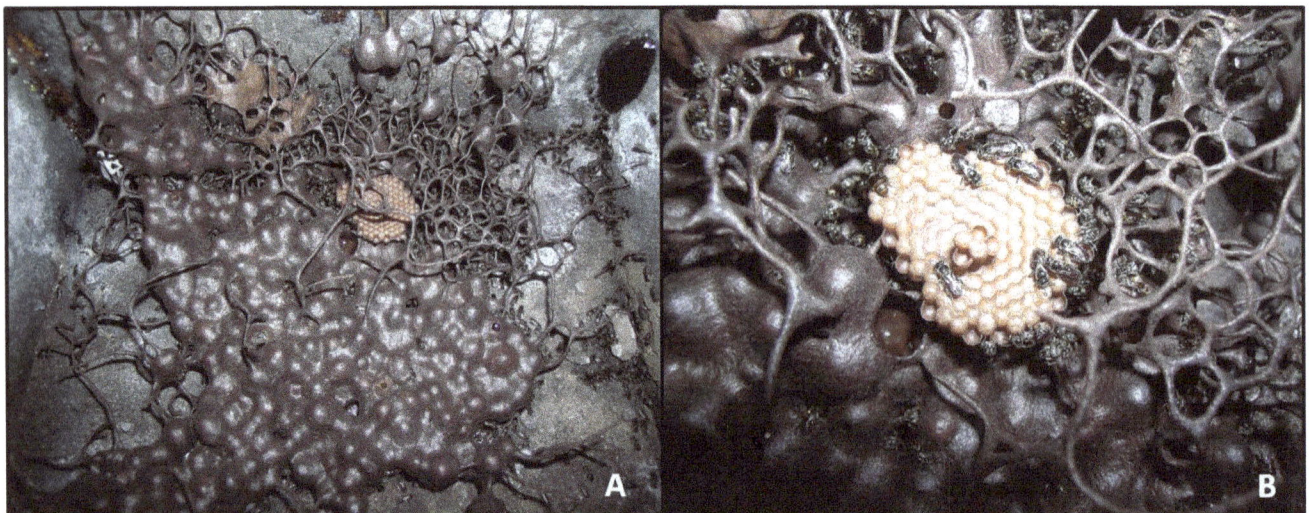

Fig. 3. Post-diapause phase: Plebeia remota (Hymenoptera: Apidae: Meliponini) nest shortly after terminating diapause. A - Note that there is a large number of food pots, several wax pillars and an incipient brood comb being built by some workers; B – Detail from A with some removed pillars.

Table 2. Overall behavioral characteristics concerning to the reproductive diapause process in stingless bees (Hymenoptera: Apidae: Meliponini). Note: diapause phases following Koštál (2006).

Pre-diapause		Diapause		Post-diapause
Induction	**Preparation**	**Initiation/maintenance**	**Termination**	
Environmental stimuli (e.g. photoperiod and temperature) achieve critical levels according to physiological sensitivity from bees.	Progressive reduction in brood cell building with a decrease in provisioning and oviposition process;	No brood cell actively built;	Some brood cells start to be built; eventually this can occur over food pots or cerumen layers.	Full removal or significant reduction in involucrum layers or pillars;
	Increased presence of cerumen involucrum layers or pillars over brood cells;	Emergence of the last brood, winter workers.		Usual provisioning and oviposition process (POP) is implemented.
	Shift in foraging pattern (1): larger input nectar and less pollen;			
	Shift in foraging pattern (2): flight activity concentrated around noon.			
	Autumn / Winter		**Spring / Summer**	

them et al., 1995; Pick & Blochtein, 2002b; Ribeiro et al., 2003; Alves et al., 2009; Nunes-Silva et al., 2010).

While some colonies may start diapause early, during the milder months (March, April), others may not start until the winter months (May, June) (Juliani, 1967; van Benthem et al., 1995; Pick & Blochtein, 2002b; Ribeiro et al., 2003; Alves et al., 2009; Nunes-Silva et al., 2010). In contrast to this variability in preparation for diapause, which occurs over a prolonged period and independently among colonies at the same site, termination of diapause among *Plebeia* colonies seems to be faster and synchronous during the first weeks of spring (in August, September or eventually in October) (Juliani, 1967; van Benthem et al., 1995; Pick & Blochtein, 2002b; Ribeiro et al., 2003; Alves et al., 2009; Nunes-Silva et al., 2010).

Despite the arrest of POP throughout reproductive diapause in stingless bees, the nest population is not entirely lethargic or immobile. Other tasks, e.g., manipulating waste, storing food (nectar, pollen), or foraging still occur (Juliani, 1967; van Benthem et al., 1995; Pick & Blochtein, 2002b; Ribeiro et al., 2003; Nunes-Silva et al., 2010). Regarding foraging, the activity of workers outside of the nest is more intense when ambient temperatures are milder and the wind is light or absent (Pick & Blochtein, 2002b; Ribeiro et al., 2003; Nunes-Silva et al., 2010).

The foraging pattern and food sources harvested by foragers also change from the reproductive phase to reproductive diapause, at least in *P. saiqui* (Pick & Blochtein, 2002a, 2002b) and *P. remota* (Nunes-Silva et al., 2010). During the reproductive phase, *P. remota* forage constantly throughout the day and both nectar and pollen are collected (Nunes-Silva et al., 2010). However, during diapause foraging does not commence until noon, when temperatures are higher, and the collection of pollen is greatly reduced compared to the reproductive phase (Nunes-Silva et al., 2010). A decrease in pollen

collection and a shift in the time of peak foraging activity between phases has also been observed for *P. saiqui* (reproductive period: peak 11:00-13:00; diapause: peak 13:00-14:00) (Pick & Blochtein, 2002a, 2002b).

It has therefore been suggested that, at least for *P. remota*, the higher nectar input during diapause affects the potential energy storage by bees, whereas reduced collection of pollen (a protein source) causes no harm because there is no egg-laying or production of new individuals during this period (Nunes-Silva et al., 2010). Nevertheless, comparative observations should also be made for other *Plebeia* and stingless bee species that show diapause. An overview of the process of reproductive diapause in stingless bees is presented in Table 2.

Factors determining diapause in stingless bees

Our knowledge regarding which environmental cues (e.g., photoperiod and temperature) are indeed significant in determining the onset, permanence and termination of diapause in stingless bees is almost absent. These social insects live inside cavities in trees or dead wood, and also nest in soil or other cavities, so how could the population in the nest, especially nurse workers and the queen, foresee the onset of adverse climatic conditions? Furthermore, how could they (or foragers) communicate and/or induce reproductive diapause in their nestmates? Unfortunately, there are few studies addressing these questions. For example, an experiment with *P. remota* showed that workers have a significant role in promoting diapause behavior (Ribeiro, 2002). By introducing *P. remota* queens from colonies in diapause into colonies that were in the reproductive phase (and vice versa), it was possible to observe that diapausal queens did not reinitiate egg-laying (Ribeiro, 2002). Moreover, it remains unknown just how individuals that have never left their nests, and thus have not been exposed to external environmental cues, are induced to enter, continue and terminate diapause.

Studies with other social insects might provide some clues. For example, it has been suggested that foraging honeybees may somehow synchronize their biological rhythms and inform nestmates about external environmental conditions through their body movements (Bloch et al., 2013). Another possibility could be some sort of chemical signaling that would induce entire nest populations to enter or terminate diapause. Evidence for this has been obtained from *Myrmica rubra* (Linnaeus) ants in which larvae remain in diapause and do not pupate, while simultaneously queens do not lay eggs (Kipyatkov, 2001). In this study, it was observed that when diapausal nests were interconnected to non-diapausal nests, and odors from the latter could enter the former, diapause was terminated in the diapausal colonies, most probably due to worker primer pheromones. However, the reverse effect was not observed (Kipyatkov, 2001). Unfortunately we do not yet know if this system could also apply in stingless bees, even though chemical communication is known to also play a pivotal role within their colonies.

Conclusions and perspectives

Possibly because nest thermoregulation in stingless bees is somewhat passive and not highly efficient, species exposed to winters that are more rigorous in southern Brazil have apparently evolved reproductive diapause behavior as a strategy for overwintering. This strategy interrupts egg laying for up to 6 months, even though this is a critical activity promoting the growth and maintenance of their nests. Stingless bees are subject to several environmental stresses (Freitas et al., 2009).

Therefore, it is reasonable to assume that stingless bee colonies could be more susceptible to strong environmental pressures during diapause, due to the lack of population resilience. Furthermore, very little is known about how climate change may affect the distribution of stingless bees which show diapause behavior. The future of stingless bee populations which show reproductive diapause will depend on how the scientific community, policy makers and society act to effectively preserve the natural populations of these pollinators.

In conclusion, we have much to learn about reproductive diapause in stingless bee species. The use of a combination of methods involving molecular, biochemical, physiological and ecological techniques could allow new investigations on the mechanisms underlying the regulation of diapause in stingless bees. We hope that this review will stimulate further studies that address these questions and that it will lead to better conservaion practices to preserve these bee species.

Acknowledgements

We thank Conselho Nacional do Desenvolvimento Científico e Tecnológico (CNPq) [Process number, 500458/2013-8], Fundação de Amparo à Pesquisa do Estado do Rio Grande do Sul (FAPERGS) and Coordenação de Aperfeiçoamento de Pessoal de Nível Superior (CAPES) for the scholarships (CFS, PNS and RH).

References

Adedokun, T.A. & Denlinger, D.L. (1985). Metabolic reserves associated with pupal diapause in the flesh fly, *Sarcophaga crassipalpis*. J. Insect Physiol. 31: 229-233.

Alves, D.A., Imperatriz-Fonseca, V.L. & Santos-Filho, P.S. (2009). Production of workers, queens and males in *Plebeia remota* colonies (Hymenoptera, Apidae, Meliponini), a stingless bee with reproductive diapause. Genet. Mol. Res. 8: 672-683.

Beekman, M., van Stratum, P. & Lingeman, R. (1998). Diapause survival and post-diapause performance in bumblebee queens (*Bombus terrestris*). Entomol. Exp. Appl. 89: 207-214.

Bloch, G., Herzog, E.D., Levine, J. D. & Schwartz, W. J. (2013). Socially synchronized circadian oscillators. P. Roy. Soc. Lond. B Bio. 80: 20130035. doi: 10.1098/rspb.2013.0035.

Blochtein, B., Witter, S. & Imperatriz-Fonseca, V.L. (2008). Adaptações das abelhas sem ferrão ao clima temperado do sul do Brasil. In D. de Jong, T.M. Francoy, & W.C. Santana (Eds.), Anais do VIII Encontro sobre Abelhas (pp. 121-130). Ribeirão Preto: FUNPEC.

Borges, F.V.B. & Blochtein, B. (2006). Variação sazonal das condições internas de colônias de *Melipona marginata obscurior* Moure, no Rio Grande do Sul, Brasil. Rev. Bras. Zool. 23: 711-715.

Bosch, J., Sgolastra, F. & Kemp, W.P. (2010). Timing of eclosion affects diapause development, fat body consumption and longevity in *Osmia lignaria*, a univoltine, adult-wintering solitary bee. J. Insect Physiol. 56: 1949-1957. doi: 10.1016/j.jinsphys.2010.08.017.

Camargo, J.M.F. & Pedro, S.R.M. (1992). Systematics, phylogeny and biogeography of the Meliponinae (Hymenoptera, Apidae). Apidologie 23: 509-522.

Camargo, J.M.F. & Pedro, S.R.M. (2013). Meliponini Lepeletier, 1836. In J.S. Moure, D. Urban & G.A.R. Melo (Eds.), Catalogue of Bees (Hymenoptera, Apoidea) in the Neotropical Region - online version. http://www.moure.cria.org.br/catalogue. (accessed date: 15 september, 2014)

Chen, C., Xia, Q.-W., Fu, S., Wu, X.-F. & Xue, F.-S. (2014). Effect of photoperiod and temperature on the intensity of pupal diapause in the cotton bollworm, *Helicoverpa armigera* (Lepidoptera: Noctuidae). B. Entomol. Res. 104: 12-18. doi:

10.1017/S0007485313000266.

Denlinger, D.L. (1986). Dormancy in tropical insects. Annu. Rev. Entomol. 31: 239-264. doi: 10.1146/annurev.en.31.010186.001323.

Denlinger, D.L. (2002). Regulation of diapause. Annu. Rev. Entomol. 47: 93-122. doi: 10.1146/annurev.ento.47.091201.145137.

Derr, J.A. (1980). Coevolution of the life history of a tropical seed-feeding insect and its food plants. Ecology 61: 881-892. doi: 10.2307/1936758.

Drumond, P.M., Bego, L.R. & Zucchi, R. (1997). Oviposition behavior of the stingless bees XIX. *Plebeia* (*Plebeia*) *poecilochroa* with highly integrated oviposition process and small colony size (Hymenoptera, Apidae, Meliponinae). Jpn. J. Entomol. 65: 7-22.

Drumond, P.M., Zucchi, R., Mateus, S. & Bego, L. R. (1996). Oviposition behavior of the stingless bees XVII. *Plebeia* (*Plebeia*) *droryana* and a ethological comparison with other Meliponinae taxa (Hymenoptera, Apidae). Jpn. J. Entomol. 64: 385-400.

Drumond, P.M., Zucchi, R. & Oldroyd, B.P. (2000). Description of the cell provisioning and oviposition process of seven species of *Plebeia* Schwarz (Apidae, Meliponini), with notes on their phylogeny and taxonomy. Insectes Soc. 47: 99-112. doi: 10.1007/PL00001703.

Drumond, P.M., Zucchi, R., Yamase, S. & Sakagami, S.F. (1996). Oviposition behavior of the stingless bees XVIII. *Plebeia* (*Plebeia*) *emerina* and *P.* (*P.*) *remota* with a preliminary ethological comparison of some *Plebeia* taxa (Apidae, Meliponinae). B. Fac. Educ. 45: 31-55.

Drumond, P.M., Zucchi, R., Yamase, S. & Sakagami, S.F. (1998). Oviposition behavior of the stingless bees XX. *Plebeia* (*Plebeia*) *julianii* with forms very small brood batches (Hymenoptera, Apidae, Meliponinae). Entomol. Sci. 1: 195-205.

Fliszkiewicz, M., Giejdasz, K., Wasielewski, O. & Krishnan, N. (2012). Influence of winter temperature and simulated climate change on body mass and fat body depletion during diapause in adults of the solitary bee, *Osmia rufa* (Hymenoptera: Hymenoptera). Environ. Entomol. 41: 1621-1630. doi: 10.1603/EN12004.

Freitas, B.M., Imperatriz-Fonseca, V.L., Medina, L.M., Kleinert, A.M.P., Galetto, L., Nates-Parra, G. & Quezada-Euán, J.J.G. (2009). Diversity, threats and conservation of native bees in the Neotropics. Apidologie 40: 332-346. doi: 10.1051/apido/2009012

Gallai, N., Salles, J.-M., Settele, J. & Vaissière, B. E. (2009). Economic valuation of the vulnerability of world agriculture confronted with pollinator decline. Ecol. Econ. 68: 810-821. doi: 10.1016/j.ecolecon.2008.06.014

Goulson, D. (2010). Bumblebees: behavior, ecology, and conservation. (2nd Ed., p. 317). New York: Oxford University Press.

Greenfield, M.D. & Pener, M.P. (1992). Alternative schedules of male reproductive diapause in the grasshopper *Anacridium aegyptium* (L.): effects of the corpora allata on sexual behavior (Orthoptera: Acrididae). J. Insect Behav. 5: 245-261.

Herman, W.S. (1981). Studies on the adult reproductive diapause of the monarch butterfly, *Danaus perxippus*. Biol. Bull. 160: 89-106.

Hsiao, C. & Hsiao, T.H. (1969). Insect hormones: their effects on diapause and development of Hymenoptera. Life Sci. 8: 767-774.

Imperatriz-Fonseca, V.L. & Oliveira, M.A.C. (1976). Observations on a queenless colony of *Plebeia saiqui* (Friese) (Hymenoptera, Apidae, Meliponinae). Bol. Zool. USP 1: 299-312.

Instituto Nacional de Metereologia [INMET]. (2014). http://www.inmet.gov.br/portal/index.php?r=home2/index (accessed date: 18 September, 2014)

Jones, J.C. & Oldroyd, B.P. (2006). Nest thermoregulation in social insects. Adv. Insect Physiol. 33: 143-191. doi: 10.1016/S0065-2806(06)33003-2

Juliani, L.A. (1967). Descrição do ninho e alguns dados biológicos sobre a abelha *Plebeia julianii* Moure, 1962 (Hymenoptera, Apidae, Meliponinae). Rev. Bras. Entomol. 12: 31-58.

Kimura, M.T. (1988). Interspecific and geographic variation of diapause intensity and seasonal adaptation in the *Drosophila auraria* species complex (Diptera: Drosophilidae). Func. Ecol. 2: 177-183.

Kipyatkov, V.E. (2001). A distantly perceived primer pheromone controls diapause termination in the ant *Myrmica rubra* L. (Hymenoptera, Formicidae). J. Evol. Biochem. Phys. 37: 405-416. doi: 10.1023/A:1012926929059.

Kipyatkov, V.E., Lopatina, E.B. & Pinegin, A. Y. (1997). Social regulation of development and diapause in the ant *Leptothorax acervorum* (Hymenoptera, Formicidae). Entomol. Rev. 77: 248-255.

Kleinert-Giovannini, A. (1982). The influence of climatic factors on flight activity of *Plebeia emerina* Friese (Hymenoptera, Apidae, Meliponinae) in winter. Rev. Bras. Entomol. 26: 1-13.

Kleinert-Giovannini, A. & Imperatriz-Fonseca, V L. (1986). Flight activity and climatic conditions and responses to climatic conditions by two subspecies of *Melipona marginata* Lepeletier (Apidae, Meliponinae). J. Apicult. Res. 25: 3-8.

Kort, C.A.D. (1990). Thirty-five years of diapause research with the Colorado potato beetle. Entomol. Exp. Appl. 56: 1-13.

Koštál, V. (2006). Eco-physiological phases of insect diapause. J. Insect Physiol. 52: 113-127. doi: 10.1016/j.jinsphys.2005.09.008

Lefevere, K.S., Koopmanschap, A.B. & de Kort, C.A.D. (1989). Juvenile hormone metabolism during and after diapause in the female Colorado potato beetle, *Leptinotarsa decemlineata*. J. Insect Physiol. 35: 129-135.

Nogueira-Neto, P. (1997). Vida e criação de abelhas indígenas sem ferrão. São Paulo: Editora Nogueirapis, 445 p.

Nunes-Silva, P., Hilário, S.D., Santos Filho, P.S. & Imperatriz-Fonseca, V.L. (2010). Foraging activity in *Plebeia remota*, a stingless bees species, is influenced by the reproductive state of a colony. Psyche, (i): 1-16. doi: 10.1155/2010/241204

Ollerton, J., Winfree, R. & Tarrant, S. (2011). How many flowering plants are pollinated by animals? Oikos 120: 321-326. doi: 10.1111/j.1600-0706.2010.18644.x.

Pick, R.A. & Blochtein, B. (2002). Atividades de coleta e origem floral do polen armazenado em colonias de *Plebeia saiqui* (Holmberg) (Hymenoptera, Apidae, Meliponinae) no sul do Brasil. Rev. Bras. Zool. 19: 289-300.

Pick, R.A. & Blochtein, B. (2002). Atividades de vôo de *Plebeia saiqui* (Holmberg) (Hymenoptera, Apidae, Meliponini) durante o período de postura da rainha e em diapausa. Rev. Bras. Zool. 19: 827–839.

Plowright, R.C. & Laverty, T.M. (1984). The ecology and sociobiology of bumble bees. Annu. Rev. Entomol. 29: 175-199.

Ribeiro, M.F. (2002). Does the queen of Plebeia remota (Hymenoptera, Apidae, Meliponini) stimulate her workers to start brood cell construction after winter? Insectes Soc. 49: 38-40. doi: 10.1007/s00040002-8276-0.

Ribeiro, M.F., Imperatriz-Fonseca, V.L. & Santos-Filho, P.S. (2003). A interrupção da construção de células de cria e postura em *Plebeia remota* (Holmberg) (Hymenoptera, Apidae, Meliponini). In G.A.R. Melo & I. Alves-dos-Santos (Eds.), Apoidea Neotropica: homenagem aos 90 anos de Jesus Santiago Moure (pp. 177-188). Criciúma: Editora UNESC.

Sakagami, S.F. (1982). Stingless bees. In H.R. Hermann (Ed.), Social Insects (pp. 361-423). New York: New York Academic Press.

Saunders, D.S. (2012). Insect photoperiodism: seeing the light. Physiol. Entomol. 37: 207-218. doi: 10.1111/j.1365-3032.2012.00837.x

Saunders, D.S. (2014). Insect photoperiodism: effects of temperature on the induction of insect diapause and diverse roles for the circadian system in the photoperiodic response. Entomol. Sci. 17: 25-40. doi: 10.1111/ens.12059

Seeley, T.D. (1985). Honeybee ecology: a study of adaptation in social life. New Jersey: Princeton University Press.

Silveira, F.A., Melo, G.A.R. & Almeida, E.A.B. (2002). Abelhas brasileiras: sistemática e identificação. Belo Horizonte: Fundação Araucária.

Sung, I.-H., Yamane, S. & Hozumi, S. (2008). Thermal characteristics of nests of the Taiwanese stingless bee *Trigona ventralis hoozana* (Hymenoptera: Apidae). Zool. Stud. 47: 417-428.

Takagi, S. & Miyashita, T. (2008). Host plant quality influences diapause induction of *Byasa alcinous* (Lepidoptera: Papilionidae). Ann. Entomol. Soc. Am. 101: 392-396.

Tatar, M. & Yin, C.M. (2001). Slow aging during insect reproductive diapause: why butterflies, grasshoppers and flies are like worms. Exp. Gerontol. 36: 723-738.

Tauber, E. & Kyriacou, B.P. (2001). Insect photoperiodism and circadian clocks: models and mechanisms. J. Biol. Rhythm. 16: 381-390. doi: 10.1177/074873001129002088.

Tauber, M.J. & Tauber, C.A. (1976). Insect seasonality: diapause maintenance, termination, and postdiapause development. Annu. Rev. Entomol. 21: 81-107. doi: 10.1146/annurev.en.21.010176.000501

Terada, Y., Garófalo, C.A. & Sakagami, S.F. (1975). Age-survival curves for workers of two eusocial bees (*Apis mellifera* and *Plebeia droryana*) in a subtropical climate, with notes on worker polyethism in *P. droryana*. J. Apicult. Res. 14: 161-170.

Togashi, K. (2014). Effects of larval food shortage on diapause induction and adult traits in Taiwanese *Monochamus alternatus alternatus*. Entomol. Exp. Appl. 151: 34-42. doi: 10.1111/eea.12165

van Benthem, F.D.J., Imperatriz-Fonseca, V.L. & Velthuis, H.H.W. (1995). Biology of the stingless bee *Plebeia remota* (Holmberg): observations and evolutionary implications. Insect. Soc. 87: 71-87. doi: 10.1007/BF01245700

Vaz Nunes, M. & Saunders, D.S. (1989). The effect of larval temperature and photoperiod on the incidence of larval diapause in the blowfly, *Calliphora vicina*. Physiol. Entomol. 14: 471-474. doi: 10.1111/j.1365-3032.1989.tb01116.x.

Wasielewski, O., Wojciechowicz, T., Giejdasz, K. & Krishnan, N. (2013). Overwintering strategies in the red mason solitary bee - physiological correlates of midgut metabolic activity and turnover of nutrient reserves in females of *Osmia bicornis*. Apidologie 44: 642–656. doi: 10.1007/s13592-013-0213-x.

Witter, S., Blochtein, B., Andrade, F., Wolff, L.F. & Imperatriz-Fonseca, V.L. (2007). Meliponicultura no Rio Grande do Sul: contribuição sobre a biologia e conservação de *Plebeia nigriceps* (Friese 1901) (Apidae, Meliponini). Bioscience J. 23: 134-140.

Wittmann, D. (1989). Nest architecture, nest site preferences and distribution of *Plebeia wittmanni* (Moure & Camargo, 1989) in Rio Grande do Sul, Brazil (Apidae: Meliponinae). Stud. Neotrop. Fauna Environ. 24: 17-23. doi: 10.1080/01650528909360771.

Zucchi, R. (1993). Ritualized dominance, evolution of queen-worker interactions and related aspects in stingless bees (Hymenoptera: Apidae). In T. Inoue & S. Yamane (Eds.), Evolution of insect societies (pp. 207-249). Tokyo: Hakuhinsha.

This article has online supplementary material at:
http://periodicos.uefs.br/ojs/index.php/sociobiology/rt/suppFiles/641/0

DOI: 10.13102/sociobiology.v61i4..s556

Permissions

The contributors of this book come from diverse backgrounds, making this book a truly international effort. This book will bring forth new frontiers with its revolutionizing research information and detailed analysis of the nascent developments around the world.

We would like to thank all the contributing authors for lending their expertise to make the book truly unique. They have played a crucial role in the development of this book. Without their invaluable contributions this book wouldn't have been possible. They have made vital efforts to compile up to date information on the varied aspects of this subject to make this book a valuable addition to the collection of many professionals and students.

This book was conceptualized with the vision of imparting up-to-date information and advanced data in this field. To ensure the same, a matchless editorial board was set up. Every individual on the board went through rigorous rounds of assessment to prove their worth. After which they invested a large part of their time researching and compiling the most relevant data for our readers.

The editorial board has been involved in producing this book since its inception. They have spent rigorous hours researching and exploring the diverse topics which have resulted in the successful publishing of this book. They have passed on their knowledge of decades through this book. To expedite this challenging task, the publisher supported the team at every step. A small team of assistant editors was also appointed to further simplify the editing procedure and attain best results for the readers.

Apart from the editorial board, the designing team has also invested a significant amount of their time in understanding the subject and creating the most relevant covers. They scrutinized every image to scout for the most suitable representation of the subject and create an appropriate cover for the book.

The publishing team has been an ardent support to the editorial, designing and production team. Their endless efforts to recruit the best for this project, has resulted in the accomplishment of this book. They are a veteran in the field of academics and their pool of knowledge is as vast as their experience in printing. Their expertise and guidance has proved useful at every step. Their uncompromising quality standards have made this book an exceptional effort. Their encouragement from time to time has been an inspiration for everyone.

The publisher and the editorial board hope that this book will prove to be a valuable piece of knowledge for researchers, students, practitioners and scholars across the globe.

List of Contributors

CF Santos
Laboratório de abelhas, Instituto de Biociências, Universidade de São Paulo, São Paulo, SP, Brazil
Laboratório de Entomologia, Pontifícia Universidade Católica do Rio Grande do Sul, Porto Alegre, RS, Brazil

C Menezes
Embrapa Amazônia Oriental, Empresa Brasileira de Pesquisa Agropecuária, Belém, PA, Brazil
Faculdade de Filosofia, Ciências e Letras de Ribeirão Preto, Universidade de São Paulo, Ribeirão Preto, SP, Brazil

A Vollet-Neto
Embrapa Amazônia Oriental, Empresa Brasileira de Pesquisa Agropecuária, Belém, PA, Brazil

VL Imperatriz-Fonseca
Laboratório de abelhas, Instituto de Biociências, Universidade de São Paulo, São Paulo, SP, Brazil
Universidade Federal Rural do Semi-Árido, Mossoró, RN, Brazil

DE Oliveira
Universidade de Brasília, Brasília – DF, Brazil

TF Carrijo
Universidade de São Paulo, Ribeirão Preto – SP, Brazil

D Brandão
Universidade Federal de Goiás, Goiânia – GO, Brazil

L Chavarría
Universidade de São Paulo, Ribeirão Preto, São Paulo, Brazil

FB Noll
Universidade de São Paulo, Ribeirão Preto, São Paulo, Brazil
Universidade Estadual Paulista Júlio de Mesquita Filho, São José do Rio Preto, São Paulo, Brazil

FC Bueno
Universidade Estadual Paulista, Campus de Rio Claro, SP, Brazil

LC Forti
Universidade Estadual Paulista, Campus de Botucatu, SP, Brazil

OC Bueno
Universidade Estadual Paulista, Campus de Rio Claro, SP, Brazil

BC Barbosa
Laboratório de Ecologia Comportamental e Bioacústica (LABEC), Universidade Federal de Juiz de Fora, Juiz de Fora, MG, Brazil

MF Paschoalini
Laboratório de Ecologia Comportamental e Bioacústica (LABEC), Universidade Federal de Juiz de Fora, Juiz de Fora, MG, Brazil

F Prezoto
Laboratório de Ecologia Comportamental e Bioacústica (LABEC), Universidade Federal de Juiz de Fora, Juiz de Fora, MG, Brazil

WM Aguiar
Universidade Estadual de Feira de Santana, Departamento de Ciências Exatas, Feira de Santana, BA, Brazil

GAR Melo
Universidade Federal do Paraná, Departamento de Zoologia, Curitiba, PR, Brazil

MC Gaglianone
Universidade Estadual do Norte Fluminense Laboratório de Ciências Ambientais, Campos dos Goytacazes, RJ, Brazil

Gabriela Castaño-Meneses
Ecología y Sistemática de Microartrópodos, Depto de Ecología y Recursos Naturales, Facultad de Ciencias, Universidad Nacional Autónoma de México, México, DF
Unidad Multidisciplinaria de Docencia e Investigación, Facultad de Ciencias, Universidad Nacional Autónoma de México, Campus Juriquilla, Querétaro, México

BK Gautam
Department of Entomology, Louisiana State University, Baton Rouge, LA, USA

G Henderson
Department of Entomology, Louisiana State University, Baton Rouge, LA, USA

AL Oliveira Nascimento
Universidade de São Paulo (FFCLRP-USP), Ribeirão Preto, SP – Brazil

CA Garófalo
Universidade de São Paulo (FFCLRP-USP), Ribeirão Preto, SP – Brazil

GA Locher
Instituto de Biociências, Universidade Estadual Paulista, Rio Claro, SP, Brazil

OC Togni
Instituto de Biociências, Universidade Estadual Paulista, Rio Claro, SP, Brazil

OT Silveira
Museu Paraense Emílio Goeldi, Belém, PA, Brazil

E Giannotti
Instituto de Biociências, Universidade Estadual Paulista, Rio Claro, SP, Brazil

ZL Chen
College of Life Sciences, Guangxi Normal University, Guilin, China

SY Zhou
College of Life Sciences, Guangxi Normal University, Guilin, China

DD Ye
College of Life Sciences, Guangxi Normal University, Guilin, China

Y Chen
College of Life Sciences, Guangxi Normal University, Guilin, China

CW Lu
College of Life Sciences, Guangxi Normal University, Guilin, China

JC Pereira
Universidade Federal do Ceará, Fortaleza, Ceará, Brazil

JHC Delabie
Universidade Estadual de Santa Cruz, Ilhéus-Itabuna, Bahia, Brazil
CEPLAC/CEPEC Centro de Pesquisas do Cacau, Itabuna, Bahia, Brazil

LRS Zanette
Universidade Federal do Ceará, Fortaleza, Ceará, Brazil

Y Quinet
Universidade Federal do Ceará, Fortaleza, Ceará, Brazil
Universidade Estadual do Ceará, Fortaleza, Ceará, Brazil

MA Nakano
Universidade de Mogi das Cruzes, Mogi das Cruzes, SP, Brazil

VFO Miranda
Universidade Estadual Paulista, Jaboticabal, SP, Brazil

RM Feitosa
Universidade Federal do Paraná, Curitiba, PR, Brazil

MSC Morini
Universidade de Mogi das Cruzes, Mogi das Cruzes, SP, Brazil

RH Scheffrahn
University of Florida, IFAS, Davie, FL, Unites States of America

LCB Martins
Universidade Federal de Viçosa, Viçosa, MG Brazil

JHC Delabie
CEPLAC/CEPEC, Itabuna, BA, Brazil
Universidade Estadual de Santa Cruz. Ilhéus, BA, Brazil

JC Zanuncio
Universidade Federal de Viçosa, Viçosa, MG Brazil

JE Serrão
Universidade Federal de Viçosa, Viçosa, MG Brazil

VO Torres
Universidade Federal da Grande Dourados, Dourados-MS, Brazil

D Sguarizi-Antonio
Universidade Estadual de Mato Grosso do Sul, Dourados-MS, Brazil

SM Lima
Universidade Federal da Grande Dourados, Dourados-MS, Brazil

LHC Andrade
Universidade Federal da Grande Dourados, Dourados-MS, Brazil

WF Antonialli-Junior
Universidade Federal da Grande Dourados, Dourados-MS, Brazil

LG Santos
Universidade Federal de Viçosa (UFV), Viçosa, Minas Gerais, Brazil

MLTMF Alves
Agência Paulista de Tecnologia dos Agronegócios, Pindamonhangaba, São Paulo. Brazil

D Message
Universidade Federal Rural do Semiárido. (PVNS/CAPES), Mossoró/RN, Brazil

FA Pinto
Universidade Federal de Viçosa (UFV), Viçosa, Minas Gerais, Brazil

MVGB Silva
EMBRAPA Gado de Leite, Juiz de Fora, MG, Brazil

EW Teixeira
Agência Paulista de Tecnologia dos Agronegócios, Pindamonhangaba, São Paulo. Brazil

AM Rákóczi
Centre for Agricultural Research, Hungarian Academy of Sciences, Budapest, Hungary

F Samu
Centre for Agricultural Research, Hungarian Academy of Sciences, Budapest, Hungary

JMS Freitas
Universidade Estadual de Santa Cruz, Ilhéus-BA, Brazil
Laboratório de Mirmecologia, CEPLAC/CEPEC/SECEN, Ilhéus-BA, Brazil
Universidade Estadual do Sudoeste da Bahia, Itapetinga-BA, Brazil

JHC Delabie
Universidade Estadual de Santa Cruz, Ilhéus-BA, Brazil
Laboratório de Mirmecologia, CEPLAC/CEPEC/SECEN, Ilhéus-BA, Brazil

S Lacau
Universidade Estadual de Santa Cruz, Ilhéus-BA, Brazil
Laboratório de Mirmecologia, CEPLAC/CEPEC/SECEN, Ilhéus-BA, Brazil
Universidade Estadual do Sudoeste da Bahia, Itapetinga-BA, Brazil

MF Ribeiro
Universidade de São Paulo, São Paulo, SP, Brazil
Empresa Brasileira de Pesquisa Agropecuária (EMBRAPA), Petrolina, PE, Brazil

PS Santos Filho
Universidade de São Paulo, São Paulo, SP, Brazil

SA Souza
Universidade Federal Rural de Pernambuco, Recife, PE, Brazil

TLMF Dias
Universidade Federal de Alagoas, Maceió, AL, Brazil

TMG Silva
Universidade Federal Rural de Pernambuco, Recife, PE, Brazil

RA Falcão
Universidade Federal Rural de Pernambuco, Recife, PE, Brazil

MS Alexandre-Moreira
Universidade Federal de Alagoas, Maceió, AL, Brazil

EMS Silva
Universidade Federal do Vale de São Francisco, Petrolina, PE, Brazil

CA Camara
Universidade Federal Rural de Pernambuco, Recife, PE, Brazil

TMS Silva
Universidade Federal Rural de Pernambuco, Recife, PE, Brazil

Mostafa R. Sharaf
King Saud University, Riyadh, Saudi Arabia

Brian L. Fisher
California Academy of Sciences, San Francisco, California, U.S.A

Abdulrahman S. Aldawood
King Saud University, Riyadh, Saudi Arabia

JR Koser
Universidade Regional de Blumenau, Blumenau, Santa Catarina, Brazil

FO Francisco
Instituto de Biociências, Universidade de São Paulo, São Paulo, São Paulo, Brazil

G Moretto
Universidade Regional de Blumenau, Blumenau, Santa Catarina, Brazil

HM Meneses
Universidade Federal de Ceará,UFC, Fortaleza, CE, Brazil

S Koffler
Universidade de São Paulo, USP, São Paulo, SP, Brazil

BM Freitas
Universidade Federal de Ceará,UFC, Fortaleza, CE, Brazil

VL Imperatriz-Fonseca
Universidade de São Paulo, USP, São Paulo, SP, Brazil

R Jaffé
Universidade de São Paulo, USP, São Paulo, SP, Brazil

TM Misiewicz
University of California, Berkeley, Berkeley, California
The Field Museum, Chicago, IL, USA

E Kraichak
The Field Museum, Chicago, IL, USA
Kasetsart University, Bangkok, Thailand

C Rasmussen
Aarhus University, Aarhus, Denmark

J Heinze
Universität Regensburg, Biologie I, 93040 Regensburg, Germany

A Schrempf
Universität Regensburg, Biologie I, 93040 Regensburg, Germany

T Wanke
Universität Regensburg, Biologie I, 93040 Regensburg, Germany

H Rakotondrazafy
Madagascar Biodiversity Center, PBZT Tsimbazaza, 101 Antananarivo, Madagascar

T Rakotondranaivo
Madagascar Biodiversity Center, PBZT Tsimbazaza, 101 Antananarivo, Madagascar

B Fisher
Madagascar Biodiversity Center, PBZT Tsimbazaza, 101 Antananarivo, Madagascar
California Academy of Sciences, San Francisco, USA

TM Nunes
Universidade de São Paulo, SP, Brazil

LG Von Zuben
Universidade de São Paulo, SP, Brazil

L Costa
Empresa Brasileira de Pesquisa Agropecuária, Embrapa Amazônia Oriental, Belém, PA, Brazil

GC Venturieri
Empresa Brasileira de Pesquisa Agropecuária, Embrapa Amazônia Oriental, Belém, PA, Brazil

CF dos Santos
Pontifícia Universidade Católica do Rio Grande do Sul, Porto Alegre, RS, Brazil

P Nunes-Silva
Pontifícia Universidade Católica do Rio Grande do Sul, Porto Alegre, RS, Brazil

R Halinski
Pontifícia Universidade Católica do Rio Grande do Sul, Porto Alegre, RS, Brazil

B Blochtein
Pontifícia Universidade Católica do Rio Grande do Sul, Porto Alegre, RS, Brazil